EVOLUTION THROUGH GROUP SELECTION

V.C. WYNNE-EDWARDS

CBE, FRS
Emeritus Professor,
University of Aberdeen

BLACKWELL SCIENTIFIC PUBLICATIONS

OXFORD LONDON EDINBURGH

BOSTON PALO ALTO MELBOURNE

© 1986 by
Blackwell Scientific Publications
Editorial offices:
Osney Mead, Oxford, OX2 0EL
8 John Street, London, WC1N 2ES
23 Ainslie Place, Edinburgh, EH3 6AJ
52 Beacon Street, Boston
    Massachusetts 02108, USA
667 Lytton Avenue, Palo Alto
    California 94301,USA
107 Barry Street, Carlton
    Victoria 3053, Australia

First published 1986

Set by Enset (Photosetting)
Midsomer Norton, Bath, Avon

Printed in Great Britain
at the Alden Press, Oxford

DISTRIBUTORS

USA and Canada
    Blackwell Scientific Publications Inc
    PO Box 50009,Palo Alto
    California 94303

Australia
    Blackwell Scientific Publications
    (Australia) Pty Ltd
    107 Barry Street,
    Carlton, Victoria 3053

British Library
Cataloguing in Publication Data

Wynne-Edwards, V.C.
Evolution through group selection,
    1. Evolution 2. Natural selection
    I. Title
    591.3'8   QH375

    ISBN 0-632-05139-X
        0-632-01541-1 Pbk

CONTENTS

PREFACE

Darwin's conception of natural selection was based on the knowledge that man, in the course of centuries, had been able to separate out many diverse breeds of useful animals and plants by 'accumulative selection'— by persistently taking the best individuals for the purpose required, as the parents for the next generation. Natural selection he presumed did very much the same, the purpose required in that context being an ability to survive through every adversity and to transmit the same faculty to one's offspring. Biologists do not doubt that he was right, and that evolution depends as he said on the existence of genetic variability within each species, which gives natural selection an array af material to act upon. Since his time we have learnt much, not only about individual variation and its inheritance but also about the ecological and behavioural interactions of organisms with one another. The workings of natural selection on the other hand have remained largely presumptive, chiefly because the process of evolutionary change tends to be imperceptibly slow and not easily amenable to observation or experimental test.

Standard evolutionary teaching makes the assumption that all forms of natural selection must necessarily fall on individuals; but in this book I have tried to show that this hypothesis is not sufficient to account for all the attributes that organisms, and particularly animals, are known to possess. In practice, some of their attributes are wholly dependent on mutual cooperation for the achievement of beneficial effects, and require that individuals conform to rule in order to promote the common good. The most conspicuous example of this among animals is their (almost) universal precaution in not overexploiting their food resources—a positive response that still goes largely unrecognized by ecologists as they watch nature's bounty renewing itself each spring. If nutritional plenty is truly the outcome of innate precaution, then conforming with such rules must be vastly beneficial; and I suggest that it could not exist in a world where individuals were set against each other, all against all, in an unregulated scramble for food and still more progeny.

My previous book *Animal Dispersion in Relation to Social Behaviour* (1962) was largely about this example and the corollaries that follow from it. I drew attention to the need for a process of selection that would allow the welfare of the group to take precedence over the self-interest of the individual: but I was unable to suggest how it might work. Now I have what seems to be a reasonable suggestion to make.

The way I have approached it may appear at first sight unnecessarily devious. The reason is that the reality has been difficult to grasp, and I believe one needs to be led along a particular path of experience in order to appreciate it. Thus the book can be divided into three parts forming a logical succession. Chapters 1 to 6 give an ecological introduction to animal nutrition, written with hindsight from twenty years of *ad hoc* research on the red grouse. Chapters 7 to 13 review the grouse research project itself and contain experimental proof that this species does regulate its own populations homeostatically, in relation to fluctuations in the food supply. Chapters 14 to 20 show first that such a capability is not peculiar to the red grouse but extends to the whole animal kingdom; and second that there are similarly vast genetic advantages, obtained in a cooperative manner from the structuring of populations in the habitats they occupy. Spatial structuring, incidentally, is a more sharply focussed concept than the 'dispersion' which appeared in the title of the 1962 book.

Naturally I hope that many readers will find the book interesting enough to read from cover to cover. I have however included summaries of each chapter in the first and third parts, and for the middle part, Chapter 13 has been planned as a general overview of the six preceding chapters. The reader who tackles the whole book may find it useful to peruse each summary as he comes to it, one reason being that it was written after that particular chapter was complete and sometimes I have included afterthoughts which are relevant to what is yet to come. Above all the summaries pick out salient points that the chapter has sought to establish. The reader who by necessity or inclination skips his way from summary to summary will find they sometimes, and inevitably, take a little too much for granted in the interests of brevity, and are not always therefore the easiest of reading; but I hope they will nevertheless convey the tenor and strength of the main argument.

The subject is so broad that I have found it necessary to enter many new fields and have spent what amounts to several years just trying to instruct myself. The relevant literature is enormous, and I regret that I have not had space to mention the many authors whose works I have read but in the end did not quote. One has to make do with enough examples to demonstrate one's point and no more. I am most grateful to three referees, two chosen anonymously by the publishers and the third by me to read the first draft; they roundly and unanimously censured it and sent me back to my study for more than another year to amend it. J. Z. Young (1978) opened the preface to his book *Programs of the Brain* with the following lines, which perfectly express my feelings now. 'Anyone who writes a book dealing with such fundamental themes as this one would be a fraud if he did not feel the need to apologise for his ignorance and temerity. Can it be right for one person to try to cover so many important topics? The result is bound to irritate those who really know about them and could mislead those who don't.' The difficulty is unavoidable; and the reader will soon see how valuable Young's concept of innate brain programs has been to me.

I acknowledge help from many people with thanks, among them the Leverhulme Trustees for an emeritus fellowship when I began the book; Professor William Mordue and my former colleagues in the Zoology Department of Aberdeen University; Professor F. M. Robertson (Genetics Department); A. Watson and R. Moss (Institute of Terrestrial Ecology, Banchory); A. Holden and A. Youngson (Marine Laboratory, Aberdeen); T. D. H. Merrie (B.P. Petroleum Development, Aberdeen); and correspondents A. V. Avery (Bristol University), A. H. Lavery (Queensland National Parks and Wildlife Service), Professor J. E. Lovelock (Reading University), W. E. Ricker (Nanaimo, B.C.), M. E. Solomon (Bristol), and H. H. Williams (Open University, Cardiff).

CHAPTER 1
AN OUTLINE OF THE
ANIMAL DISPERSION HYPOTHESIS

1.1 The regulation of animal numbers

Group selection has been much debated in recent years. Its advocates are a minority of evolutionists who believe that some of the attributes which living organisms possess could not have evolved if selection had only been able to work on differences between individuals. Under individual selection, advantage depends solely on the number of progeny that an individual contributes to the breeding stock of the next generation, the fittest contributing most. This book on the other hand is much concerned with the cooperation of individuals in local populations to protect the welfare and survival of their own stock; and that may depend on their having a social organization, imposing a way of life with which they have to conform. For cooperative systems like this to evolve it seems necessary to invoke a selection process acting between group and group, with the groups persisting as semi-permanent units, giving time for the better integrated ones to prosper and supplant those that are less vigorous.

On the other side of the debate a much larger body of opinion defends the neo-Darwinian assumption that selection at the individual level is the only credible process. A still larger number of interested biologists no doubt reserve their judgment and keep an open mind. The most popular contentious issue has been how, or whether, altruism can evolve or exist in the animal world—altruists being individuals that act for the benefit of others at some cost to themselves, in contrast with selfish individuals who gladly accept the benefits of altruism but do not reciprocate.

In a previous book, *Animal Dispersion in Relation to Social Behaviour* (1962), I emphasized the necessity of group selection, but as already stated in the Preface I had to leave the basic mechanism unresolved. I had no reply to give to critics who were not prepared to accept even massive circumstantial evidence without being given some hint of how mutualists and cooperators would actually get the better of the selfish individuals in their midst. The answer, now that I think I have found it, has turned out to look very simple; but it took me a long time to realize what it was, and

1

that implies a need for retracing the path that led to it, step by step, if it is to be made simple and credible to others.

I shall start therefore where my first suspicions were aroused about the ability of orthodox natural selection to account for all the facts of animal adaptation; and that was while I was absorbing the message of Lack's then new and afterwards very influential book, *The Natural Regulation of Animal Numbers* (1954). I had been interested in population dynamics in a general way for years, but there had never before, to my knowledge, been a treatise that set out to cover it. Lack had digested a large and scattered literature and analysed it in an evolutionary perspective, enabling one to grasp and think constructively about the underlying problems. His own speculations about the nature of the regulatory processes—natality, mortality, dispersal, and competition—were guided by his personal conviction that natural selection works solely at the individual level, and that any alternative conclusion would be misconceived.

The subject lies close to the centre of animal ecology. To lead into it I cannot do better than to summarize his own introduction, in the following simple terms. Most wild animal populations that have been repeatedly censused have been found to fluctuate irregularly from year to year. The majority go up and down about a constant average, and for birds it generally appears that the highest recorded population of any species in a locality is between two and six times, but rarely as much as ten times, the lowest. These are narrow limits, considering the powers of increase that birds possess. There are many examples, such as the rapid multiplication of house sparrows and starlings when introduced to North America, to show that, given the chance, they are capable of exponential increase. The fact that an equilibrium has been reached in the majority of well-established populations shows that there are stabilizing processes at work, though there is no obvious indication of what the controlling forces are.

An important conclusion could however be drawn, namely that whatever they are, the forces must be density-dependent in their action. This means that the birth and death rates must vary with the population density itself, and change when the density changes. Crowding must somehow cause a population to stop growing, and make it decline; and should it fall below the equilibrium level, natality must rise or mortality fall, or both, to push the density up again. The further the density moves from the point of balance, the stronger must the restoring forces become.

Darwin (1859) was probably the first person to suggest an answer. In Chapter 3 of *The Origin of Species* he broadened Malthus's assumption that human populations tend to increase exponentially, to include all other organic beings; and it implied that if they are not being destroyed more or less as fast as they are produced, the earth would soon be filled with the progeny from a single pair. What the 'checks to increase' are that suppress their natural tendency to keep on multiplying seemed, he said, 'most obscure'. Common sense suggested there might be four kinds of checks, the first being the amount of food (or, for plants, of nutrients). It must set an extreme limit to the increase of each species. A second was, not the obtaining of food, but serving as food for others. The third he chose to call climate, assuming it to cover transient physical events in the environment that could endanger life, including wind, flood, drought, heat, cold, volcanoes, landslips, and the like. His fourth was 'epidemics', that is to say infective lethal diseases. On epidemics he commented that the spread of infection would be favoured when a species had inordinately increased in numbers in a small area; and that 'a sort of struggle between the parasite and its prey' enters into their relationship.

He did as well as anyone could, 130 years ago, in making a reasonable analysis of a novel subject; and in anticipating that overcrowded conditions would favour the spread of disease he came close to recognizing the essential property of density dependence. Two of his other checks—starvation and predation—also contain an element of density dependence, since both tend to increase when the affected populations become more crowded; and of the four, only climate is wholly density-*in*dependent.

Darwin was deliberately guarded in putting these suggestions forward, but his successors just took it for granted he must be right, and had given the only conceivable explanation of how a natural balance could be struck between exponential increase and the forces of mortality. It was almost another century before alternative possibilities began to be taken seriously, although much earlier than that it had been suggested, by more than one lone voice, that the animals themselves appeared to be taking part in preserving the balance between population and resources; and that at least some of them could, on their own initiative, make density-dependent responses to changing conditions, for instance by means of their territorial systems. Dividing the habitat into a set of property holdings each owned by a separate individual or family, might enable surplus individuals to be kept out when the need arose, and thus help to forestall the exponential increase that would otherwise occur.

1.2 My own hypothesis

I began writing *Animal Dispersion in Relation to Social Behaviour* (1962) the year after Lack's book appeared, initially with the idea of bringing together the plentiful evidence already at hand, that the higher animals were capable of adjusting their own population densities. As just mentioned, many of them have territorial systems or other forms of property tenure, and it had been shown for various species that when a habitat becomes full of individual claims, no more conspecifics can gain admission. The owners quickly reached the point at which any attempt to compress their territories further is implacably resisted. Alternatively, or even simultaneously, individuals living in a group often form hierarchies in which they each eventually become ranked in a linear order as if they were placed on the rungs of a social ladder. Ideally they have an alpha individual at the top and an omega one at the bottom. When food and other desirable prizes are in short supply, those standing higher in the scale have a prior right to take them, and those nearer the bottom, when their turn comes, may find nothing left. Hierarchies consequently have the same general effect as property holdings, of giving a sufficiency of goods to as many individuals as the resources will support, and obliging the remainder to look somewhere else or go without.

The simplest kind of territorial system results in spreading the population out in a mosaic over the habitat. In birds, territories are very conspicuous in the breeding season, when each is occupied by a mated pair. Hierarchies on the other hand are most characteristic of gregarious animals, living together in herds or schools, and separated at larger intervals from similar concentrations. In many gregarious species each flock owns a group territory. There are also more or less solitary animals like foxes that have, instead of individual exclusive territories, 'home ranges' which overlap with one another; and overlapping individuals may at the same time establish personal precedence relationships with close acquaintances (Figure 1.1). It was because of this fundamental characteristic, present in so many animals, of adhering to one or other of these systems or a mixture of both, that I called the book *Animal Dispersion*. 'Dispersion' in this sense was taken as applying to an established, *static* pattern of individuals within a habitat, even if the pattern persists only for a relatively short time. 'Dispersal' was given a different meaning, to denote the *active movement* of individuals that leads to their re-arranging themselves and establishing new dispersion patterns.

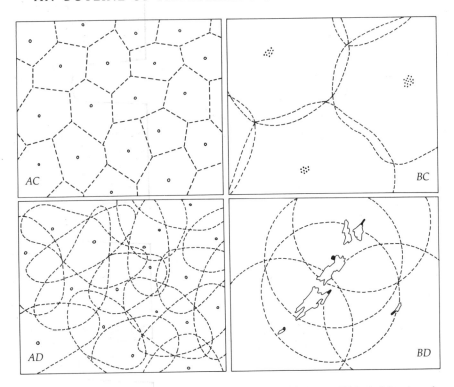

Fig. 1.1. *The four basic types of 'home range', each implying established rights to seek food. AC, the solitary and exclusive, frequently all defended. AD, solitary and overlapping; the base, if any, usually defended. BC, gregarious and exclusive (the diagram actually shows a fractional overlap as in rooks). BD, gregarious and overlapping (shows a group of islands with five seabird colonies from which residents fan out in all directions up to a maximum radius). From Wynne-Edwards (1962).*

There had been a succession of authors before my time who had observed that territorial systems existed in birds and were capable of limiting population densities. One of the first, the German naturalist Altum (1868), even suggested that birds 'need a territory of a definite size, *which varies according to the productivity* of any given locality' (quoted from Mayr's translation, 1935; my italics). An even clearer-sighted pioneer was the Irishman C. B. Moffat (1903). In a modest paper of 15 pages he postulated that possessing a territory is, for many species of birds, a necessary prerequisite to breeding; and that there must be many individuals that do not succeed in winning one, as indicated by the speed with which accidentally widowed birds tend to acquire a new mate. He made other equally impressive suggestions—that a male bird when it sings is giving notice that it is holding a plot of ground and is prepared to defend it; and similarly that male adornments have evolved as 'war paint', more for warning rivals than for captivating females. He concluded that the holding of territory puts a prudential limit on the size of breeding populations, and accounts for their more or less constant density from year to year; and he took issue with Darwin on the latter's insistence on the high rate at which all organic beings tend to increase.

Darwin (1859) had in fact been led to reject the idea that the Malthusian remedy of 'a prudential restraint from marriage' could have any application to the natural world. His basic assumption was that organisms can be said to strive to the utmost to increase in numbers; and in the struggle for existence that results, the fittest ones survive to reproduce. It must follow therefore that natural selection operates by evoking ever fitter, more productive individuals; and that the good of the group or the species, in so far as it results at all, must arise incidentally from the production of fit individuals, and cannot be subject to selection in its own right. He came to this conclusion not without misgivings, which remained unresolved; but it has become the standard interpretation now applied to his theory by evolutionists.

The subject of territoriality was brought to the general attention of ecologists by Eliot Howard (1920); but, in the absence of any clear lead from him, its central function of limiting the numbers of occupants per unit area of habitat continued to be obscure and controversial. In *Animal Dispersion* I suggested that the common characteristic of territoriality is the competition it entails for wanted items of real estate, which as a rule are in artificially short supply, so that not all the competitors can succeed in obtaining one. The essence of territoriality is to regulate the number of winners, and divide the population into haves and have-nots. Food territories in particular provide an artificial method of dividing up the food supply into shares. Living foods of all kinds can be regarded as crops, produced each year from a permanent capital resource, like interest on investments. In exploiting them, so long as the consumer only uses the crop, the capital will be safe and will produce another crop next year. The important principle is obviously to keep the capital intact. The smaller the food territory, the higher the density of consumers and the greater the demand for food. By programming the consumers to take territories rather larger than the critical minimum size, therefore, over-exploitation can be prevented and the capital safeguarded, and the function of the system fulfilled.

The outcome of the competition has profound effects on the participants, in as much as the status and security of the winners of property is increased and in some cases almost assured, whereas losers are generally made much less secure than they were before, and exposed to greater risk. The winners qualify themselves to obtain one or more of the three basic rights of citizenship, namely to belong, to feed, and to breed, and the losers are excluded from these rights. In some species and circumstances all three rights are won at the same time in a single contest, but in others a succession of status contests are required.

When one turns to examine the parallel intraspecific contests for dominance rank in a hierarchy, it is not surprising to find that their consequences in granting or withholding rights are exactly the same. Hierarchies were first recognized in relation to personal feeding rights in domestic ducks and poultry by Schjelderup-Ebbe (1922), but as far as their biological function was concerned they remained in the limbo, even more than territoriality. Before the *Animal Dispersion* hypothesis offered this joint explanation for both in 1962, their close connection had been noted by only one behaviourist, the late Professor W. C. Allee (1938, 1949) of Chicago, the man whom we ought to revere, in retrospect, as the founder of sociobiology.

Hierarchies are abstract relationships between individuals and can often function without reference to individual property possessions. They are particularly appropriate to animals living gregariously and sharing a collective territory. Typical hierarchies are usually found in relatively small groups of individuals, all personally known to each other. The members have all had one-to-one confrontations with each other, usually involving threats and menaces but causing little physical damage; hence the use of the term 'pecking order'. In a small enough group the order is linear, and no ambiguity exists as to who takes preference to whom. Dominants can displace subordinates from their feeding positions at will, and normally do so without incurring opposition or retaliation. In their nature and characteristics, hierarchies are thus even more obviously an artificial contrivance than territories.

Occupying individual territories, or living in a flock, are often alternative forms of dispersion, both occurring in the same species. Birds such as finches and buntings (and numerous other vertebrates and invertebrates as well) can be owners of parental territories in the breeding season, but live in flocks (with hierarchies if small enough) at other seasons. In summer, white pelicans (*Pelecanus onocrotalus*) may associate in parties of up to 20 when feeding, and cooperate in driving small fish into shallow water. When they have fed, each will fly back to its own jealously guarded nest-site in the breeding colony. In winter, starlings return at night to a personal, defended roosting site, often a small perch in a large communal roost, but they feed in flocks during the day. Thus in practice, property ownership and abstract rank are often mixed together; and indeed the property owner is in effect responding as a dominant against all intruders as long as he stays on and is defending his rightful possessions; but as soon as he trespasses on another's ground he becomes a subordinate.

The effect of the hierarchy on the allocation of scarce food resources is to grant preferential access to higher ranking individuals, who are able to satisfy their appetites and remain in good health while their subordinates starve. The results, as already stated, are virtually the same as those derived from a territorial system which affords owners a secure food supply, but excludes non-owners.

Dominant adults in a collective group or herd can equally well inhibit subordinate adults, or younger recruits, from breeding—once again imposing a disqualification that is equivalent, in a territorial species, to not possessing a territory. The truth of these statements can be demonstrated experimentally, as we shall see (p. 115), by killing or otherwise removing dominant adults from a group, or territory owners as the case may be, whereupon a corresponding number of subordinates will normally step into their places and proceed to breed. The two types of exclusion mechanism are complementary to one another, and their function is the same.

One of the best illustrations of the interplay of the two methods of exclusion is the communal display of males in a 'lek', exemplified by the black grouse (*Lyrurus tetrix*), or the birds of paradise, or the antelopes known as Uganda kob (*Kobus kob*), or (perhaps) the dancing aerial flocks of many flying insects. Their leks are tournaments, held before and during the breeding season, day after day, when the same group of males meet at a traditional place and take up the same individual positions on an arena, each occupying and defending a small territory or court. Intermittently or continuously they spar with their neighbours one at a time, or display magnificent plumage, or vocal powers, or bizarre gymnastics, like so many virtuosi, and sometimes doing them all at once. Though they have territories, yet they have a hierarchy with the top-ranking males typically placed in the middle and ungraded lesser aspirants ranged outside. Females come to these arenas in due course to be fertilized, and normally they make their way through to one or other of the dominants in the centre.

Though much less dramatic, another interaction of the same two phenomena is seen in the orderly flocks or schools of many vertebrates, especially when all the members are orientated in the same direction. All seem to display equal rank and close cohesion, but each nevertheless maintains a characteristic 'individual distance' from its neighbours, and resents encroachment.

1.3 Social systems

In the course of writing *Animal Dispersion* I came to realize that the behavioural world I have just been describing is in fact the social world; and to perceive moreover that there is some underlying unity of function in sociality wherever it is found (Wynne-Edwards 1959, 1964). It appears in part to provide a kind of theatrical stage, or make-believe way of life, where individuals play instructed parts, instead of being tossed out into a free-for-all in the rough world outside. Nevertheless it is a competitive world and the struggle for existence remains keen; but the members of social groups cooperate in civilizing it and, so far as the competition is concerned, they enact it according to rules. Everything the social code decrees is done for the common good, and it is typical that the rules for social competition virtually eliminate bloodshed in the pursuit of personal advantage. After describing a lek, it is needless to say that social competition is conventionalized; and the prizes the competitors seek are equally artificial substitutes for the true requisites of life. In a lek the males are contesting for position and status. Those that emerge at the top, holding the highest rank and the best sites, will later engage in reproduction, while the rest will be inhibited.

The social code is absolute and binding. It applies in exactly the same way to food as it does to reproduction. Access to either of them depends on obtaining the qualification or right required, through social competition for either property or status or both. The underlying, primary function is to regulate the population, by controlling reproduction, or thinning the numbers down to a density the food resources can support. Later we shall see that there are important subsidiary benefits that arise from sociality as well.

When I wrote *Animal Dispersion* I was in no doubt that these social adaptations existed in the higher animals and brought great benefits to social groups and species, by supplying a mechanism for conserving and husbanding food resources and ensuring their efficient use. It looked as if the restraints that the social programs imposed must demand some sacrifice of fertility and survivorship on the part of the individuals concerned; and if so, the programs could not have evolved through the operation of individual selection alone. If groups rather than individuals benefitted, then a process of group selection must be necessary to elicit the result. But how could it possibly be effective in a natural world where free-lance individuals could apparently enjoy higher fitnesses than the social conformers (Wynne-Edwards 1963)? This was a dilemma I could not resolve, and for over 20 years it has invalidated my hypothesis of sociality as a general adaptation for population control, in the eyes of most biologists.

Fortunately the problem has now evaporated, with the realization that it was due to a wrong assumption. Deeper reflection on the subject has revealed that the individuals in species that regulate their food consumption optimally do not make any personal sacrifices at all. On the contrary they find themselves born into a world of plenty. The fitnesses they enjoy are far higher than those attainable by a population of independent non-cooperators, exploiting similar habitats for personal gain without regard to the consequences. These self-seekers would live in a ravaged world by comparison; whereas social programs are good for all qualified individuals, since the group-benefits they bring stem largely from the increased fitnesses of the members.

This principle, on which a theory of group selection can acceptably be founded, is discussed at greater length in Chapter 19 (p. 313).

It may be necessary for some readers to adjust their concept of what sociality or being 'social' actually implies, in this functional and evolutionary context. Being social has nothing primarily to do with being gregarious, but is founded instead on conventionalizing the relationships and interactions of individuals. As stated earlier, solitary foxes can develop the same kind of social bonds and rivalries with other individuals to whom they are personally known, as gregarious animals can. 'The cat that walks by itself' acknowledges hierarchies, and the rights of others, that are just as powerful, ritualized and topographical, as those found among the mice in the barn or the sparrows in the yard (cf. Leyhausen 1965). Personal acquaintance and recognition are essential to the more intimate manifestations of sociality, although they can become submerged in large conventional assemblies.

Another mistaken concept, widely held and much in need of revision, is that sociality is a phenomenon that occurs sporadically, even in vertebrates or insects, and has evolved more or less from scratch and in parallel in different species. The evidence shows on the contrary that it is an extremely ancient phenomenon, which may have originated as few times as twice, once in the vertebrates and once in the arthropods. The members of each of these phyla probably inherit the basic elements of social cooperation from their respective Cambrian ancestors, although the particular features of it differ much, even from species to species. Some of the products, such as the male adornments involved in social selection, and the local dialects found in bird song and human language, are especially labile. Sociality is not apparent in the molluscs (except perhaps the cephalopods), nematodes, platyhelminths and coelenterates, nor in any of the minor lower phyla of the Metazoa; and it is doubtful if it exists in any of the annelids. Nevertheless, as stated above, these organisms all probably have homeostatic populations and are able to react prudently in exploiting their food supplies (see Chapters 17 and 18).

The phenomenon of sociality thus appears capable of unification, as an adaptation primarily for promoting population homeostasis, and hence the optimal use of food resources. Seen in this light it bears little resemblance to the disorderly tangle of threads it presented 30 years ago.

Summary of Chapter 1

1 The natural forces that regulate animal numbers are not readily apparent, and many population ecologists have been content to accept Darwin's presumption that the checks on unlimited increase must be either food shortage, or predators, or parasites, or climatic factors, or some combination of these, which remove whatever excess of population results from reproduction.

2 In an earlier book (Wynne-Edwards 1962) I developed the hypothesis that, on the contrary, most population regulation is effected by each species for itself. Two main mechanisms were identified for expelling surpluses. One is provided by territorial or other systems of property rights, which afford space or sites for a finite number of owners per unit area and keep surplus individuals out. The other is the social hierarchy which, especially in gregarious species, allows the more dominant individuals preferential access to feeding and reproduction, and again squeezes subordinates out. Thus competition for property or social status can, when necessary, stratify a local population group into satisfiable and unsatisfiable members, and lead to the exclusion of the latter.

The supposition is that these adaptations have evolved because animal populations, as consumers, benefit from controlling the exploitation of their food supplies (and secondarily use the same mechanisms for purposes of genetic management). Species that kill their prey or feed on living tissues are restrained from taking more than a renewable crop from their food resources. If they did take too much, future crops would be reduced and their successors would suffer accordingly, whereas prudent consumption tends to maximize food production (see Chapter 5).

3 Holding a parcel of ground or a personal sleeping or breeding site, or achieving accepted membership and status in a peer group, are artificial substitutes for the ultimate objects of competition; when won they give an adjustable number of winners access to the basic prizes—food, living and procreation. As part of the same substitution, crude strife has virtually been replaced by artificial (i.e. conventional) forms of competition, which frequently recur at set times or in artificial arenas, or both. Inherited brain-programs govern the rules, and the normal outcome of one-to-one contests is bloodless, decisive and binding. It determines the two individuals' mutual status for the season ahead, so saving much future time and energy. Losers may or may not be carried by the group as subordinates until the next contest, but sooner or later they will be ejected, exposed to Darwin's checks, and thus eliminated.

Resource levels usually change with time, and recurrent competitions allow the consumers' density to keep pace with the changes, in a process of population homeostasis. This presumably requires accessory brain-programs that integrate feedbacks from the individual's current nutritional experience and from its perception of crowding, and make it respond appropriately, e.g. by taking a territory of suitably adjusted size on each occasion, or by cooperating to expel more subordinates from, or accept more recruits into, its flock.

The chief generalization to emerge from the hypothesis was that this artificial regime of cooperation plus conventional competition, constitutes the social phenomenon, or sociality in its basic biological form. General demographic functions were also assigned, as above, to property tenure and the social hierarchy.

In 1962 I wrongly assumed that, by cooperating to regulate their density and reproductive output, individuals would have to forfeit some of their potential fitness, as compared with their freeloader non-cooperative rivals. I omitted to note that, while the cooperators would be able to hand on to their successors well cared-for, productive habitats, the freeloaders would be unable to prevent their own numbers and conflicts from escalating, with the result that their habitats would be stripped of renewable assets in a ruthless pursuit of personal fitness. It has become obvious now that cooperative populations in well-conserved habitats would speedily and easily attain higher mean fitnesses than self-seeking individualists could maintain.

As originally presented, the hypothesis rightly insisted that adaptations for the common good could not evolve through individual selection, pure and simple, but only through selection acting between groups. I was however unable to suggest how social cooperators would be able to equal, much less exceed, the fitness of freeloaders. Consequently most evolutionists rejected it, notwithstanding the weight of circumstantial evidence in its favour. The present work sets out to rectify this defect, and establish the explanation just given, namely that cooperators, bequeathing productive resources as well as genes to posterity, have developed an unsurpassed strategy for survival.

CHAPTER 2
FOOD AS A LIMITING RESOURCE

2.1 Food production

Animals are by definition heterotrophic, that is to say dependent on other living organisms for producing the foods they eat. Some have varied diets, but the majority are specialist feeders, many of them normally relying on a single species of plant or animal food, and perhaps only on one of its parts or products. Specializing allows the perfection of skills and mechanisms for finding, handling, and digesting the food which increase the efficiency of nutrition. It also tends to segregate consumer species and reduce the number that share the same resource. When carried to the extreme it has the disadvantage of linking the welfare of the consumer population inseparably with that of a single producer or host; and for that reason the majority of specialist feeders can probably fall back in an emergency on some alternative resource. It is remarkable that almost every substance or material produced by living organisms can be put to nutritional use and metabolized by some other organism.

Green plants and green microbes alone possess the ability to absorb radiant energy from the sun and store it in chemical form, that is, to make fuels. All other types of organisms ultimately derive their energy supplies from plant production. Plants use carbon dioxide, water and inorganic materials for the synthesis of numerous carbon compounds; and again the animals are almost entirely dependent on obtaining plant-produced amino acids, as the raw materials for synthesizing proteins. Like the fuels, amino acids are passed along the food chains from consumer to consumer, and residual amounts are present even in organic detritus.

All organisms produce excretions and dead matter in solid form, and it is these that constitute the organic detritus characteristically present in many habitats. It accumulates chiefly on soil surfaces on land and at the bottom of bodies of water. There are many animals that depend in part or whole on detrital materials for food. The inter-relations of the three great biotic 'industries', consisting of detritivores (or decomposers), primary producers (or synthesizers), and vivivores (consumers of living matter), form the subject of Chapter 4.

The largest category of consumer animals are the herbivores, which feed by ingesting living plant materials. It is also common to find one kind of consumer feeding on another, but all food chains have to start with plants as the primary producers and proceed through a herbivore or detritivore to a second, third or a higher trophic stage. These later stages may comprise (1) predators that feed by eating other animals, (2) parasites that take tissues or body-fluids from living hosts, or (3) detritus feeders. At each trophic level more of the original stock of chemical energy is dissipated by oxidation, and more of the organic nutrients are metabolized and lost as dead matter. Thus the biomass of herbivores in any ecosystem is many times that of the top predators and parasites. In Chapters 7–12 special attention is paid to the nutrition and population regulation of a representative vertebrate herbivore, the red grouse (*Lagopus lagopus scoticus*); and in Chapter 18, on a smaller scale, to host-parasite relationships in flatworms (Platyhelminthes).

Food resources are renewed as a result of growth and reproduction, and are all produced at a finite rate per unit area. Primary production rates change with the seasons in many parts of the world and in many types of habitat, and may consequently have to be reckoned in quantities per year rather than per day. Consumers generally become adapted to the logistics of food production, and many invertebrates have life cycles that divide the year accordingly, into a period of growth and reproduction and a period of dormancy. A minority of vertebrates do the same. It often happens that the primary producers undergo a surge of growth at a particular season, which creates food materials faster than they can be consumed, and yields a standing crop which can be made to fulfil the consumers' needs for weeks or months; but eventually the consumers must either go into suspended animation until the following year, or switch to an alternative food source. Larger animals, particularly those that are active all the year round and live for a number of years, frequently depend on a regular succession of crops, which may consist of either plant or animal foods.

Our special interest lies in viable relationships that can exist between producers and consumers. Any given food resource can be seen as having a characteristic carrying capacity for consumers, which is theoretically measurable in consumer-days per unit area. If it is a standing crop it will support either more consumers for a shorter time, or fewer for longer; and if, say, it contains 100 daily rations of food and has to last for 20 days, the average number of consumers will have to be restricted to five. For its efficient use, therefore, the consumption rate may need controlling.

A consumer population of insectivorous birds might exploit several different food crops in succession. In any given habitat the series of crops that different insect species yielded for them would be unlikely to provide identical carrying capacities. If there were no re-dispersion of consumers between crops, their population density would have to be limited in relation to the crop with the lowest carrying capacity, and the rest of the time their food supplies would be superabundant.

On a longer time-scale there are usually differences in food productivity from year to year. Consumer species can often respond to such changes by adjusting their population density to match. But some of the long-lived, slowly reproducing species are incapable of doing so and are obliged to adopt, as a compromise, densities that are below the current ceiling capacity most of the time.

Any population that inhabits the same ground in perpetuity (the normal situation) has to depend on the renewal of its food resources year by year. There are some kinds of resources that require conservation in order to provide for this, and some that do not—a distinction that is important. Examples of the kind requiring conservation are the prey species eaten by a predator; and of those that do not, the most prevalent examples are the various kinds of organic detritus, such as carrion. All kinds of detritus can be consumed in their entirety without detriment to their future production; but live plant and animal foods generally require that a stock of seed or its equivalent be left to produce the next year's crop. Some kinds of live foods are better shielded against consumer depredations than others; but succeeding chapters will show that whatever type of food resource is consumed, whether it benefits from conservation measures or not, there is normally an optimum density of consumers that brings them the highest collective return (or fitness). So great are the benefits of controlling consumption that we shall be led eventually to expect that no consumer species (i.e. no animal) could afford to ignore them, although some are considerably more adept than others in tracking the constantly moving optimum.

One can, in fact, state as a general hypothesis that there are two vital activities of organisms in which great rewards accrue from mutual collaboration between individuals. One is in exploiting nutritional resources, and the other is in exploiting the resources of shared genetic systems. *Food resource management* necessitates the density regulation (homeostasis) of populations; and *genetic management* necessitates the structuring of populations. In the higher animals, *sociality* creates an *ad hoc* regime, and mechanisms, for collaboration in both these enterprises alike.

This book provides evidence intended to substantiate these three propositions. Insight into the respective roles they play came to me gradually. It started with speculations about controlling consumption, which have subsequently been tested by many years of research. The structural/social aspect had scarcely emerged from the shadows when the first draft of this chapter was written, and the nutritional/social aspect consequently predominates throughout most of the book. My generalization about this latter aspect can perhaps be put into a somewhat better perspective by giving some illustrative examples, three of which follow.

Dunnet (1955) showed that rural starlings (*Sturnus vulgaris*) in the Aberdeen area depend for much of their diet, from midwinter till June and the end of the breeding season, on the larvae of a crane-fly (*Tipula paludosa*), commonly known as leather-jackets. The larvae live in the soil, feeding on grass roots in pasture land, at a shallow depth within reach of the starling's probing bill. They develop from eggs laid in July–August, and become large enough to attract starlings about the time that winter sets in and slows their further development. In May they put on a great spurt of growth and reach a maximum biomass per square metre about the time the first young starlings hatch. After pupation, the adult crane-flies eventually emerge a couple of months later, and within a few days they mate and launch the next year's generation.

In two consecutive years the average number of larvae present in soil samples differed by a factor of six. But the supply was evidently super-abundant on each occasion, because only 2 and 7% of the available stocks were actually carried to the nestlings; and the birds' population density and breeding success were similar in both years, notwithstanding the difference in food abundance.

Dunnet concluded from this that the starling density was not related to the carrying capacity of their food supply in the breeding season. One reason for citing this example is to illustrate that it is common for vertebrates to be living below the potential carrying capacity, because they face a need for restriction at some other period of the year. On the other hand we rarely find them living even briefly above the carrying capacity, and imprudently running into difficulties that could have been avoided. There was possibly an alternative explanation of the low breeding density here, which was that breeding sites for this cavity-nesting species were in short supply, and constituted a limiting factor.

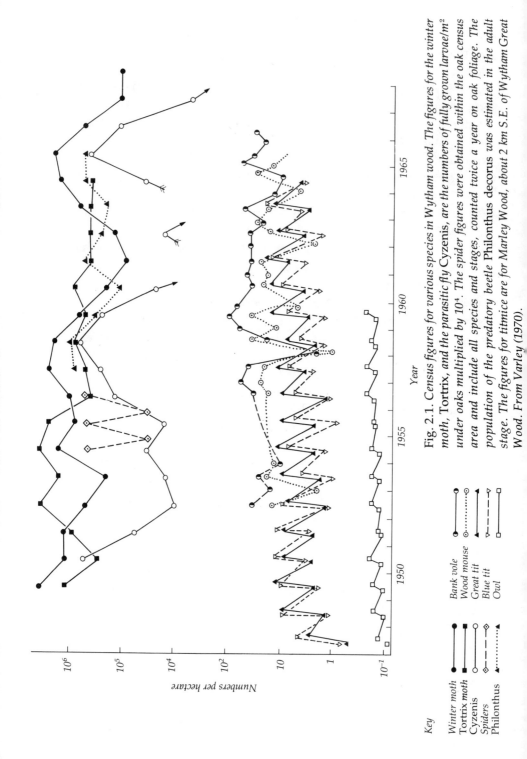

Fig. 2.1. Census figures for various species in Wytham wood. The figures for the winter moth, Tortrix, and the parasitic fly Cyzenis, are the numbers of fully grown larvae/m² under oaks multiplied by 10⁴. The spider figures were obtained within the oak census area and include all species and stages, counted twice a year on oak foliage. The population of the predatory beetle Philonthus decorus was estimated in the adult stage. The figures for titmice are for Marley Wood, about 2 km S.E. of Wytham Great Wood. From Varley (1970).

In Figure 2.1 are shown a set of 20-year runs of population estimates, for ten diverse animal species, living in Wytham Wood, Oxford, chiefly in the oak-wood (*Quercus robur*) community (from Varley 1970). The species fall roughly into three groups as to body-size. The small, numerous arthropods appear at the highest density level; four small vertebrates are at an intermediate level, and the large predatory owl is alone at the lowest. This reflects an obvious fact that there is a broad correlation between body-size and population density (characterized as the 'pyramid of numbers'). In most years the two herbivorous (leaf-eating) moth larvae are the most numerous arthropods in the assemblage. Their arthropod predators (spiders and beetles) and the parasitoid *Cyzenis* (a tachinid fly) require more individual feeding space, and occur at lower densities.

The insectivorous birds (*Parus major* and *P. caeruleus*), which feed their nestlings largely on the moth larvae, come much further below their prey on the density scale. The two mainly herbivorous rodents are on a different food chain, and are two orders of magnitude more numerous than their main predator the owl. The saw-tooth shape of the graphs pertaining to the three bird species results from counting them twice a year, the lower point being the breeding population and the higher one the post-breeding or autumn number.

All the species included were no doubt capable of regulating their own populations in some degree. Actually the moth larvae defoliated many of the oaks in May 1948, and did considerable damage again in 1957 and 1965 (Varley 1970: 391). That this occurred three times in 20 years suggests that normally the populations are controlled. The larvae are in a race against time, because as soon as the oak leaves attain full size they become charged with tannins, making them toxic to most insects and protecting them from further depredation. The larvae hatch as the first leaf-buds burst, and float themselves from tree to tree, dangling on silk threads and wafted by the breeze, in a search for early shoots uncrowded with larvae. When they too have grown to full size they drop to the ground and pupate. The oak trees are able to repair the damage, at least in part, by producing a secondary crop of leaves.

Taking the whole run of years over which each specific study lasted, and comparing the highest population figure with the lowest, it can be seen that some of the insects varied as much as 100-fold within the 15–20 years. The rodents *Apodemus* and *Clethrionomys* fluctuated up to 50-fold; whereas the great tit, which entered the study period at a very low breeding density after an exceptionally severe winter in 1946–47, showed a 12-fold range of variation (Perrins 1965).

The same winter may also have caused the uniquely low initial population of tawny owls. For the first eight years their numbers climbed slowly up from 17 to 30 breeding pairs, after which they remained steady. Some 15 years after the study had ended their territories were checked again, and found still at the same ceiling level (Southern 1970, and verbally). In fact, though their density tends to be so constant, their reproductive output is not so. Over the period it varied between zero, in a year when none attempted to breed though they still occupied their territories, and an average of 1.3 flying young per pair in the best year. The main proximate cause of this is the fluctuating abundance of their prey; but Lack (1966: 141–3) showed that Southern's data also suggest an inverse feedback from their own density. He compared years with similar mouse numbers, both in the initial period when owl numbers were building up, and at the later stage after they had levelled off, and found the owls had raised consistently higher averages of young per pair while their own population was below the ceiling density.

Most of the demographic variation that occurs between years can be attributed ultimately to chance climatic fluctuations. It so happened that two other great tit populations were being monitored at the same time as the one at Wytham (Perrins 1965: 605). One of these was also in the south of England, 80 km west of Wytham, and the other near Arnhem in Holland, 500 km to the east. All three showed synchronies in their fluctuations, particularly in the last eight years when they were most accentuated. It may not always be possible to characterize the components and combinations of the weather that set off these demographic changes, but there is no obvious alternative cause to which one can ascribe such widespread effects. The actual changes in the great tit population, however, would turn out to be almost entirely homeostatic, due to their own responsive behaviour to perceived environmental change.

The two rodents in the survey undoubtedly use social competition in the regulation of their populations at Wytham (Watts 1969; Flowerdew 1974, 1978). However, the subject of rodent population dynamics figures prominently in Chapter 14 (p. 177), and discussion of it seems best deferred till then.

For the great tit, as for the tawny owl, it is known that the Wytham population is directly controlled by the territorial system. When a number of territorial males were experimentally removed from the wood in March, before the breeding season had begun, they were promptly replaced by a similar number of substitutes, which were birds that had not previously held territories in the area. Average territory size varies from year to year (much as in the red grouse as described in Chapters 8 and 9); but the critical resource that presumably elicits the response has not been identified (Krebs & Perrins 1978). The larvae of the winter moth (*Operophtera brumata*) form a large part of the diet of small young tits in early broods, but, as Dunnet found with the starlings, only a few percent of the stock available is fed to the nestlings (Perrins 1979: 225).

The winter on the other hand is a more difficult season nutritionally for tits, as for many other birds and mammals. A large part of their food then consists of overwintering and often dormant insects and spiders, many of which do not reproduce or grow before the spring. Nevertheless there are a few that do increase, especially in February and March in certain winters. Gibb (1960) made a four-year study of the subject, working in pine plantations near Thetford, Norfolk, which were chosen because their invertebrate fauna is much simpler and more uniform than that of oak woods (Lack 1966: 80). The predominant bird species there was the coal tit (*Parus ater*), though other tits and the goldcrest (*Regulus regulus*) were also present.

Gibb found the food stock varied in quantity from year to year, and diminished substantially through the winter. In the foliage of Scots pine (*Pinus sylvestris*) and Corsican pine (*P. nigra*) where the tits foraged, there was an exceptionally heavy stock in November–December 1953. The results of laborious sampling showed an average of 0.56 g of invertebrates per square metre of forest. By March it had fallen to 0.12 g (Gibb 1960: 187 & 189). One of the commoner insects was selected for special study because it offered a unique opportunity for estimating the numerical proportion of the stock eaten (Gibb 1958). It was the larva of the moth *Enarmonia conicolana*, which grows from an egg laid singly in June on a young green cone of *P. sylvestris*. The larva soon hatches and burrows into the cone to get at the seeds, on which it feeds. In September, fully fed, it works its way out again almost to the cone surface, where it makes a cell in which to hibernate. If it survives it pupates in March, and the moth emerges in June. The larvae are found by the tits (it is not certain how, but perhaps by the hollow sound of the cone when tapped), and dug out. In the following summer after the cones have fallen to the ground, large samples can be examined. Those that were infected either have a neat round hole cut by the pupa to let the moth eclose, or a hole torn by a tit to extract the larva.

The infestation rate varied according to locality within the same year, and also from year to year. On the average the birds took 45 to 50% of the larvae in the course of the winter; but where the infestation was high and searching consequently more profitable, they might take 70% or more, and vice versa. The rewards in relation to the time and effort spent are an important factor to foragers, and there is a threshold below which they do not repay the costs (cf. Royama 1970).

In addition to the gradual disappearance of the food stock there is also a parallel tendency for the bird density itself to decline as the winter progresses. In the main study area, Gibb (1960: 167) made thorough counts of the birds once or twice a month. They were then in roving flocks consisting of three or more species of tits and the goldcrest. In the four winters the average September–October and March–April counts per 10 ha were as follows—30 & 19, 42 & 13, 31 & 16, and 48 & 29—indicating an overall average reduction between these months of close to a half. When the survival rate of the coal tit was considered in isolation, winter by winter, its reduction was found to be closely and significantly correlated with the lowest level to which the food biomass fell. In fact, from the four winters with their different states of food stocks and weather, Gibb (1960: 194) found 'abundant evidence that the density of the tits and goldcrests in the pines varied closely with the stock of their invertebrate food'; in other words, as shown most clearly in 1953–54, the stock of food per bird remained roughly the same throughout the winter.

How this regulation was brought about he did not discover, though it was not visibly due to predators or disease. He accepted that some sort of intraspecific competition is an essential element in density-dependent mortality. Some kind of exclusion mechanism must have operated, and forced individuals to leave the habitat and die elsewhere, weakened by losing access to their normal diet, and hence overtaken by one or other of Darwin's checks.

In rather less natural conditions, the progressive removal by tits (mainly *P. caeruleus*) of cocooned larvae of the codling moth (*Cydia pomonella*), which had hibernated in crevices in the bark of apple trees in an orchard, was found to amount on average to 95% over three years (Solomon *et al.* 1976; Glen *et al.* 1981). It was again found that the larval mortality rate decreased markedly as the winter went on.

It may be connected with the need for adjusting the number of consumers, that the various tits, together with goldcrests and sometimes other insectivorous birds, spend so much of their time in mixed-species flocks at this season of year. Their diets, though not identical, more or less broadly overlap, and they all undergo similar density-dependent adjustments of their numbers. Flocking is beneficial, apart from anything else, in the deterrence of predators; but it has long seemed likely to me that such mixed bird flocks (which occur in many parallel situations) serve an epideictic function; that is to say, they show the members of the assembled company what their own local consumer guild consists of and how numerous it is. If this is so, it would feed information into their brains that could be coordinated there with the other relevant variable, namely their own current state of nutrition. My interpretation is that these two factors put together must stimulate a response in the homeostatic program of each individual, and that collectively these responses will bring about an appropriate density change. This is a theme that will continually recur throughout the book, and there is no need to insist on its immediate acceptance here. Whether or not it is credible can be deferred for later decision. It appears self-evident, however, that the tits and their followers do not blindly 'muddle along' till they find they have reduced the food stock to the threshold of unprofitability. Preventive action is taken in time, probably by raising intraspecific aggression just enough to make the lowest-ranking members of the company unwanted and resigned, and ready to opt out. The survivors will thus be enabled to reach the spring in good health; and enough 'feedstock' will also survive to renew the invertebrate food crops.

To conclude this section it seems desirable to introduce one insect species whose populations unquestionably show social, self-regulatory powers. Others will appear later in special contexts. The example chosen here is the sycamore aphid, *Drepanosiphum platanoides* (Hemiptera, Aphididae), and the data are from Dixon (1970) unless otherwise stated. It feeds by inserting its stylets into the phloem tissue in leaves of the sycamore tree (*Acer pseudoplatanus*), and extracting the sap. The sap contains high concentrations of sugars but little protein. Consequently the latter is the limiting nutrient for the aphids' growth and reproduction. The insects have to process large volumes of sap to obtain the amount of protein they require, and the excess of sugar they cannot avoid ingesting is excreted as honeydew. The highest proportions of soluble nitrogen are found in the phloem of young growing leaves, and similarly in senescent leaves. The amount falls sharply as the leaf matures, then rises gradually again between July and October.

The aphids' life cycle is geared to this sequence. They multiply rapidly when conditions are good, and there are typically two population peaks, one in June and the other in September. In between, numbers fall to perhaps a third of the peak density, and both the body-weights and reproductive rate of the intervening generations are much reduced.

The aphids are sensitive to crowding, and space themselves out when feeding on the undersides of leaves (Figure 2.2). Especially during the mature-leaf period their reproductive rate has been shown to respond inversely to their own population density and the number of individuals in their group. Their awareness of density depends on sensory preception (probably tactile, through antennae, legs and wings; see Figure 2.3). Though the aphids maintain individual distances, Kennedy & Crawley (1967) showed they are mutually attracted and aggregated at the same time, and consequently leave many leaves, and spaces on leaves, unoccupied at any given moment. We have met this typically social phenomenon already (p. 9), and it is one which is also very familiar to us in gregarious vertebrates. Though the authors (1967: 162) may have sensed the fact, they did not actually remark that the aphids' behaviour is easily accounted for by the simple *Animal Dispersion* hypothesis; namely that it serves to prevent them from exceeding the density at which they would, between them, extract sap at a greater rate than the individual leaves, or the tree as a whole, could afford. Were overtaxing to occur, they could chronically stunt or even kill the plants on which their posterity depends. It seems a reasonable presumption therefore that this

Fig. 2.2. *Natural distribution of adults of* Drepanosiphum platanoides *on a sycamore shoot seen from below (traced from a photograph). From Kennedy & Crawley (1967).*

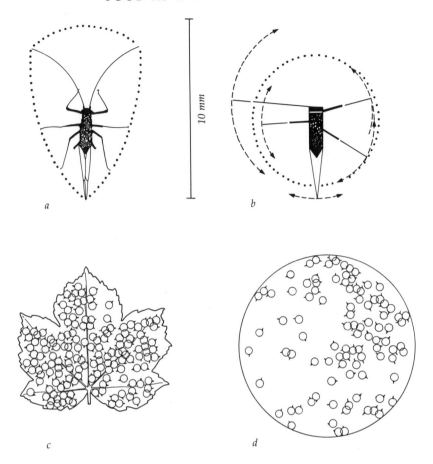

Fig. 2.3. (a) 'Tactile envelope' of a settled, immobile Drepanosiphum platanoides. (b) Approximate 'reactive tactile envelope' (dotted circle) allowing for the movements of the appendages (broken lines) when a settled aphid is touched. (c) 'Reactive envelopes' of aphids settled at maximum density on a leaf. (d) The same for aphids settled at maximum density on a glass disc. From Kennedy and Crawley (1967).

is the purpose of their homeostatic adaptations; and that Dixon was mistaken when he rejected Kennedy and Crawley's suggestion that the aphids' population regulation is self-induced rather than food-induced. Dixon (1970: 283) did incidentally make it quite clear that there is no evidence for supposing that predators or parasites are capable of regulating the aphid numbers. In his view they are held in check by what he appears to regard as unavoidable or non-adaptive crowding effects among individuals.

2.2 The overfishing phenomenon

A distinction has been made (p. 16) between food resources that are open to damage from overexploitation and those that are not. Far more than half the species in the animal kingdom are dependent on the overexploitable kind, and their problems seem properly the ones to consider first. The *Animal Dispersion* hypothesis proposes that all the species so exposed to risk are normally prevented from harming their resources by genetically coded brain programs, and that these programs consequently have a very high survival value. Under natural conditions, damage so rarely comes to our notice that we might fail altogether to appreciate the nature of the problem, or even to realize a problem exists, but for our repeatedly adverse experiences whenever we try to gain a living ourselves by exploiting similar resources.

Scientific insight into the subject developed largely between about 1920 and 1960, as a result of research into the difficulties being encountered by commerical fisheries. Fish lend themselves particularly well to population research for two main reasons. First, it is technically easier to take large, standard and repeatable population samples in the water than it is on land; and second, fish grow throughout life, and in temperate zone climates where many of the most valuable fisheries have occurred, their growth-rates change with the seasonal cycle of temperature. Consequently the scales and bones acquire annual growth-rings like those in the wood of trees, which record the exact age of each individual, and even its approximate linear dimensions in former years.

Most commercial fishing is done with nets that tend to be selective in the sizes of fish they catch. The meshes are large enough to allow at least some, and sometimes all of the small ones to slip through and escape. Mesh size thus determines the size at which young fish are recruited into the commercial catch. This is often the age at which they first mature, because many fisheries are actually conducted on spawning grounds to which fish return, and where they are known to congregate at the same season year after year.

Only one major group of sea-fish, the sharks and rays (elasmobranchs), have evolved reproductively in the same direction as the land vertebrates, towards greater parental care and lower fecundity. They copulate in pairs and the females in most species lay small numbers of large yolky eggs with horny shells. Some female sharks have gone further, and gestate the eggs in their 'uteri' until the young are ready to be born alive.

The reason why the majority of the bony fishes, and many marine invertebrates as well, retain the practice of spawning huge numbers of eggs appears to be that, in the sea, green plants are confined to the daylight zone near the surface, mostly in the top 50 m. All the primary production of bio-fuels and organic nitrogen originates there, and nutriment is carried to greater depths and to the sea floor only in the bodies of mobile animals, or in detritus falling under gravity. Broadly speaking, therefore, young fish must either be born large enough to catch and eat sizeable prey from the start, as young sharks and rays are; or they must float up to the surface in the egg (or when they hatch) and be left to look after themselves among the unicellular algae and small animal life of the plankton.

Mortality in the plankton is evidently high, and variable from year to year. In some fisheries, for example those for herrings (*Clupea harengus*), there tend consequently to be big recruitments in occasional years, and much of the commercial catch at any time may consist of one or two particular year-classes that happen to have been recruited in unusual numbers. The same classes reappear annually thereafter, their members becoming bigger but scarcer with advancing age, until they fade out, leaving younger supernormal cohorts to replace them.

The fisherman plays the same role as other predators, except that his equipment enables him to catch enough in a day to feed not only himself but scores or hundreds of other people as well. Nowadays it is theoretically possible, after doing the research necessary for measuring the population parameters (recruitment rate, age-specific growth rates and mortality rates), to plan a regulated, sustainable exploitation of a fish stock. In practice, however, fish population dynamics are full of 'noise', caused by unpredictable variations in the parameters from year to year; and the commercial landings, measured in terms of catch per unit effort, are often so variable that it may take years to discover whether overfishing is really occurring or not. Its reality was first convincingly demonstrated by the Great War of 1914–18, which interrupted commercial fishing in the North Sea. The industry had undergone a revolution in the previous 50 years. Railways had created markets for fresh fish in growing cities far from the coast. Boat-owners had converted from sail to steam and designed larger vessels equipped with much more efficient gear. But the war quickly turned the Dogger Bank and European continental shelf into battlefields, fraught with mines and U-boats, and the fishermen went into naval service.

When peace came and the industry resumed in 1919, the catch per unit effort rose by 80% compared with 1913. However in a very few years the high catches were over, and the fish stocks were back in their pre-war state.

The same enforced experiment was repeated in 1939–45, lasting six years instead of five. This time the catch per unit effort on resumption was more than double its pre-war rate (Figure 2.4). This is a hard fact, and the reader might do well to recall it when he comes to Chapter 19, and needs

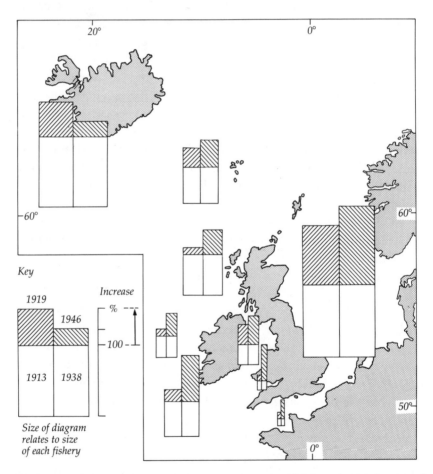

Fig. 2.4. *Histograms showing the increase in the catch of fish in various regions round the British Isles immediately after the two World Wars (i.e. in 1919 and 1946) expressed as percentages of the catches just before the wars (1913 and 1938). The effect of the reduced fishing during the wars is clearly seen. The size of each diagram gives a measure of the relative yield of each fishery. From A.C. Hardy (1959) The Open Sea, Collins. (Figure after M. Graham).*

to weigh up in his mind the advantages of a regulated, cooperative exploitation of living food resources by animals as consumers, as an alternative to free enterprise and each individual for itself.

By 1946 the experts understood that proper management was potentially able, not only to revive and perpetuate high stocks indefinitely, but to produce higher landings per unit effort and higher total yields. The management principle was epitomized not long afterwards in the western United States in the slogan, 'fish less and catch more'. But for nearly 40 years since that time the political and practical difficulties have stood in the way of progress towards efficient management in Europe, especially about how to divide the cake between nations, and how regulations could be enforced. Although everyone concerned knows that management would bring in more fish and make them cheaper to catch, it spells an end to the traditional freedom of the seas, and sets limitations on the public right to fish and the quantities to be caught. Only the nations that have unilaterally declared their sovereignty over a sufficient territorial zone offshore, notably Iceland and Norway, are yet taking proper care of their most precious resource. Unfortunately fishing can remain profitable even when costs are driven up by overexploitation. If we are to learn nature's lesson, we should realize that the only practicable principle for managing such a renewable resource is to subdivide it territorially, assigning territories to communities of people who willingly cooperate in conserving their own property, in their own interests and those of their heirs to come. It will be hopeless as long as exploiters are trying to get the better of rivals in international waters.

A previously unexploited animal resource such as a fish population in a steady state can almost always stand a certain amount of human exploitation; but how much is not predictable by rule of thumb. The first effect of exploitation is usually to deplete the oldest, largest individuals which take longest to produce; and this brings a compensatory homeostatic response from the remaining stock, which reacts to replace the components that are being lost, and restore the equilibrium. Younger fish grow faster than before, taking the food their missing elders would have eaten had they been alive; and, being younger and further below the asymptote to growth, they convert more of it into flesh and less into respiration than their seniors would have done. We shall see later that exploited populations generally attempt to repair the shape of the population, no matter on what age group the perturbing mortality falls, and extra depletion at any level tends to stimulate extra production to compensate for it. It can be predicted that populations of fish being prudently exploited should have a higher productivity, coming from a slightly reduced biomass, than populations not so exploited.

Increasing the predatory pressure leads also to increasing the recruitment rate. Many species of vertebrates, when their populations are steady, put in several years of adolescence before they attain sexual maturity for the first time, but can shorten the period if circumstances require. Very slow breeders such as the albatrosses and petrels (Procellariiformes) lay only one egg when they do breed, and may not make their first attempt before they are 10 years old. The main function of their long adolescence is probably to provide a massive standing reserve of recruits, any required number of which could mature and breed in a matter of months if one of their parent breeding colonies were wiped out by, say, a volcanic eruption. They appear to provide the emergency cover needed to allow these species to resort to very low, though normally sufficient, reproductive rates.

Dr Janet Ollason (personal communication) has run a computer program for the fulmar (*Fulmarus glacialis*), a species of petrel on which a demographic study has been running continuously since 1950 in the Orkney Islands off the north of Scotland (Ollason & Dunnet 1983). Her model was designed to estimate the proportions of prebreeders and breeders in the adult population. It incorporates known parameters on the average age at first breeding (10 years), and the survival rates of fulmars from fledging up to age 10 years (0.90 per annum), and from 10 years onwards (0.97 p.a.). But it has been necessary, in order to estimate the proportions of breeders in the population, to make reasonable guesses about the maximum lifespan, which cannot yet be derived from the ringing data. Three alternative estimates are listed in Table 2.1.

Table 2.1. *Estimation by Dr Ollason of the proportions of prebreeders and breeders in a Scottish population of fulmars*

If maximum life-span (years) is:	Proportions (%) of full-grown birds that are:	
	Prebreeders	Breeders
50	38.5	61.5
60	36.0	64.0
70	34.4	65.6

Not all 'breeders' necessarily return to breed every year (Wynne-Edwards 1939; Ollason & Dunnet 1983: 189), but no estimate can be given of the proportion of these absentees, which constitute an extra category of non-breeders.

Some insects, if homeostasis so directs them, can speed their progress towards maturity by omitting one or two larval instars in their development. Under other conditions additional instars may be interpolated.

These and similar adaptations allow most species to accommodate to unusually high incidental mortality. If the chronic pressure of exploitation imposed on them is progressively increased, however, all must eventually reach a limit when an already reduced population biomass can no longer make good the toll imposed. Under natural conditions, hunting no doubt ceases to be profitable to the predator at some level during the progressive overexploitation of a prey stock, and this may save some heavily persecuted populations from extinction.

Market forces may tend to do the same with species exploited by man. For example, the first modern 'industrial' fishery sprang up off the coasts of the New England states in the 1870's. Astronomical quantities of a plankton-eating herring-like fish, the menhaden (*Brevoortia tyrannus*), were caught in seine-nets, hundreds of thousands being occasionally taken in a single haul. They were not considered palatable as food, but were converted into oil, soap, fish-meal and fertilizer. The small town of Boothbay, Maine, for instance, is credited with having lifted 50 million fish in one year. At the outset the menhaden was the most abundant fish in the region; but within ten years the industry had collapsed. Since then a century has passed without the species either becoming extinct or resuming its former abundance.

The two right whales, *Balaena mysticetus* and *B. glacialis,* have rather similarly failed to recuperate from former exploitation, and they remain rare animals notwithstanding the many years of international protection they have supposedly received. When one appreciates the importance of demographic structuring and genetic balances in the survival of populations (see Chapter 15), one begins to fear that these patterns may have been disrupted, and cannot be restored from the genetic codes of the survivors without long (perhaps millennial) processes of trial and selection. The alternative explanation that the survivors have forfeited their niches to competing species, seems less probable, especially for the whales.

Man has completely exterminated a number of animal species as a result of exploitation pursued to the bitter end; and several other useful

or comestible animals now survive only in domestication. Among the
former were Steller's sea-cow (*Hydrodamalis gigas* = *Rhytina stelleri*, a
sirenian), the great auk (*Pinguinus impennis*) and the passenger pigeon
(*Ectopistes migratorius*). According to the chronicler of the last, Schorger
(1955: 199), 'no other species of bird, to the best of our knowledge, ever
approached the passenger pigeon in numbers'. When Europeans reached
North America there are supposed to have been three to five billion (3 or
5×10^9)of them; but being intensely gregarious, laying only one egg in a
clutch, and being good eating as well, they were exceptionally
vulnerable.

Living plant resources are little if any less prone to damage by overexploitation than animal prey. The most familiar illustration is
overgrazing by livestock, which probably began as soon as man
domesticated oxen, sheep and goats. Agriculture is highly productive
only when farmers are territorially dispersed, with hereditary tenure of
their land; and 'common land' by contrast is notoriously less productive,
simply because no one holder of common grazing rights can prevent
other holders from overstocking the ground. But just as happens when a
fishery is prudently exploited, a moderate amount of grazing by
herbivores can elicit an increased rate of production by the plant sward
(Hutchinson 1971, Vickery 1972).

It may be a necessary condition for maintaining stability in an
ecosystem, that the demands the various kinds of vivivores make on their
staple food resources be self-regulated, at levels the resources can sustain.
It is also apparent that the future productivity of vulnerable resources can
generally be optimized by restricting demands to appropriate limits.

Summary of Chapter 2

1 Animals are consumers of nourishment produced by other organisms. Some have broadly varied diets; more are narrow specialists. Only
plants can convert solar energy into fuels (high-energy chemicals), and
synthesize the amino acids that animals require but cannot make.
Naturally the largest trophic category of animals are the herbivores,
which ingest these materials at first hand. Carnivores and parasites
obtain them indirectly by feeding on other animals, at second, third or *n*th
hand. The biomass that can be nourished necessarily diminishes at each
successive stage. Except for the germ-cells and other propagules, living
matter is eventually either metabolized, excreted or dies; and part of it

forms solid detritus, still capable of use as animal food.

Foods have finite production rates per unit area which commonly vary with the season. Some foods appear as crops (e.g. seeds or insect larvae) which consumers eke out, often until a different crop is ready. A habitat thus has a variable carrying capacity, measured by the number of consumer-days for which its standing crop would last, or the number of consumers its daily produce would satisfy.

Consumer populations commonly perpetuate themselves on the same ground, which passes from parents to progeny.

Not all food resources need conserving to ensure their future productivity. Detritus, both animal and vegetable, is the most important exception. Consuming it in fact helps rather than harms its continued supply. It is again a finite resource, and detritivores can benefit by limiting their density to the level that enables the largest number to complete their life-cycles before it is all gone, or to be carried until new resources appear.

Some quantified relationships between producers and consumers are quoted as illustrations. Fast reproducers like many insects can stand far heavier cropping by predators than slow-breeding vertebrates. Homeostasis occurs in arthropods as well as vertebrates; e.g. aphids normally limit the damage they inflict to levels their food plants can sustain.

2 Man's commercial fisheries are large-scale predations on naturally renewed fish stocks. They show how vulnerable such stocks can be to overfishing, which reduces both the stock and the crop it produces. This was convincingly demonstrated by the two world wars during which fishing was suspended for several years in the North Sea. Each time when it resumed the catch per unit effort was found to have doubled; but the stocks were soon being overfished again. If all parties could agree to cooperate, management could sustain the double (or nearly double) catches indefinitely. Experiments have shown that controlled exploitation of an animal resource may actually stimulate its productivity, through a homeostatic compensation for the extra mortality being imposed. Food-plant resources, which are every bit as liable to overexploitation and its consequences, can perhaps sometimes be similarly stimulated.

It seems probable that the self-limitation of demand by consumers to sustainable levels is a necessary condition for ecosystem stability; and also that controlling consumption will generally help to optimize future food production.

CHAPTER 3

FOOD IS NOT ALWAYS THE LIMITING FACTOR

3.1 Alternative limiting resources

Food is not the only resource that animals cannot provide for themselves, and it is not therefore necessarily the limiting resource. Animals need habitats in which they can obtain protection from avoidable risks, and some of them need access to drinking water. Some birds for example can find sufficient supplies of food in areas that offer no safe or suitable nesting sites. It was shown by experiment in a tract of rather treeless, intensively cultivated farmland north of Aberdeen that the breeding of carrion crows (*Corvus corone*) is limited in some places by a lack of nesting trees. These birds have large and very permanent territories, each centred on a traditional nesting tree. Though many of the territories do contain alternative trees that could be used if necessary, Charles (1972, unpublished thesis) found there were some parts of his study area where suitable trees were separated by up to 1 km; and by erecting single small dead trees, each tied to a fence post, in these treeless sites and placing a pile of nest materials from and old crow's nest in the vicinity (with additional sticks etc.), he induced the establishment of several new territories, in which the owners nested and bred. Rather like the tawny owls at Wytham mentioned earlier, which have similar semi-permanent territories, the crows were found to increase their brood sizes and breeding success when supplied with extra food; but extra food alone would not induce the interpolation of extra territories (Yom-Tov 1974). In this species the populations are normally stratified into territorial pairs and 'flock birds', the latter forming another of these reserves of inhibited adults and adolescents. The flock birds were found to possess a well-defined hierarchy in which the senior birds included betrothed pairs. Shooting the male of a territorial pair was predictably followed by a take-over of the territory by a pair (presumbaly the alpha pair) from the flock (Charles 1972).

An apparently parallel 'experiment', dating from time immemorial, is the attraction of white storks (*Ciconia ciconia*) to a human settlement by erecting a nest platform, as can be witnessed in various north European countries. Yet another is the more familiar practice of erecting nest-boxes for hole-nesting birds, in habitats where natural forest or its old dead trees have been removed (cf. Bruns 1960).

Hippopotamuses need water in which to submerge their bodies during the heat of the day, although they feed as herbivores on land. Like beavers (*Castor* spp.), which obtain food by felling trees, they may also gain some protection from predators by being aquatic. Neither of these mammals can range far from suitable waters although food is not lacking elsewhere; but in habitats where the presence of suitable streams and lakes allows them to enjoy their amphibious life, food normally remains their limiting resource. Because the production rate of beaver food is so slow, they need a family territory of large size, only part of which is under exploitation at any time, the rest being left fallow on very roughly a 20–40-year rotation. Some species of bats (Chiroptera) similarly need caves in which to sleep by day, to hibernate, or bring forth young. Such examples as these could be extended indefinitely.

In the warm-blooded vertebrates (homeotherms), evaporation of water from the body is used to prevent dangerous rises in body temperature. It is a simple consequence of the surface/volume ratio that overheating presents more danger to larger species than to smaller ones; and overheating thus tends to become most acute in large animals in hot environments. The fact that the oxidation of carbohydrates (e.g. sugars) produces equal numbers of molecules of carbon dioxide and water (and the oxidation of fats produces more water than CO_2) means that all organisms routinely produce substantial amounts of metabolic water, enough in fact to satisfy most needs of small homeotherms, so that they drink only rarely if at all, especially in cool environments. For the majority of terrestrial birds and mammals, therefore, access to permanent drinking water is not a common limiting factor, and they can no doubt find what free or extra water they require in most habitats that are capable of supplying them with food.

A fascinating analysis of the social ecology of a water-drinking homeotherm, namely the Aborigine people living as hunter-gatherers in the arid interior of Australia, has been given by Birdsell (1978). Before the white man intruded they had populations structured into territorial tribes and bands. Depending on the mean annual rainfall in their region, a tribe's territory was either centred on a permanent water hole, surrounded by a reserve hunting zone and set aside to provide a refuge in time of drought; or, in still more arid regions, it had a radial corridor or projection leading to a similar water hole, but shared with other contiguous tribes. The tribes had rather uniform populations of about 500 persons, and the average size of tribal territories was strongly and inversely correlated with mean annual rainfall in the same region. The rainfall measurement gave of course an index of the food productivity per unit area, but the dispersion patterns were primarily determined by the distribution of water holes as a limiting resource.

These Aborigines practised population regulation by infanticide, and especially by limiting the recruitment of girls, because 'women are the funnels through which fertility is poured'. During the birth the mother secluded herself with a 'granny woman', respected by the band and representing their communal interests; and she was empowered to decide on seeing the baby whether to let it breathe and live, or whether to hold her hand over its face so that it did not ever enter into independent life—a definition that by custom absolved the granny from murder.

More information from Birdsell about the Aborigines' population structuring and genetic (mating) system can be found in Chapter 16.1. Water holes are of course typically shared by many species of large animals, for example in the Kruger Park, South Africa.

3.2 Alternative limitation by predators

Two of Darwin's other checks—predators and parasitic infections—are capable in some circumstances of directly limiting their victims' numbers in a partly density-dependent manner. The third, 'climate', can inflict direct mortality, but the mortality is catastrophic in its incidence, and density-independent, which prevents it from serving as a reliable population regulator. Predation is the alternative check that appears most often in a regulatory capacity. A good illustration of its potential for limiting numbers of prey emerges from the description of overfishing given in the last section. That showed that sustained predation by commerical fisheries could hold an exploited population in check indefinitely, at a density below the carrying capacity of the food resources on which the fish depend. Thus when predation pressure was withdrawn as a result of the world wars, the biomass of the North Sea fish stocks rose in response.

If my hypothesis is true, however, we should not expect a natural predator to act as prodigally as the fishing industry, but rather to conserve its prey stocks at the level that yields the largest annual crop. That should be produced when the prey is close to the carrying capacity of its own food resources. The predator is not then regulating the prey population, but only removing a dispensable surplus. However, there is an independent source of evidence on the question, arising from man's successful experiments in achieving 'biological control' of agricultural and other pest animals by introducing predators or parasites capable of keeping pest densities down to an innocuous level. Although this type of biological control is difficult to procure and failures have been far more common than successes, there are instances in which such excellent and effective results have been obtained as to present something of a challenge to the hypothesis.

Many of the plant and animal pests of agriculture and forestry are species introduced from abroad which have multiplied so enormously in their new home that they cause serious damage to crop production, although they do not behave in this way in the country of origin. The usual reason is that they have left their own predators behind; and in some cases when these have been sought out and introduced into the pest-ridden neighbourhood, successful and permanent biological control has been obtained.

A classic example, in which the pests were plants, pertains to the prickly-pear cacti (*Opuntia inermis* and other species) in Australia. *O. inermis* came from North America about 1870, and in the arid parts of Queensland especially it spread on an extraordinary scale, until it dominated the vegetation over more than 20 million hectares of range land. The growth became dense enough to be physically impenetrable over about half this area. Prolonged searches were made for some organism that would destroy it, and many possible ones tried in vain. In 1925, at what proved to be the turning point, a borer moth, *Cactoblastis cactorum*, native to Uruguay and northern Argentina, was found to be capable of killing the plants. Its caterpillars feed gregariously, and exclusively on cacti so far as is known. Having two generations a year, their reproduction is very rapid. Thousands may be found on a single plant, consuming large parts of it. The lesions they cause allow access to bacteria and moulds which finish off the demolition. In 1928–30 three billion artificially-reared larvae were distributed and released at points all over the affected region; and although the adult moths seldom disperse beyond the nearest living cactus to lay their eggs, by 1933 it was reckoned that nine-tenths of all the prickly-pears in Queensland were dead (Dodd 1936).

When Dodd wrote, neither the cactus nor the moth had become extinct, though both were scarce. Cactus seeds are spread by birds which eat the 'pears', and some new stands spring up at a distance from their parent plants, sufficient to delay their discovery by the moths long enough to allow the plants to reach maturity and keep the species going. As of 1983, the cacti were still present throughout the area that had been overrun at the time the moth was introduced 50 years ago, and an unstable, cyclic relationship exists between the two. The cacti have been in an increasing phase in the last decade, but are now expected to give way as the moth and ancillary destroyers build up again. Only in coastal areas and on some offshore islands, where control is ineffectual, do some cactus stands exist at densities that exclude agriculture (Dr Hugh Lavery, Queensland National Parks and Wildlife Service, personal communication).

Evidence is discussed in Chapter 19.3 of the beneficial mutual integration of species to form ecological communities, the existence of which has long been intuitively evident to community ecologists. Wilson (1980) has indicated how communities are probably able to accept new species that contribute to mutual productivity, and squeeze out those with a detrimental effect. One of the most ancient coded rules for cooperative animals must be to 'do no murder', presumably because indulging in murder somewhat resembles playing with fire, which can suddenly grow into a conflagration. It is safer to ban it altogether. There are exceptions, as when parents or other adults habitually kill and then often eat their young (see *Animal Dispersion*: 531). It occurs in some spiders, fishes, birds and mammals, and probably in other groups, but presumably only when the young are surplus to the local group or cannot be fed. Generally of course this particular act is inhibited along with all other killing of conspecifics; and in the higher vertebrates the inhibition often seems to be reinforced by the disarming beauty and trustfulness of the defenseless young.

The red grouse, as will appear later (p. 98–9), expel a large surplus from their populations each autumn, but do not actually kill them. Instead the predators conveniently find the outcasts easier to catch than the birds securely established on their territories; and in effect they provide the established grouse with a useful clean-up service. They do not directly control the grouse population. The controlling is normally all done by the grouse themselves through social competition, which selects how many and which individuals are to be exposed to risk before the predation actually takes place. This discovery of what is very likely a common relationship between predators and their prey gives one a new insight into interspecific mutualism, as far as predators are concerned.

What looks like a parallel example is suggested by the famous incident of the Kaibab population of mule deer in Arizona (Leopold 1943; Allee *et al.* 1949: 706), which erupted after their predators the pumas and wolves had been suppressed by shooting; and as a result they devastated their food resources to the point where they starved and died by scores of thousands. In a totally different environment—a marine intertidal community or 'food web'—the systematic removal of the top predators led to a chain reaction in the habitat, at the finish of which an original community of 15 species had become reduced to seven (Paine 1966).

Populations of insect species no doubt tend to be able to endure far heavier predation rates than those that vertebrates can stand because their potentials for recovery are much greater; and in their homeostasis they too must often have become mutually dependent on one another in complex ways. Harking back to the phenomenon of biological control, one should not therefore be unduly surprised if, when one species in a community is singled out and let loose in an alien continent, it consumes its food resource with such apparent abandon as actually to exterminate itself.

Man, incidentally, is not the only species that uses predatory insects to combat pests. There is a remarkable symbiosis between the bull's-horn acacia trees (*Acacia cornigera* and others) in Central America, and the ant *Pseudomyrex ferruginea*, in which the tree provides nesting cavities inside its swollen thorns, and foliar nectaries or food hand-out points of two kinds, for the fierce watch-dog ants, in return for being kept free of herbivorous insects. Janzen (1966) showed experimentally, by excluding the ants from a sample of young acacia suckers, and comparing their growth and survival with those of normal suckers on which the ants were present, how great a benefit the ants confer. In fact he found the symbiosis was obligatory on both sides, and neither tree nor ant can exist without the other.

3.3 Alternative limitation by parasites and disease

In the long term, parasite populations are no doubt most secure when they live in hosts that have well regulated populations and dwell in stable habitats. Individual hosts provide them with discrete habitat units, each having a certain carrying capacity. There are, broadly speaking, two sustainable options or strategies they can develop for exploiting a host population. Either they can load up an expendable proportion of the hosts to levels that eventually kill the infected ones (or in some cases emasculate them); or alternatively the infection can be broadcast widely through the stock of hosts, while the damage the parasites inflict is kept down to what an individual host can tolerate and still fulfil its own life cycle. It may actually be possible to combine both strategies, by permitting the host a lead a fairly normal life and at the same time limiting the proportion of hosts parasitized.

In the first option the parasites resemble predators in killing the animal on which they feed; and as with some predators, it may often be possible for several parasites to share in exploiting one individual host. However, parasites normally defer killing the host until their own parasitic stage is complete and they have no further use for it. This is the strategy followed by, among others, the 'parasitoid' insects. In their life cycles a free-living adult female hunts for and finds a host, always another arthropod and most often a lepidopteran caterpillar. She lays one or more eggs in or on it, and her offspring go through their larval development while selectively consuming the living tissues inside the host's body. There is sufficient food to allow them to complete their growth, after which they generally emerge and spin their cocoons beside the dead shell of the victim's body.

This is a very efficient and successful type of carnivorous parasitism, which does not involve much modification of the parasite's anatomy and physiology, compared with what is required in parasitic flatworms for example. Arthropod host species tend to have high fecundities, as previously stated, and it is usually safe to parasitize a substantial proportion of them without diminishing the prospect of finding future supplies of hosts. Furthermore the adult female insect can identify suitable hosts and lay her eggs in them with great economy, so that she requires to have no abnormal and extravagant genital system. It seems predictable that adult parasitoids can regulate homeostatically the proportion of hosts that become 'infected', and ensure that the survival of the host populations is not under threat. The hosts in their turn are likely to have their own demographic programs for adjusting their densities to the capacity of their habitats, and for compensating if the parasites by accident impose excessive mortality.

The second kind of infection strategy, in which the burden on the individual host is moderated to enable it to complete its own life cycle, is typical of the parasitic flatworms (among others), which are the subject of Chapter 18. The majority of these worms have two parasitic stages in their lives, spent in different species of hosts, with brief free-living stages in between. In the very numerous digenean trematodes, for example, the egg usually hatches outside the host in an aquatic environment, and releases a minute mobile larva with a short life. It has to find a member of the first host species, perhaps a water snail, in order to survive, or die in the attempt. If successful it gives rise asexually to a numerous clone of larger larvae in the snail, and these when liberated have similarly to find a final host, usually a vertebrate but not necessarily an aquatic one.

Once inside the final host they grow into adults and become almost wholly committed to converting nutriment derived from the host into prodigious numbers of small shelled eggs. The adults are normally hermaphrodite, and their eggs pass into the host's gut and so to the outside world. So slender are an egg's chances of completing such a complex life history that it may need millions of eggs to ensure that one adult worm replaces itself on average by one adult daughter in the next generation.

An alternative common route from the first to the second host is for the latter to be a predator which feeds on the former. When this type of life history evolves, and the first host is, say, a nimble fish rather than a sluggish mollusc, the parasites may place a sufficient burden on the fish to slow its movement and make it easier for the predator to catch. Even when molluscs are the hosts there are species of parasite known that contrive to make them visually conspicuous and catch the predator's eye, to stimulate its curiosity,

The saturation tactics of helminth reproduction are probably second to none in the animal kingdom, and appear to provide an unpromising start for a program of homeostasis. Nevertheless the evidence reviewed in Chapter 18 indicates that there is an auto-regulation not only of the number or biomass present in any one host, but also of the egg output itself. If the success rate of the larval stages is sufficiently high, it appears likely that a signal does get through to the adult population, buried though they are in the host's dark interior, and devoid of sensory receptors other than physico-chemical and tactile ones. Its message may be that egg-production should be reduced or stopped altogether in order to ease up on the hard-pressed hosts.

Microbial parasites are not infrequently lethal to their hosts, or at least to some of them. They usually multiply prolifically soon after infecting a naive host, and then quickly achieve transmission to other hosts, before dying out. Hosts can also be reciprocally lethal to invading parasites (both animal and microbial ones). They have probably all evolved some kind of defensive 'immune system', which in vertebrates takes some days to bring into action after a first infection. The immunity reaction is specific to the type of invader, and once activated it can be put into storage and rapidly retrieved if it becomes necessary to repel a subsequent invasion by the same kind of parasite.

If parasites kill their hosts too quickly they may fail to get transmitted to others in the time available; or conversely they may spread too fast and decimate the host stock. In either event natural selection will turn against them, and if any of the parasite population survive, it will be the less virulent individuals. The same process will automatically continue until, after many generations, the host is able to tolerate the parasite and a stable, even though burdensome, relationship is established. Parasitologists have long assumed that this is the normal mature relationship to be expected between parasite and host; but their conclusion has recently come under criticism for its implicit dependence on the principle of selection 'for the survival of the species' (cf. May & Anderson 1983). This objection does not arise in the context of group selection (see Chapter 18).

Parasites by definition live at the host's expense. If in the course of further evolution the balance of advantage moves into a mutual partnership of advantage to both, it ceases to meet the strict criterion of parasitism. Such symbioses have evolved, for instance, with the bacterial floras found in the guts of many animals, where the microbes dissolve cellulose, and synthesize enzymes, vitamins and other useful proteins, in return for the nutritive habitat the host provides.

Being nocuous, parasites are persistently sought out by the immune response their hosts mount against them, and they can only survive by contriving to hide from its effects. Many viruses, for instance, mutate at intervals, with the result that they cease to be recognizable to the hosts' defenses, and render the hosts' previously acquired immunity obsolete.

Virulent parasites might be expected to cause density-dependent mortality among their hosts, as Darwin suggested, because of the greater ease of transmitting infections which denser crowding gives. But such parasites cannot function satisfactorily as external agencies for controlling their host populations. Their dispersal is haphazard and they cannot be relied on to appear when required; and moreover their virulence and ability to kill are all the time being whittled down by natural selection.

Like predators, the parasites generally stand to lose by reducing the stock of their hosts, and any excessive tendency they showed to do this would be counter-selected.

It seems safe to conclude therefore that parasites and the disease they bring are not only a burden to their hosts, and best eliminated from the community if possible, but are also too chancy and incompetent as agents on which to rely for population limitation.

3.4 Climate has a destabilizing effect

The last of Darwin's checks was the climate, taken to mean the physical elements in the ambient environment which vary through time at any place, and produce a succession of conditions capable of affecting the lives of organisms. Such elements are found in both terrestrial and aquatic habitats. Their variability is in part regular and predictable, as between night and day or summer and winter, and in part irregular, as between one week and another or one year and another. They can have large effects on metabolism and performance; for instance on the availablity and internal transport of oxygen and other substances; on water loss, heat loss, energy consumption, reproductive success, and capability of locomotion; and they can affect the rate of food production and the carrying capacity of habitats. A considerable part of the total adaptability that individual organisms possess serves for accommodating to climatic change, in the broadest sense. But sudden changes and extremes may nevertheless contribute substantially to the 'uncontrollable mortality' (see next section) that populations suffer. Most of the seasonal and year-to-year variation that occurs in population densities, though it may ultimately be attributable to climate, nevertheless originates spontaneously in the organisms themselves, as they adapt their numbers to the environmental changes they experience.

Climate imposes geographical limits on the ranges of some species, particularly those that pioneer into the barely habitable fringes of deserts and polar or alpine zones. Populations living in these environments often exist more or less as perpetual colonists, advancing into no-man's-land when conditions ameliorate, only to be exterminated when harsher climes return. 'Their struggle for life is all against the elements', in Darwin's words, and their temporary extinctions can be wholly and directly due to climate. Catastrophic impacts are typical of climatic mortality wherever it occurs. The mortality is however density-independent, since the proportions of those that get killed or escape may be little affected by whether the population density is high or low; and there seems no justification for thinking that direct climatic effects can ever assist in maintaining the stability on balance of animal populations.

This view has not invariably been accepted by ecologists. There is a classic example where the climate was first identified as a population regulator by the original researchers, who were only to find their conclusion reversed by later workers re-analysing the same data. The subject of the study was the minute Australian insect *Thrips imaginis*. It lives in flowers of many kinds but has been called the apple-blossom thrips because in 1931 it appeared as a plague over the southeastern quarter of the continent, and virtually destroyed the apple flowers. The apple crop failed in New South Wales, Victoria and South Australia; and this led to an investigation of its population dynamics by Andrewartha and his colleagues, which extended over many years (Davidson & Andrewartha 1948 a&b). Looking back, Andrewartha (1961) guessed that at the height of the 1931 outbreak there might have been 20 million thrips per acre of apple orchard, with similar numbers in gardens, fields and other habitats. In normal years they were of course much less in evidence.

The vicinity of Adelaide where the research was centred has a Mediterranean climate, with a moist mild winter in which much of the vegetative growth of plants takes place. Flowering comes to a peak in spring and early summer; and in high summer the weather gets hot and dry, bringing vegetative processes nearly to a halt. The female thrips needs pollen to eat before she can lay eggs, and the population builds up to a high density during the flower season, culminating in November, after which it declines. It was the factors determining the peak height that interested the investigators most. It varied much from year to year; and in the flowering seasons that began earlier there was a longer run up to the climax, and the peak tended to climb higher. But analysis showed that the variable most closely correlated with the subsequent peak height was the relative warmth of the winter, presumably because it affected the profusion of bloom afterwards produced, and thus the peak size of the food supply.

Data on thrips numbers were actually obtained by counting a daily sample from 20 cultivated rose flowers, the species being chosen because roses bloomed all the year round. After the November peak there was a rapid fall in numbers each year; but the insects remained active and reproduced all the year round in favoured spots. Over most of the wider habitat invaded during the favourable months of early summer, the thrips died out in the dry season. From then on, they became restricted to small protected refuges, from which they colonized the extended habitats once more the following spring. In a normal year the number of flowers available in spring increased so fast as to become superabundant, with the thrips population unable to keep pace; and the insects appeared not to catch up with the carrying capacity until the decline in blooms began with the onset of drought.

Andrewartha saw the peak population sharply cut off by the drought. The spring increase was to him a race against time, dramatically ended by 'the weather'. But more recent workers who re-analysed his 14 years of data have shown that there were density-dependent forces at work at all times (Smith 1961; Fretwell 1972). Numbers were regulated in the refuges in which the thrips survived the off-season, and they did not increase to the point of destroying the available flowers. Even during the population growth in spring there were signs of an inverse correlation between increase rate and density.

What the control mechanisms were or are has not been established, nor have the actual carrying capacities been experimentally determined at any season or in any habitat. The likelihood seems to be that there is a coded, self-operating program in the insects' central nervous systems, modulated by crowding and nutrition and comparable with that of the sycamore aphid (p. 23).

Many short-generation insects have rather similar annual cycles, except that it is commoner to include a diapause (hibernation or aestivation) that carries the dormant insect through the least favourable season, as an egg, larva/pupa, or adult. Fretwell (1972: 53), in one of his 'Asides', recalled the irreconcilable conflict of minds that occurred in the 1950's, between the vertebrate demographers headed by Lack (1954), who championed density-dependent regulation, and the insect ecologists like Andrewartha and Birch (1954) who clung with equal fervour to density-independent 'control'. The reason for their opposing points of view, Fretwell suspected, may have been that the former's attention was aesthetically directed towards song-birds on their breeding territories, and at their annual minimum density; whereas the economically-minded entomologists had their eyes fixed on the density peaks. Density-dependent effects were far more conspicuous to the former than the latter.

3.5 The equation of recruitment and loss in a population

As we have just seen, the carrying capacities of habitats in a seasonal environment may vary very greatly from the best season of the year to the worst (when it may be zero). In general, very short-lived animals tend towards prolific reproduction in the former, sometimes with telescoped generations (e.g. among freshwater crustaceans living in temporary ponds), and they can thus contrive to utilize their ephemeral resources to somewhere near their full capacity. Medium-lifespan vertebrates such as small birds produce one generation annually, which may temporarily double or treble their baseline density in the season of plenty. Migration may also add substantially to consumer concentration when and where resources are abundant; nevertheless some food resources may remain under-exploited during part of the year. Long-lived non-migratory vertebrates seldom depart from their baseline by as much as 1.5 times, and often by less then 1.25, with at least half the population remaining immature throughout the breeding season. A large proportion of their summer or rainy season resources consequently go uneaten.

The existence of this spread or spectrum of resource utilization strategies has given rise to the notion that natural selection favours fast reproduction in strongly seasonal habitats and slow reproduction in those that remain nearly constant round the year. The extremes have been polarized under the designations of r-selection and K-selection, where r is the rate of population growth (natality minus mortality),and K is the carrying capacity of the habitat. Selection is expected to increase r where fast multiplication has an advantage in attempting to 'track' fast-growing resources. In species that do not at any time fall far below the limit set by K, in contrast, individuals will supposedly be disadvantaged by having large families, and their reproductive rate will be optimized at a level sufficient only for maintaining a steady population (MacArthur & Wilson 1967: 149).

In practice, however, animals representing the whole strategic range are normally found in all climates; and the most prolific animals of all, the helminth parasites, live in one of the most constant habitat types known—the interior of warm-blooded vertebrates. Some of the animals that live in tropical rain-forests and other unvarying habitats maintain steady populations by spreading their reproduction out evenly all through the year. Body-size and length of life-span (which are loosely correlated) appear to have much more to do with determining what is practicable in the selective manipulation of generation time, than seasonality of resources has.

If a population remains steady, or returns to the same density at the same season year after year, it is clear that the annual rate at which existing members are lost must be equal to the annual rate at which new ones are gained ($r_{annual} = 0$). The majority of animal populations normally approximate more or less roughly to this steady state. However, births and deaths are quite unconnected events; and if their rates turn out to be the same they cannot both be independent variables. One must be matching the other; and it is not immediately obvious which is the independent and which the dependent variable.

Darwinists have tended on logical grounds to assume that *recruitment* is maximized by natural selection, and must therefore be the independent variable. Consequently, losses must automatically, somehow or other, remove whatever excess is produced. We have seen already that this assumption is untenable, in terms of practical ecology, and inconsistent with numerous proven facts showing that some animals at least can program their own reproductive rate for homeostatic purposes.

In *Animal Dispersion* (p. 486) I broke down the two sides of the equation into their component parts, and pointed out that that four out of the five terms are capable of regulation by homeostatic programs, as follows.

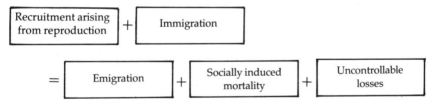

Recruitment arising from reproduction can be controlled internally in various ways, chiefly by regulating the quota of adults per unit area that qualify to breed, or by varying their individual fertility rates or the standard of parenthood they provide. Immigration from outside can be resisted, or accepted, by those already in occupation of a particular habitat; and the same residents can force some of their number to emigrate, if necessary, by stepping up the intensity of social competition. Social competition and exclusion can lead to socially induced mortality among rejected or low-ranking members of the group. Only 'uncontrollable losses', due to mortality caused by accidents and outside agents, are incapable of mitigation from within; and they are thus the only completely independent demographic variable, which the other terms in the equation must, if possible, combine their efforts to match.

It is important to notice that homeostatic controls are able to influence either or both sides of the equation; that is, to produce falls as well as rises in population density.

Summary of Chapter 3

1 Food is not always the ultimate factor to which consumer density is adjusted. Some habitats well stocked with food are defective in other necessities, e.g. in possessing both feeding and safe nesting sites for birds, or acceptable combinations of land and water for amphibious animals.

2 Predators also may keep some prey species below the carrying capacity of their food supply for long periods, although perhaps seldom permanently at any locality, or everywhere at any time. Homeostatic adaptations can probably seldom be dispensed with as a safeguard when predation pressure happens to relax.

3 Parasites, in their own interests, tend to inflict mortality on their hosts either sporadically, or proportionally, on a scale for which both they and their hosts can readily compensate. They are probably seldom capable of holding down their host population densities on a dependable, permanent basis.

4 Climate cannot regulate populations although its irregularities may perturb or exterminate them.

5 The concept that species with high reproductive rates are favoured in strongly seasonal climates, and those with low rates in equable ones, because of r or K selection, cannot be supported. Reproductive rates are necessarily and inversely correlated with mean lifetimes, and more loosely with body-size. Species with both high and low rates are present in most environments; and, for example, many species of the most fecund type of animals, the parasitic flatworms, are anomalously found in one of the most equable of all habitats, inside the bodies of homeothermic vertebrates.

An equation of recruitment and loss in a typical vertebrate population shows an income arising from reproduction plus immigration, which is balanced by losses from emigration, socially-induced mortality, and uncontrollable mortality. The last term is the only independent variable. The other four can all be homeostatically manipulated, and can normally compensate for the last one, and hold or restore the balance.

CHAPTER 4
THE THREE TROPHIC INDUSTRIES

4.1 Their interdependence

The existence of three biotic 'industries', differing in their sources of nutrition, was mentioned on p. 14. They are (1) the primary producers, consisting of green plants and green microbes, and nourished by inorganic nutrients and photosynthesis; (2) the decomposers, consisting of microbes and animals that break down dead organic matter and waste; and (3) the vivivores, which are animals and microbes that feed on the living tissues and body-fluids of other organisms.

The producers and decomposers were probably the first to become differentiated from one another in the early stages of life on earth, not long after growth, reproduction and mortality became programmed processes, and nutrient cycling was evolved. It is conceivable that a single type of organism could have existed previously which was simultaneously phototrophic and saprophytic and completely self-sufficient. That is to say it would have been able to photosynthesize, to ingest and digest the dead bodies of its own species, and recycle its own excreted nutrients. Some of the present-day euglenoid Protozoa appear to approach this factotum condition. But specialization no doubt soon led to the divergence of phototrophic producers and saprophytic consumers, and the first herbivores would have emerged not far behind. All three are well represented among our contemporary one-celled organisms.

Among the multicellular organisms one can broadly identify the primary producers with the plants, and the vivivores with the majority of animals. The decomposers are a mixed bag. They include animals such as earthworms, millipedes and beetles that perform the useful task of grinding up large tough structures in detritus; and these animals are backed by a huge army of fungi and bacteria. The decomposers as an integrated industry have established a complex mutual-benefit relationship with the other two industries. They provide a comprehensive recycling service of the elements picked up from the environment and used by green plants, such as carbon, hydrogen, oxygen, nitrogen, phosphorus and sulphur.

(The carbon cycle involves all three industries, beginning with the plants which take in atmospheric carbon dioxide and reduce it during photosynthesis, thereby releasing free oxygen into the atmosphere. The subsequent oxidation of organic carbon occurs through respiration and energy release in organisms belonging to all the industries). In all but a few restricted ecosystems organic wastes are disposed of at the same annual rate as they are produced. This incidentally implies that organisms are debarred from synthesizing substances that none of the decomposers can liquidate. Any one that did evolve impossibly refractory or non-detoxifiable materials might eventually be smothered in its own litter.

It is important to realize the vastness and significance of the mutualism that has evolved between the plants and the decomposers, neither of which can get along without the other. Some animals, as I have just said, contribute usefully to the cycling process, but most of the conspicuous animals are vivivores, members of the third industry, which live at the expense of the other two and contribute little but detritus in return.

The feeding habits of members of the animal kingdom are obviously diversified. While some are detritivores and many are herbivores, there are also carnivores which, in the widest sense, can be identified with species that kill to eat, plus others that feed on the tissues and body fluids of their animal prey. The latter comprise blood-sucking and ecto- and endoparasitic animals. Herbivores too sometimes kill whole plants when they eat; for example, elephants pull down trees, deer and rodents eat seedlings, and countless birds, small mammals and insects eat seeds. Some animals including many insects are internal parasites in plant hosts.

It is often useful to think of consumers as living at different trophic levels, with herbivores or detritivores as a first level, lower carnivores or parasites as a second, higher carnivores and hyperparasites as a third, and so on. It has previously been noted that though many animal species have extremely specialized diets, there are at the other extreme a more limited number of others that are relatively omnivorous, and may for example eat carrion one day and kill live prey the next, or alternate between insects and fruits.

4.2 What controls the numbers in each industry?

The majority of consumers of living foods would be capable of diminishing the rate of food production if they were to overexploit their resources, in the manner already discussed in Chapter 2.2; and in this respect they stand in sharp contrast to organisms in the producer and decomposer industries, neither of which are generally capable of diminishing the rate of accumulation of resource supplies. This fundamental difference was first brought into the open in an illuminating paper by Hairston, Smith and Slobodkin (1960; but see also Slobodkin et al. 1967). These authors considered the means whereby the populations of species that belong to each of the three industries are controlled in numbers, and they pointed out for a start that the decomposer industry as a whole must be food-limited, because of the fact that dead organic matter does not accumulate in the biosphere; or if it does the phenomenon is exceptional, as the relative scarcity of fossil fuels shows. When the supply of dead organic matter increases, the normal consequence is that the decomposers increase to cope with it. It is especially relevant in our present context that consuming every morsel of litter in no way diminishes the prospect of future litter supplies; on the contrary, decomposition of litter aids primary production. Decomposers do of course often face seasonal fluctuations in their supplies, and where this happens they tend to evolve seasonal life cycles with some form of diapause during the unfavourable part of the year; or at those times they turn to consuming live food. The three authors incidentally point out that some decomposers may of course have their populations limited by predators or by competition and exclusion, in ways that I have already discussed in Chapters 2 and 3. If that results in their being too few in numbers to deal with all the waste available one must conclude that there are normally alternative decomposer species, ready to consume what they have left unprocessed. It is the industry as a whole that is food-limited.

The authors apply the same logic to the global limitation of all organisms taken together, in saying that 'if virtually all of the energy fixed in photosynthesis does indeed flow through the biosphere, it must follow that all organisms taken together are limited by the amount of energy fixed' (1960: 421).

Their next deduction is particularly significant, namely that the producers as a whole must be limited by their own exhaustion of a resource. That could mean, for example, running out of living space, or water, or the chemical nutrient in shortest supply. The mass destruction of vegetation by meteorological catastrophes can be ruled out as a general limiting factor because, once more, it is an exceptional phenomenon. Similarly the authors rule out the limitation of plant growth by herbivores, on the grounds that 'cases of obvious depletion of green plants by herbivores are exceptions to the general picture, in which plants are abundant and largely intact'. The fact that temporary exceptions do occur 'when herbivores are protected by man or by natural events . . . indicates that the herbivores are able to deplete the vegetation whenever they become numerous enough, as in the cases of the Kaibab deer herd, rodent plagues, and many insect outbreaks. It therefore follows that the usual condition is for populations of herbivores *not* to be limited by their food supply'.

What limits herbivore populations if it is not food? The authors again dismiss the vagaries of the weather as a general factor (a conclusion I have already endorsed), and they turn to predation (including mortality due to parasitoid insects), as the only general method of limitation remaining. They admit there is no positive proof that herbivores are controlled by carnivores, but point to numerous instances where the removal of predators by man has allowed herbivores to increase and seriously deplete their food plants.

If indeed predators and parasites were controlling the herbivores, that would mean that the populations of predators and parasites are food-limited. The authors do not overlook the fact that some carnivores are regulated by territorial systems which provide an internal check; but taking all carnivores together, their argument goes, the carnivore guild as a whole must be food-limited; for if it were not so, the herbivores would multiply to the point of depleting the vegetation, and that only rarely and transiently happens.

Thus the conclusions of Hairston *et al.* are, in brief, that decomposers are food-limited, herbivores are limited by serving as food for carnivores, and carnivores are food-limited.

Important as their arguments were for spot-lighting the essential problems, they came in for criticism. One critic (Murdoch 1966) pointed out that the arguments for and against limitation by food appeared to have been used in different ways with reference to herbivores and carnivores. If it is claimed that herbivores cannot be food-limited because the green plants they eat are not depleted, can it not equally well be claimed that carnivores are not food-limited either, because they seldom deplete their stocks of prey? The reply of Slobodkin *et al.* (1967) did little to dispose of this criticism. They pointed out that whereas the standing stock of plant protoplasm vastly exceeds the biomass of herbivores in terrestrial ecosystems, the biomass of herbivores exceeds that of carnivores by a very much smaller multiplier; and 'even allowing for the superior quality of animal protoplasm . . . as a food, it is evident that the food per predator is a much smaller number than the food per herbivore' (1967: 118). Why then, one may ask, can the predators not deplete their food resources even more easily than the herbivores deplete theirs?

The facts that are available to us fail to support the view that there is a fundamental difference between herbivores and carnivores in the logistic relationships that obtain between them and their respective food species. Live animal food may be more evasive than live vegetable food, but plants are much more often protected by being unpalatable or completely inedible. It is estimated that some 90–95% of the global production of terrestrial plants goes straight to the detritus feeders and only 5–10% is actually eaten by herbivores (Ricklefs 1973: 663). This can be compared with an estimate made for the Serengeti region in Tanzania, where on 25 000 km², some 40 kilotons of mammals (mostly herbivores) perish in an average year from one cause or another, this being equivalent to the mean annual production. Thirty-six per cent by weight (14.3 kt) are killed and largely consumed by carnivores, and the remaining 64% (nearly 26 kt) die from 'other causes' and are consumed as carrion by decomposers, chiefly vultures, fly-larvae and bacteria (Houston 1979 and personal communication).

In these comparative estimates at least, the toll that mammalian carnivores take from their largely herbivorous prey (36% of the annual production) turns out to be not vastly greater than the 5–10% that herbivores take from plants. At these overall consumption rates there must be many instances where the consumers, whether carnivore or herbivore, cannot be the key factor that limits the abundance of their respective food producers.

4.3 Homeostasis benefits animals whatever their trophic industry

The possibility of prudential limitation of populations by the consumers themselves was largely discounted in Hairston, Smith and Slobodkin's discussions and arguments. What emerges now is that, whereas the autotrophs (green plants) and necrophages (decomposers) are in no danger of harming their prospects of future sustentation, even though their populations grow big enough to use up all of their limiting resource that is currently available, the vivivores are on a different footing. Their industry depends for its future on husbanding vulnerable food stocks.

We shall appreciate more clearly later (Chapter 19) that if their demands were to exceed the sustainable yield, food production would fall below the high level at which their husbandry sustains it, and under extreme pressure it might even become insufficient to support a consumer population at all. There are just two ways of avoiding that catastrophe: either the vivivores' numbers must be held down below the carrying capacity of their food resources by predation or disease, or they must exercise homeostatic controls for themselves to produce the same effect.

A feature that has tended to obscure the second alternative and prevent the wider recognition of it by ecologists as a normal means of population control, is that territorial systems and hierarchies do not suggest any self-evident *raison d'être* to the observer, for the simple reason that their purpose is a negative one, to prevent and not to cause. The *ultimate* limiting factor is still food, but the function of the proximate factor (homeostasis) is, in essence, *to prevent food from appearing to be the limiting factor.* In some and probably many species—herbivores and carnivores alike—food territories change in size from year to year according to the quantity and quality of the current food crop. In high-yield years and habitats, the consumers' territories are smaller and their population densities higher, and in low-yield years and habitats the opposite is true. As a result, consumers are seldom obviously in want. Slobodkin *et al.* (1967: 110), in their general definition of limiting resources, refer to the classic paper of F. F. Blackman (1905), and epitomize Blackman's axiom by saying that 'if the supply of the limiting resource is increased, there will be an increase in the limited phenomenon'. However, though Blackman was right enough, considering the state of knowledge when he wrote, he had little inkling of the cybernetic systems that have become familiar since in the physiology of organisms, involving the mediation of internal, proximate control mechanisms which shape the individual's responses to external, ultimate events in the environment.

Thus in a territorial population, as will be shown in Chapters 8–10, densities can respond in exactly the manner described. When the food supply of the red grouse was experimentally improved their density went up; but grouse are never food-limited if they can help it, in the sense that decomposers show themselves to be food-limited by habitually cleaning up the entire supply. The grouse populations are limited by their own precautionary responses, based on the present state of the heather as food; and that means they will not tolerate densities that would exceed the carrying capacity of the moor and overtax the heather. There were two successive years, 1958 and 1959, in which nature performed the opposite experiment. The heather suffered abnormal winter frost damage, and the ravages or each winter were not made good in the succeeding summers. In the autumn of 1957, before it happened, the grouse on a study area of 56 ha took up territories of 1.33 ha in average size. As the heather productivity was successively damaged and reduced, the mean territory size rose to 2.8 ha in the autumn of 1958 and 3.5 ha in that of 1959; and the corresponding number of territorial cocks fell from 42 to 20, and then to 16.

Food production for detritivores follows very much the same patterns as it does for vivivores. It will always benefit them to keep their density within the available carrying capacity, whether they are vultures, earthworms, or burying-beetles. The latter (e.g. *Necrophorus*), for instance, collaborate to bury the corpse of a small bird or mammal safely underground before their competitors the blowflies (e.g. *Calliphora*) arrive. But the brood of beetle larvae produced, which are the progeny of a single female, must be limited in number so that they will not run out of food before they are ready to pupate (p. 324). Such animals as these show the fine control they can exercise in tailoring their demands to the amount of food available, by being able to dispose of the whole of it without any obvious mismatch in their own population dynamics.

The main difference for detritivores is that their homeostatic programs do not need to make allowance for conserving the food producers. Most interestingly in this context, it has been found that some earthworms do actually produce plant growth hormones and incorporate them into the soil, thus increasing the eventual crop of litter on which the earthworms feed (cf. Syers & Springett 1983). We shall see other examples of mutualisms such as this which result in exchanges or the sharing of benefits between species in a community. In many of them, animals make substantial amends for the burdens they place on the plants.

Summary of Chapter 4

1 Living organisms can be grouped according to their sources of nourishment into three 'industries': (1) the primary producers, engaged in photosynthesis, i.e. green plants and algae; (2) the decomposers, i.e. animals and microbes that consume dead organic matter; and (3) the vivivores, i.e. animals and microbes consuming or destroying whole organisms or their living parts and tissues. The first two are mutually indispensable because, between them, they operate the nutrient cycles. Animal species are all ultimately dependent on the other two; most of them are vivivores, and many of these contribute nothing in return except detritus. A minority are detritivores themselves and assist the decomposers. A still smaller number can live on both live and dead foods.

2   Since first being aroused by Hairston, Smith and Slobodkin (1960), interest has been attached to the possible factors that control the global numbers in each industry. Taking detritus first, it can be assumed that because the detritivores normally consume it all so that it does not accumulate indefinitely, they must be limited by the amount that becomes available. Second, the green plants as a whole are seen to thrive, and suffer only minor or occasional damage from consumers; clearly therefore they are not being limited by the herbivores. Instead their limiting factor must also be some resource, e.g. living space, water, or the nutrient in shortest supply. Last, among the vivivores, the herbivores only exceptionally endanger or exterminate their food plants; and the carnivores and parasites equally rarely do the same to their prey and host populations. Thus as long as they are occupying their normal habitats none of the vivivores appear to be limited by the amount of food, and few of them by any other type of resource. For the great majority this leaves us with no convincing alternative to set against the hypothesis that their numbers are self-regulated. In short, detritivores taken as a whole appear to be food-limited, plants to be variously resource-limited, and vivivores to be self-limited. All organisms taken together are limited by the amount of energy fixed by the primary producers.

3   The hypothesis that animals normally regulate their own numbers in relation to their current food resources, is explored in subsequent chapters. It postulates that, as far as possible, vivivores adjust their densities to the changing levels of food available, so that their resources do not become chronically overtaxed; and, especially in hard times, so that the individual residents remaining in a habitat obtain enough to carry them through until conditions improve.

It postulates that detrivores similarly adjust their densities so as to allow individuals a prospect of obtaining a share of the food available sufficient to enable them to complete their life cycles or pre-diapause stages. Thus in both the consumer industries, member species can benefit from population homeostasis.

CHAPTER 5
EXPLOITING SEEDS, FRUIT, AND NECTAR

5.1 Seed-eating

Slobodkin, Smith and Hairston (1967: 115) divide the herbivores into two major groups, according to whether they feed on the producer itself, or on its products, such as seeds, nectar and pollen. They say 'many seed-eaters (ants, birds, mice, for example) are also carnivores, especially during the breeding season when demand on the general food supply is greatest'. They had previously come to the conclusion that carnivores must be food-limited, whereas herbivores were limited by the carnivores that prey on them; and they now considered that seed-eaters, like carni-vores, often reveal evidence of being similarly food-limited, 'or else show an evolutionary response (e.g. territoriality) to what must have been intraspecific competition for resources in the past'. Whether or not seed-eaters turn alternatively to animal foods, they 'seem not to harm the producers' and to 'have little direct effect on the vegetation'.

These comments were intended to show that consumers of plant products such as seeds could be exempted from their rule that herbivores are kept in check by predation; and that the rule in fact applied only to those that destroy 'living vegetation'. Thus the clear distinction between food-limited and predator-limited consumers was retained, and no species were left with a foot in each camp.

However, it is no easier to accept that seed-eaters are thus auto-matically immune to the risk of overexploiting their food plants, than it is that carnivores are immune from overexploiting their prey. In fact D. H. Janzen has rebutted the concept that seeds are plant products in the same sense as nectar and fruit are. Seeds are functionally identical to the eggs of animals. He studied for many years (Janzen 1969) the relationship between the seeds of leguminous trees in tropical America and the Bruchidae, a large family of small beetles whose larvae specialize in eating legume seeds. *Acacia cornigera* is a swollen-thorn acacia, one of the species that are policed by fierce ants (see p. 39); a representative tree in lowland Mexico produces some 60 000 seeds in its lifetime, 'and it is doubtful if more than 600 escape from the parent tree' (Janzen 1969: 11). The survivors avoid the bruchid attack largely because, before the beetles have got at them, birds (*Saltator* grosbeaks and the brown jay *Psilorhinus morio*) split some of the pods open to get at the pulp inside, incidentally swallowing the seeds which they later excrete somewhere else, where bruchids are scarcer than they are round the parent tree.

Janzen showed that the leguminous trees have two main methods of protecting enough seeds to perpetuate themselves. They either resort to chemical defence and make the seeds toxic or unpalatable, in which case they produce fewer, heavier seeds; or they rely on predator satiation, with more abundant, smaller seeds. The heavier seeds give rise to sturdier seedlings; but the strategy of predator satiation consists of producing a huge crop irregularly, unpredictably, and preferably very locally, which floods the capacity of the existing specialist predators to consume it, and allows some survivors to germinate.

It is a strategy that has evolved many times and is found far beyond the confines of the Leguminosae. In southeast Asia, for example, the forests contain many trees in the family Dipterocarpaceae, which, as the name implies, bear two-winged seeds. It has been shown (McClure 1966; Janzen 1974) that in some species of the large genus *Shorea*, seed crops ('illipe nuts') are produced at intervals of roughly five years, and in a given locality all the individuals of the same species, and sometimes of several related species, simultaneously flower and set seed. From a platform 43 m up in the forest canopy (on which I spent a day with him in 1962), McClure observed that, of 60 trees belonging to 35 species, kept under surveillance from 1960 to '65, there was a particular set of six that all flowered and seeded together in 1963, but at no other time; three were *Shorea curtisii*, and the others were single specimens of three other *Shorea* species, all being the only ones of their kind visible from the platform. Five of them next matured seeds together in 1968 (Medway 1972). The specialist seed-predators presumably have great difficulty in latching on to the particular trigger mechanism that activates the trees, and thus in being ready in force when the big event occurs. The same adaptive purpose may also account in part for the irregular seed production found in many temperate-zone trees and shrubs (see Janzen 1971), although the irregularity there extends to species bearing edible fruits for which, as will be seen shortly, the plant-animal relationship is just the opposite and predator satiation a misfortune.

Predator satiation is a suitable strategy for a tree because the tree has a long enough life and a big enough storage capacity to make infrequent and prolific reproduction possible. But it first gained recognition among evolutionists as an explanation for the remarkable synchronized appearances of the periodical cicadas (*Magicicada* spp.) in North America, at intervals of 17 or 13 years (Lloyd & Dybas 1966a, b). These large homopteran insects pass their exceptionally long nymphal lives in the soil, and because they have a densely patchy distribution, millions may appear in a single night when the emergence finally takes place. Again more than one species are associated, and each species tends to have separate 13-year and 17-year stocks which are distributed in allopatric (regionally exclusive) associations. The predator-satiation hypothesis accounts very convincingly for these extraordinary synchronizations. Individuals that emerge (or set seed) in the wrong year, or even the wrong month, are devoured.

It might be asked how, if tropical trees and root-eating cicadas have been forced to evolve such spectacular anti-predator devices, one can believe that predators do really have homeostatically controlled populations, in order to prevent them from exterminating their prey. If the latter is true, what need can there be for this elaborate self-protection on the part of the prey? The answer is important and easily given.

Consumer pressure inevitably results in selection of the prey, as long as some prey individuals are better adapted to avoid being attacked than others. This is what induces defensive armour, distastefulness, self-concealment or fleetness in animal prey, and prickles or toxins in plants. *Whether the predator is prudent or not, he cannot avoid exerting such pressure.* The individual dipterocarp trees whose intermittent seeding first became triggered by identical environmental stimuli and so rendered simultaneous, presumably gained an advantage over the individuals that seeded at random; and thereafter the more the genes for 'gregarious' reproduction increased in the population, the stronger became the selection against the remaining individuals that did not conform. Anti-predator adaptations are all maintained in this way; and a responding selection goes on in the predators, favouring those best able to overcome the producers' defences. At the same time group selection will of course always favour the consumer populations that are most successful in keeping their numbers down to the producers' carrying capacities; and in fact, some of the first insects in which the self-regulation of populations was demonstrated, namely *Sitophilus granarius* (MacLagan 1932), *Rhizopertha dominica*, *Oryzaephilus surinamensis* and *Acanthoscelides obtectus*

(Crombie 1943) were seed-eating beetles, the last actually a bruchid. Janzen (1969) stressed the enormity of the damage done by the bruchids that infest the susceptible kinds of leguminous seeds; but the likelihood is that these bruchids are all adapted to conserve their food-giving trees.

5.2 Fruit as a food resource

It is only a short evolutionary step from seed-eating into the utopian world where plants enclose their seeds in attractive fruits (using the word in the popular sense) and where legitimate consumers are not destroyers but dispersers of seeds instead. Seeds ingested incidentally with a meal of fruit are voided some time afterwards from the mouth or anus, unharmed and at a distance from the place where they grew. The consumers are birds principally, and less often monkeys or bats; but many mammals are attracted to fruit they can reach. Some of the consumers are omnivores that turn to fruit in season when it suits them to do so; others, mostly living in the humid tropics, are more or less total fruitarians. They all benefit from the symbiotic adaptations that have been evolved, especially on the part of the plants, though sometimes mutually.

Nutritious or succulent fruits are produced at considerable cost, often very abundantly. The strategy has obviously worked well for many kinds of plants. Fruit production has evolved many times independently, and there are now whole genera and families of plants that have been sustaining partnerships with animals for millions of years. Bearing fruit can entail much waste, however. This may come from pirate consumers that crush and then digest the seed as well as the pulp, or from a shortage of consumers which leaves some fruit unpicked, or from seed being dumped by consumers in unsuitable places, for example by various fruit-eating bats in both the Old and New Worlds, which feed at night and void their guano where they sleep by day, hung up in dry caves. Until recently the co-evolution of fruits and their consumers was taken entirely for granted by naturalists, and it is only in the last 20 years that it has come under the ecological spotlight largely because of the pioneer work of the Snows (1971–76) in the forests of Trinidad and tropical South America. A large and interesting literature has followed.

D. W. Snow contrasted fruit and insects as common alternative bird-foods in the neotropical forest, where a rich variety of both are found, and of birds as well. Insects tend to be scattered and hard to find, and some insectivorous birds have to spend most of the day looking for food in order to get enough. Fruits are the complete opposite: they are aggregated, immobile and as conspicuous as possible, and selection by consumers favours the ones that are easiest to see and most satisfying to eat. They are intended for eating, and in the long course of time neotropical fruits as a group have become highly nutritious, providing oil-birds (*Steatornis*), bellbirds (Cotingidae) and manakins (Pipridae), among others, with a balanced diet, suitable even for nestlings. Selection by frugivores has effectively lengthened and separated the fruiting periods of the different producer species so that between them they maintain a supply all the year round. Frugivorous birds consequently live on Easy Street, able to keep informed about where there are trees in current production; and when they get there, able to tell at a glance which fruit are in prime condition for eating. For them, feeding is quickly accomplished, and, as D. W. Snow has pointed out, in South America it is the bird species subsisting on fruit, and similarly on honey and pollen, that have had leisure enough to evolve lek displays, which characteristically demand that the males be present, and performing, many hours a day for most of every year.

Insectivore evolution appears to have diverged from that of frugivores in another respect. Snow drew attention to the large numbers of species that characterize the neotropical families of insectivores. Each species tends to become specialized in the insect food it takes, no doubt as a result of competitive exclusion between sympatric species, which forces them into separate niches; and the members of any one species are generally found to be scarce. By comparison, species diversity in the frugivorous families is low, and the individual species are generally found living at higher densities. This reflects the much smaller diversity of fruit types than of insect types, and the fact that fructuous plants often have to compete among themselves for dispersers. Thus trees bearing fruit commonly attract a variety of animals, and occasional instances are recorded of more than 20 (once over 50) different species taking fruit from a single tree.

Morton (1973) has shown, however, that as well as advantages there is one major disadvantage in making an exclusive diet of fruit. Although fruits, especially in the tropics, often have a sizeable protein content, that of insect foods is normally something like ten times higher. He compares a tropical house-wren (*Troglodytes aedon musculus*), which feeds its nestlings on insects, with the thick-billed euphonia (*Euphonia laniirostris*), which is a mistletoe bird and exclusive frugivore; both live in the same habitat and the house-wren sometimes actually usurps a euphonia nest. The growth of the wren chicks is rapid and sigmoid, whereas that of euphonia chicks is slower and almost linear, resulting in a longer fledging period. This confirms Lack's tentative earlier conclusion that fruit is a less satisfactory diet than insects for nestlings (Lack 1968: 171, footnote, & 193). Morton considers that the more numerous visits the parents have to make to the nest, and the slower development, increase the risk of nest predation. He uses the Snows' results in contrasting the higher clutch size of the oilbird, which has a safe nest-site in an inaccessible cave, and lays four eggs, with the one-egg clutch of the bearded bellbird, laid in an open nest in the forest (see next page). The feeding of nestlings on fruit is in general a rare and sporadic habit, which suggests that total frugivory has had a checkered career in bird evolution and carries a considerable extinction risk; and unless there was such intense competition for insect food as there is in the tropics, fruit would never be 'chosen' as food for nestling birds.

For adult birds, however, frugivory is a completely viable way of life in the tropics; and for opportunistic omnivores in most parts of the world, fruit is a welcome, quickly gathered extra, requiring little or no structural or physiological adaptation, or special skills, of the consumer. The fruit harvest can always be wholly consumed if required without detriment to future crops: indeed maximum seed dispersal can only promote their production. Like nectar, fruit is thus a true plant product in the sense of Hairston *et al.*

5.3 Frugivores must limit their numbers

Even in the humid tropics where food production continues throughout the year and supports perennial frugivores, seasonal changes in the volume of the standing crop are likely to occur; and at least at times when production falls to a low level, consumption will need to be limited if the supply is not to run out. We should therefore expect to find that resident frugivores limit their population densities in the normal way, through territorial or other social exclusion mechanisms.

Three examples, two of them taken from the Snows' researches in Trinidad, and the third deferred until the next section, suggest that this expectation is fulfilled. The first comes from David Snow's study (1961–2) of the oilbird, a nocturnal species whose nearest relatives are the nightjars (Caprimulgidae). Most remarkably, it has become a total frugivore. As already stated, it nests colonially in caves; and in Snow's main obser-vation colony in the Arima gorge the ledge-space for nesting was very limited and nearly full. The dozen or so nest sites were occupied all the year round by permanent owners. At the beginning of the breeding season prolonged disputes occurred, when apparently unestablished birds tried to gain a foothold near an established site and were eventually repelled. There was a small amount of extra space which if occupied might have brought the colony total up to 15 or 16 nests. In Trinidad, oilbirds are confined to the Northern Range, and in 1960 they had a total breeding population of about 1500 birds, in five colonies spread out along some 60 km of steep hills and sea-cliffs. The largest colony was sub-divided between four adjacent caves. Regulating the breeding population by confining the right to breed to the owners of established sites in traditional colonies, and excluding the surplus that fails to get estab-lished, is the normal practice of colonial birds, and *Steatornis* appears to be no exception.

The second example is the bearded bellbird (*Procnias averno*), an unrelated frugivore also found in the Arima valley, which was studied by Barbara Snow (1970). She found its distribution in Trinidad was similarly confined to areas where forest still remains and carries a varied flora of fruit-bearing trees. In fact the bellbirds' diet was largely the same as the oilbirds'. The bellbirds' population was rather sparse, and structured and patterned round a set of traditional display centres, which were well separated and each dominated by a top-ranking male. In effect the males in a local area form a vastly expanded lek. Their extremely loud, explosive *bock*! and repeated *tonk - tonk - tonk - . . .* calls, when made from a perch projecting above the forest canopy, can carry up to a mile, and are presumably audible from one display centre to another.

At most, there were five or six dominant-male sites in some 5 km of the Arima valley, and each site typically attracted to itself two or three subordinate males as well, which called from their own fixed stations within 50–200 m of the dominant bird's. One closely watched dominant was confined to his post by calling and other displays for 59–87% of the day, for 9½ months of the year. Thereafter, from August till mid-October, he was silent, performing his annual moult, while one or more of the subordinate males temporarily took over his position. The spatial pattern thus remains perpetually in force, and presumably ensures that each dominant has a successor in readiness when his reign ends. The system is inherently flexible, should the need for density change arise.

Females come to the dominant male's display site to be inseminated; and while he has a female (or any other conspecific) in his company he remains silent, except for a *bock*! at the moment he jumps to mount the female. The meaning of the signal is not likely to be lost on other bellbirds within earshot.

Some further clarification is desirable at this stage of the general function of male competition and property holding, as a means of regulating population density, especially in birds; the subject was earlier mentioned in Chapter 1 (p. 6). It will be shown in Chapters 7–10 that the red grouse is one of many bird species in which the males hold individual feeding territories prior to and during the breeding season. In an average year their competition to obtain a territory results in cutting down the successful candidates to less than half the number that initially entered the contest. The average size of territory claimed by a successful male changes from year to year in accordance with annual variations in the state of the food supply; and it has been possible to prove by field experiments that the territorial system regulates the population density, and adjusts it to a level that the food resources can withstand without detriment.

There is much evidence that species with similar territorial systems are similarly self-regulating in relation to food productivity. But not all foods are evenly spread over the habitat like the heather plants on which the herbivorous grouse subsist, and some birds rely on more or less aggregated food sources which are productive enough to allow many of their conspecifics to feed together at the same time. Thus swallows often assemble to feed on insects flying over inland waters. Similarly terns in close proximity circle over and plunge into shoals of small coastal fish, and the oilbirds in Trinidad gather around particular fruiting trees. Such species as these and others with comparable feeding habits generally nest in colonies at traditional places, in which the property possessed by each established pair is reduced to the crowded site of the nest itself. Still others, such as vultures and eagles, may have to range individually over areas much too large for them to defend in their entirety, hunting for food items that are scarce and scattered; and these species either breed in colonies as well, or, if there are enough safe and isolated sites available, they may occupy traditional eyries, where only the near vicinity of the nest is defended, somewhat as it is in a colony. Again the evidence is strong that the numbers of eyries and colony sites are normally limited by tradition, and similarly serve to regulate the breeding populations of their owners. It is typical of these nest-site-holding, distant-feeding species to have the two sexes alike, so that either member of a pair can 'hold the fort' while the other is away hunting.

The lek birds form a third distinct category, much fewer in numbers of species than the previous ones. In them the males characteristically hold small token display-territories or stances on a traditional arena (lek), which the females need to visit at least once if they are to copulate and breed. As previously said, these males may spend a sizeable part of their lives actually on the lek, engaged in conventional displays and bouts with their companions. There are several similarities between lekking and the more familiar territorial systems. One of them shows particularly clearly in the bearded bellbird, where the arena appears to have expanded to fill the whole habitat. The males still spend long periods at traditional display stances, where females visit them for copulation, and where the males continue to challenge their now-distant rivals with resounding calls; but it is display posts and not nest-sites or feeding territories that are defended. It has never been proved that leks can limit the local quota of breeding females, though circumstances make the presumption very strong. All three male-contest systems have evolved in parallel in numerous orders and families of birds, and in other non-avian groups as well. Intermediate stages, like the one just mentioned, are found linking the three together and making them appear to be true alternatives in function. Moreover no species to my knowledge makes use of more than one of the three contest types, except for some birds (e.g. gulls and terns) that tend to be colonial in coastal and solitary in riverine habitats.

5.4 Food gluts, defendable food sources and undefendable ones

The rate of fruit production in the Arima valley was not uniform all the year round. D. W. Snow (1962: 214) says that in the largest family of fruit producers, the Lauraceae, there were several times as many species ripening their crops in April–June as there were in August–February. The breeding seasons of the frugivores were timed to make their own peak food demand coincide with the greatest plenty. The synchronization may be partly mutual; but consumer demand often failed to rise as high as the peak of fruit production, even though various omnivores might be temporarily attracted to share it. Fruit often went to waste (Snow 1976: 74), which is the test of superabundance. In temperate and even arctic latitudes a similar wastage can be observed. There, fruit production is more strictly seasonal for climatic reasons, but many consumers, from foxes and gulls to thrushes and regular herbivores like grouse and mountain hares, are still drawn to the berry harvest. Especially in bumper years, much soft fruit is destroyed by the frost and never consumed, although tree-borne berries that resist the frost are generally all taken by birds before the spring.

The gluts and transience, so characteristic of fruits, emphasize the intra- and interspecific competition that can often exist between individual producers, all trying to obtain the right sort of consumers at once. Nevertheless there are occasions when fruit is less plentiful and antagonism is to be seen among the opportunists that compete for a share in a transient food-crop, which may be resolved by territorialism and the exclusion of rivals. As a third illustration of population regulation among frugivores, I can use the highly specialized Indo-Australian bird family Dicaeidae or flowerpeckers, especially the members of the genus *Dicaeum*. They are very small birds, the size of goldcrests (= kinglets, *Regulus*), and their principal food is mistletoe berries (Loranthaceae). Mistletoes are hemi-parasites, numerous in the tropics and growing in the crowns of forest trees, whose tissues they penetrate in order to withdraw nourishment. Individually they are small plants compared with the trees that are their usual hosts. Their fruits are enclosed in a skin, and contain a translucent, sweet and extremely viscous pulp and a single large seed.

Some flowerpeckers swallow the seed with the pulp, and have a specially reduced and modified stomach that allows the bolus to pass immediately from the gullet to the intestine. The seeds add materially to the weight of the small bird, and it is adapted to rush them through the intestine and get rid of them within 5 or 10 minutes—an act that looks to the observer as if it needs concentration and effort. The seed on emerging is still viscous, and the bird turns longways to its perch and lowers its belly to deposit it on a twig or bough, to which it adheres. It germinates *in situ* within a couple of weeks (cf. Ashworth 1895; Docters van Leeuwen 1954; Keast 1958). Other flowerpeckers manipulate the fruit in their beaks (as the mistle thrush, *Turdus viscivorus,* also does) to separate the pulp from the seed, which is then wiped off the beak and adheres similarly to the tree. Remarkably enough, a parallel adaptation of the stomach for the rapid transit of the seed has evolved independently in *Euphonia musica* and its allies, the mistletoe-birds of the New World (cf. Wetmore 1914), and yet another in the honeyeaters (Meliphagidae) of Australasia (see below, p. 69.

One of the flowerpeckers in India (*D. erythrorhynchos*) has regular feeding beats or feeding territories, and in the non-breeding season this is true of single birds. As they feed they utter an almost incessant *chik, chik, chik,* occasionally varied by a twittering song. They test each berry for ripeness before picking it, taking them usually at the stage when they have only a slightly yellowish tinge, that is, as soon as they are edible. Only in January–February when the berries are in their greatest profusion do any remain long enough on the plant to become rosy red and completely ripe (Ali 1932). This suggests that competition is strong and that it is normally impossible for the birds to defend fruit long enough to let it ripen. The same can be true in temperate latitudes of domestic cherry trees, fruiting in summer.

The eating of unripe fruit appears to indicate an example of food resources being subject to mismanagement. For consumers in seasonal climates, however, the fruit is merely a bonus to their normal food supplies, and there are no penalties for exploiting it in a multispecies scramble. There is not at present enough evidence to show whether food-fighting takes place only, or significantly more often, over totally consumable resources such as fruit, and carrion (cf. Kruuk 1967), than it does over living foods that need husbanding. Lockie (1956) noted among rooks, jackdaws and crows (*Corvus* spp.) that a concentrated source of food for hungry birds, such as a stack yard in hard weather, would break down the individual distance they normally maintain and force them into close contact, with increased restlessness and fighting; and there is no doubt that the combination of fierce demand and urgency in exploiting a short-lived supply is a frequent cause of failure of conventional restraints. Birds as a class may be particularly prone to food-fighting because of their great mobility and powers of discovery of point sources of food.

It was the Snows' (1971) observations in Trinidad that prompted Lack (1971: 155) to review the question, first raised by E. O. Willis (1966), of whether or not the sharing of food crops between many species, sometimes including several members of the same genus, contravenes the 'competitive exclusion principle', namely that one species tends to exclude other species from its own niche. Willis accepted that sharing normally occurs when fruit is superabundant and there is nothing to be gained by excluding competitors; but there are neotropical fruits such as the berries of *Conostegia* which occur irregularly, and are usually eaten by more than one species. It seemed to him possible that such foods were always superabundant when they were discovered, and that there might be an 'irregularity principle' which allowed a succession of irregular food resources to be shared by several species, without ever leading to competition. As long as there were intervals in which food was not superabundant and the various species withdrew to habitats where competition and exclusion kept them apart, and determined their several population densities, there would be no conflict with the principles of ecological isolation. Lack, who added much to our understanding of competitive exlusion (cf. Lack 1944, 1971), had already accepted this view; and he used the Snows' data to show that, of the ten species of tanagers (Thraupidae) in Trinidad, all but one were seen to consume fruit, at least occasionally; the exception was a rare species that appeared to eat nothing but insects. One other species, *Euphonia violacea*, was exceptional in that it ate fruits almost exclusively. The remaining eight were mixed feeders taking both fruit and insects, and their habitats showed that as regards insect hunting, there was little overlap between them.

It seems likely then that opportunist fruit-eaters whose staple diet is arthropods or other small animals, generally have their population densities regulated in relation to their animal food supplies, and can then afford to regard fruit as a superabundant extra, free to all comers. This appears to be the normal relationship, especially in the tropics.

5.5 Nectar as food

Nectar is a near but not exact parallel to fruit as an attractive food for animals, and it is similarly bartered in return for sevices. Though it is manufactured solely for the purpose of being given away, the amount produced is restricted by individual flowers, so as to induce the bees or other pollinators to visit numerous blooms before they are replete; but the flowers tend also to grow in dense enough masses to make repletion possible and economic for pollinators, in terms of their energy balance (Heinrich 1975). Plants need to attract specialists that bring pollen from other plants of the same species, and similarly take pollen away. The plants have of course evolved many structural devices of great elegance, to restrict the availability of their honey to legitimate pollinators, and thwart the robbers that fail to give the service required. Many also have circadian rhythms that narrow the hours of scent and honey-flow to match the feeding times of, say, specialist moths. Competing with other plants to retain the services of a good pollinator tends to coerce the producer into providing a satisfying diet, which in some species actually contains appreciable amounts of amino-acids and fats as well as sugars (Baker & Baker 1975). The parallel with fruits and frugivores is obviously close; and nectar is just as liable to pilfering, for example by bees that bite a hole into long-tubed nectaries, which their tongues are too short to reach by the proper route; consequently they come and go, again and again, without ever touching the stamens or pistil.

Many mellivorous animals eat pollen as well, and there are numerous kinds of flowers including poppies and roses that manage to attract pollinating insects in exchange for liberal supplies of pollen, without offering any nectar at all. Larger pollinators such as fruit-eating bats and flying foxes (Pteropodidae) may commonly be drawn more by the copious pollen than by the honey; and some honey-eating birds also take small insects attracted or trapped by the nectar. If honey and pollen are to succeed in promoting cross-pollination, they must avoid attracting as many casual omnivores as fruits do. Several families of nectar-eating birds are nevertheless polyphagous (varied in their diet), including the lories, a subfamily of parrots found chiefly in Australia, the honeyeaters (Meliphagidae) of the same region, and the honeycreepers (Coerebidae) of the West Indies, Central and South America. Within each of these groups there has been an adaptive radiation. Some Meliphagidae for instance have forsaken nectar altogether, and there is a common tendency in all three groups to supplement nectar and pollen with insect-catching, fruit-eating or both. Much convergent evolution has occurred between them mutually, and also with some of the frugivores in other bird groups. Thus the Meliphagidae include yet another mistletoe specialist, and as a family have evolved the same kind of gizzard by-pass as we noted in the flowerpeckers and euphonias. The flower-piercers (*Diglossa* spp.), among the coerebids, prick a hole into the long corolla tubes of some flowers to get at the honey from the outside, in the manner of short-tongued bees; and the majority in all three families have brush-like or otherwise specialized tongues for withdrawing nectar. (For references see articles on 'honeycreeper' by D. W. Snow, 'honeyeater' by F. Salomonsen & H. A Ford, and 'parrot' by D. G. Homberger, in Campbell & Lack 1985).

Lastly it should be noted that, being a finite resource, nectar is capable of supporting only a finite number of consumers per unit area. Animals dependent on it may therefore again be expected to have self-limiting populations. The eusocial bumblebees (*Bombus*) and honeybees (*Apis*) show especially advanced examples of this capability.

Summary of Chapter 5

1   Vivivores of many types use seeds as food, and plants have reacted by evolving defences, such as making their seeds hard-shelled or toxic, and producing them in transient superabundance. Especially in some tropical tree families, predator satiation is achieved by synchronizing the seed crops in local populations of one or more species to fall in the same year, followed by a run of barren years.

Why should these devices have evolved if vivivores really possess the resource-conserving adaptations I am postulating? The answer applies generally to producers of living foods (excluding fruit and nectar). If individuals in any of these species differ among themselves in palatability, conspicuousness, catchability etc., the consumers will automatically tend to take a heavier toll of the ones less well protected, and this will perpetually evoke better counter-measures, whether the consumer populations themselves are self-regulated food-conservers or not. It is in fact an escalating two-way process, in which the consumers have to keep ahead if they are to retain the use of the resource.

2 & 3   Fruit (understood in its non-technical sense) proffers an easily found, abundant bait for animals to eat, and thereafter to carry away the seeds it contains, disposing of them elsewhere. Again in the equable tropics, different producer species evolve to avoid each other's fruiting seasons and they thus come to share and sustain the same guild of consumers the whole year round. Frugivorous bird genera and families run to few and generalist species, as compared with insectivorous ones, many of whose species have had to segregate as expert narrow specialists in order to find enough of their often cryptic prey. Fortunately for them, insects are many times richer than fruits in protein. For the same reason tropical frugivores need to spend little time hunting and feeding, and the females of many species can rear one-parent families, freeing the males to stay much of the day and year in attendance at a lek; whereas the majority of insectivores spend most of their lives hunting, and have seldom if ever evolved leks.

Though never proved, there is a strong presumption that leks provide a third alternative density-regulating mechanism, in parallel with the territorial and the traditional breeding-site mechanisms. A study of the dispersion of a frugivorous neotropical bellbird has shown how effectively the usually-distinctive features of the three mechanisms can be recombined.

4 Generally speaking, some resources are uniformly dispersed in the habitat and can readily be divided into territorial units. Some, including many fruits, are locally clumped and can provide food for many consumers at once in a small space. Still others (e.g. big-game carcasses for vultures) are so sparsely scattered that no consumer could defend a large enough exclusive territory. Where feeding territories are impracticable, traditional breeding-colony nest-sites, traditional solitary nest-sites, or leks, usually take their place, at least among birds.

5 Nectar is a food product evolved by flowering plants, along with showy petals and fragrance, to entice animals to serve them as cross-pollinators, and in several respects it is a parallel type of food to fruit. Pollen, being a richer source of protein, is commonly collected at the same time as nectar. The production rates of both are finite, although at least the honey can all be consumed without detriment to future supplies. It can therefore benefit consumer populations to match their densities to its carrying capacity, and so allow the individual members a chance of obtaining an adequate ration. Honey is more susceptible than fruit to pilfering by consumers that evade the service required by the plant, and this weakens the selective advantage gained by the plant, and the mutualism the advantage sustains.

CHAPTER 6
GENERAL RELATIONSHIPS AFFECTING CONSUMERS

6.1 Food resource conservation

We have seen that fruit and nectar eaters share with the necrophages a freedom to consume the whole of their resources without detriment to future crop production. Nevertheless populations that make these their staple diets stand to gain in fitness and security when their members are programmed to cooperate, in holding the density down to a level that allows all the accepted residents to get the food they need. That means that supernumeraries will have to be excluded from the habitat as necessary. For vivivorous species—a much larger number—there is the additional condition that population density should not only allow adequate rations to residents, but should be low enough also to prevent their collective consumption from exceeding the amount that the living resources can sustain, and duly renew.

To gain these advantages, both categories of consumers must take the initiative and regulate themselves; and the device all the higher animals appear to share is to divert their intraspecific competition into artificial channels and make the contestants compete for conventional, and appropriately limited, rewards as substitutes for the food itself. As resource levels change, cybernetic programs in the consumers' brains usually lead to responding changes in population density which preserve a prudential relationship between the two. Food territories are the easiest of the conventional rewards to understand because they are very close substitutes for the resources they contain and protect; and territory size can obviously be adjusted, and made as large as necessary to relieve the resources from any insupportable pressure. But as we have seen, individual food-territories are impracticable in many circumstances and types of habitat, and other less direct and more abstract susbtitutes have therefore evolved as well, such as the restriction of breeding pairs to the use of traditional sites and colonies, or the admission of individuals to the rights of feeding, residing or breeding in a habitat on the basis of personal rank alone.

For vivivore populations, limiting their demands to a dispensable portion of the food available has deeper implications. The presumption is that they are able to attain their highest sustainable density and prosperity where and when the food organisms on which they depend are, in their turn, similarly procuring the highest productivity from their own trophic resources; and there seems no reason to doubt that in obviously flourishing natural communities an approximation to these relationships does commonly exist. If so, vivivores must normally require their members to exercise restraints on population growth when they are surrounded, figuratively at least, with forests of food. This must be literally true of beavers (*Castor* spp.) which cut down slowly regenerating trees, or of species that 'milk' trees like many sap-sucking or defoliating insects. Incidentally not all vivivores are animals; there are parasitic and carnivorous plants, and microbial parasites, which share the biophages' risk of harming the species that nourish them. Presumably they too need to possess equivalent prudential adaptations.

The most important implication is that genetically coded food-resource conservation of the kind depicted above, ought to be an invincible strategy for group survival, because, so far as lies within the vivivores' power, it keeps their resources at or near their top productivity; and by so doing it allows them to maximize their own mean individual fitness, and their population productivity per unit area. Such advantages would be denied to non-cooperative genotypes; and if such genotypes were to arise and infiltrate a cooperative group they would lack the normal prudential discipline and would exploit the resources for personal gain instead, regardless of the common good. Temporarily their genes would multiply, but so would the resource destruction. The productivity of their local group would fall, bringing a frequency-dependent selection down upon them. It is quite possible that this actually happens in nature; but because of their penalizing effect on themselves and other members of their group they can never rise above a low frequency (see Chapter 19).

6.2 Syntrophism, or the exploitation of the same population of food organisms by different species of consumers

Allusions have already been made to the sharing of a food resource by two or more species of consumers. One can deduce that many pairs of species, especially closely related ones, must in the past have competed for food, with the result that one of them has gained local supremacy over the other and has excluded it from sharing the resource. For example, some species pairs are found to be allopatric, each occupying a separate geographical range; still more have become differentiated in their ecology so that they occupy exclusively different habitats. Thus the red grouse and the ptarmigan (*Lagopus lagopus scoticus* and *L. mutus*) are altitudinally separated in Scotland by the upper limit of tall heather, at the borderline of the infra-alpine and alpine zones. Species pairs may actually occupy the same broad habitats, but be separated by their feeding stations and specialist foods. The subject has been extensively explored and reviewed, especially by Lack (1944, 1971) and Diamond (1975).

If potentially competing species do manage to avoid conflict by subdividing the resources under dispute, so that each becomes a specialist exploiting different facets or constituent elements within it, the effect is to constitute two food niches where previously there had been only one. Possibly sometimes as a result, many herbivorous insects, for example, have specialized on a single species of food plant. Specialization that averts interspecific competition could conceivably also simplify the consumers' problems in conserving their resource. But being mono-phagous must also increase the risk of catastrophe if the food supply fails for extraneous reasons, or some other superior competitor arrives on the scene. That could bring selection to bear in the opposite direction, tend-ing to increase the breadth of the food niche, and favour alternative methods of avoiding interspecific competition. In fact, overlaps in food-resource requirement are much more commonly observed than mono-polies.

The essential prerequisite for food sharing is the elimination of con-flict. Within a group of conspecifics an established personal hierarchy or a territorial system generally serves this purpose. Interspecifically it can be achieved if the competitors feed at different seasons or times of day. Thus the red grouse and mountain hare (*Lepus timidus*) in the Grampian Region of Scotland share the same moorland habitat and the same staple food plant, *Calluna vulgaris*. All three are abundant in the most productive habitats, but the grouse feed by day and the hares mostly by night (cf. Hewson 1962); and the consumers may gain some additional segregation from the fact that hares like young heather swards up to 10 cm deep, whereas grouse prefer them taller, up to 30 cm (Hewson 1976 b; Savory 1978). In the same region, the intertidal mudflats in the Ythan estuary support enormous populations of two small invertebrates—a snail *Hydrobia ulvae,* and an amphipod *Corophium volutator.* In autumn and winter they are eaten when the tide is out by large numbers of redshanks (*Tringa totanus*), and when it is in by flounders (*Platichthys flesus*). In spring and summer the main ambulant consumers are shelducks (*Tadorna tadorna*); and under water the flounders, though fewer in number, are joined by gobies (*Gobius minutus*) (Milne & Dunnet 1972).

When consumers sharing the same resource are thus segregated from each other by living in different media, or feeding by night and day, they cannot compete. What must happen is that each species of sharer exploits the food crop in ignorance of any opposition, and has to spin it out for a certain season, or perhaps for the whole year round. The crop must be given a proper chance to build up during its growth period, and it will then decline at a certain, perhaps variable, rate. Whether there are other hidden populations of consumers is beside the point, as far as crop management is concerned. Each consumer species must adjust its consumption independently, in relation to the level and rate of change of abundance in the food stock, by altering its own density as required.

Segregation of consumers can be much more complex than in the examples just given. Pearson (1968) showed that nine species of seabirds nesting on the Farne Islands, off the coast of northeast England, all took substantial quantities of sandeels (*Ammodytes*) as food during the breeding season. Five were surface feeders, fishing by hovering and plunging from the air, namely the lesser black-backed gull (*Larus fuscus*), kittiwake (*Rissa tridactyla*), and three species of terns (*Sterna*); and four were underwater swimmers, the puffin (*Fratercula*), guillemot (*Uria*), shag and cormorant (*Phalacrocorax* spp.). There seemed ample food for all—a normal but not necessarily diagnostic sign of successful management— and although at times Pearson could see some members of nearly all the species at once, fishing in sight of the islands, the majority of each (except the common and arctic terns) were segregated, by the distance they went out to sea, by the water stratum in which they hunted, and perhaps by the difference in the size of fish they pursued.

A yet more complex example of shared food resource is the krill, *Euphausia superba*, a planktonic shrimp that abounds in some parts of the Antarctic seas and is the main food in austral latitudes for four species of *Balaenoptera*—the blue, fin, sei and minke whales. The natural historian of *E. superba*, J. W. S. Marr (1962: 240), has left us a picture of the Weddell Sea as it was in the 1930's. The vast expanse of open water in summer was populated by compact swarms of krill feeding within a few metres of the surface. An average-sized patch might measure '60×40 yards in surface area and 1 yard thick' and might contain, 'at 40 000 to the cubic yard, 96 million euphausians'; and there would be other patches of different sizes and shapes nearby. The krill grazed upon and virtually removed the phytoplankton 'with enormous local destructive power', leaving a patchy distribution of algae behind them. From time to time a questing herd of a hundred or more whales could be seen, which, when they happened on food-rich water by chance (or possibly echo-location), would in their turn proceed to cruise around, sifting out the krill, and departing when satiated, or perhaps when further feeding in that locality was inhibited.

There are so many species of predators on krill that one can only guess how far the different species manage to keep out of one another's way. Marr (1962: 127–136) adds the following list of consumers to the four rorquals already mentioned: two other large whales, abundant in years gone by (the southern right whale and the humpback); crabeater seals (*Lobodon*), of which Scheffer (1958) estimated there were 2–5 million haunting the Antarctic pack-ice; three very abundant species of penguin; 13 species of albatrosses and petrels, which 'represent a very substantial part of the huge oceanic bird population of Antarctica'; and an unknown variety of fish. The list is not exhaustive; Bierman and Voous (1950, quoted by Marr) concluded that *Euphausia* is the basic diet of all antarctic seabirds. Marr concludes (page 240): 'When one contemplates the hosts of its predators and the vast weight of whale flesh, seal flesh, fish, penguins and other birds that is built up on its wholesale destruction, one is left wondering as to how in its seemingly inexhaustible myriads it continues to survive, and above all at its astonishing capacity year after year for making its staggering losses good . . . . . The onslaught of predators upon this species is a phenomenon perhaps without parallel in marine ecology'.

The possibility cannot be ruled out that the krill is a superabundant summer resource for the fish, birds, seals and whales that feed on it then, but have to limit their populations in relation to scarcer resources and lower carrying-capacities available to them at other seasons; that is to say that in spite of the multifarious pressures of its consumers, *E. superba* manages to achieve predator satiation. With a 2-year (or longer) life cycle it does not have a particularly fast turnover, though it is clearly extremely prolific. It grows rapidly through the summer months, and makes itself increasingly vulnerable to predation as it matures by its densely gregarious behaviour, close to the surface by day and lit up by phosphorescence by night: vulnerable that is to whales that plough through the swarms, although shoaling exposes only a 'surface' to small predators and may be protective against them. The whales are migratory summer visitors; the pack-ice gives partial protection from them and other warm-blooded predators in winter. The *Euphausia's* metabolism must slow down then, as must that of the fish that may or could prey on it all the year round. There seems no obvious reason why the consumers, between them, should not have built up a capability of using its full carrying-capacity. Predator satiation is usually coupled with a life-history that produces a short-lived abundance but starves out the potential predators at other times; and no such alternation exists for the krill, which offers a dependable feast to vertebrate predators.

It may be even less credible, at first sight, that 50 or more species of exploiters could share 'syntrophically' in the management of the same resource. The fact that the exploiters are many and diverse does not, however, necessarily magnify the problem, provided each species, and indeed each individual, is separately and reliably programmed with a similar scale of responses for resource conservation. As long as individuals do not have to compete except with their own conspecifics, density-dependent feedbacks will operate quite normally and their respective social hierarchies will press, when the feedbacks so dictate, for the exclusion of conspecific supernumeraries from their midst. In fact it is difficult to see how such a productive 'fishery' for the consumers could ever persist, were the consumers not all adapted to conserve it.

Segregating many different exploiters in order to prevent interspecific conflict may require some elaborate dove-tailing, by local geography and distance from shore in the land-based seabirds (chiefly penguins), by fishing methods, speed and body-size. Thus lesser predators no doubt voluntarily give way to whales—a relationship with many parallels elsewhere.

In most circumstances where segregation cannot be achieved, inter-specific competition can be expected to lead ineluctably to competitive exclusion. It seems possible, however, that there is another way out, at least for closely related species. It is to merge potential competitors of two or more species into a single social union. It is not common, and perhaps not often successful indefinitely. In *Animal Dispersion* (1962: 428) I drew attention to a number of such instances where related species look alike and appear to unite into a common social organization, at least in the non-breeding stages. The herring and the sprat (*Clupea harengus* and *C. sprattus*) are one example. Their juveniles, when of the same size, form mixed shoals and are extremely difficult to tell apart. Another pair are the common and arctic terns, mentioned earlier in this section, which often nest in the same colonies and take similar foods in similar though not identical proportions. Their resemblance in ecology, voice and looks is closer than usual in birds of the same genus, though probably not close enough to deceive the birds themselves. The most striking examples I know among vertebrates are found in a number of African species of weaverbirds (*Ploceus*) and bishop- and widowbirds (*Euplectes* and *Coliuspasser*), belonging to the family Ploceidae. The males in breeding plumage are mostly colourful and in some species 'bizarrely different'; but in each genus the postnuptial moult levels most of them down to resemble females and juveniles, and the different species become difficult or impossible to distinguish in life. At this time look-alike flocks form and roam the plains, feeding on seeds. This contrasts sharply with the majority of bird species that form mixed flocks, but retain their distinctive recognition characters. A rather different type of interspecific union, affecting the participants' combined dispersion, is interspecific terri- torialism (cf. Lack 1971: 127–8).

The generic resemblances in non-breeding ploceids are presumably adaptive. If their purpose were to allow an interspecific social integration, the flocks would have to adopt a common scale of social dominance, otherwise individuals could not operate the normal process of excluding subordinates as a means of reducing population density. It is likely that closely related species share large portions of their homeostatic codes through common descent, and this may facilitate their ability to respond in unison to the same external feedbacks. Judging by appearances, some real advantage has been gained by the retention of nearly identical 'gregarious phase' plumages. An alternative, perhaps less likely explanation is that the resemblances serve as a kind of Müllerian mimicry, and that larger crowds of seemingly uniform individuals are less vulnerable to birds of prey than smaller single-species groups would be.

Elaborate brain programs exist in all the higher organisms, for purposes quite other than conserving their food resources. One may recall that in any of a score of small species of warbler (Sylviinae or Parulidae, it makes no difference) an individual can guide itself on a migration of several thousand miles each spring, stopping to refuel itself on the way and often travelling by night; and eventually it arrives at a particular temperate latitude, and even a precise locality, in order to breed. It can then select a site suitable for its territory, or choose an established mate, before there is any direct evidence of the insect food that will appear a month later, when its young are hatched. The perceptions and inputs of information that are required, and the programs that integrate them and direct the bird's complex responses, are as varied as they are remarkable.

The programs required for population homeostasis do not seem any more difficult. The average red grouse cock in autumn receives a neural feedback from its day-to-day feeding experience and its state of nutrition. Its brain, after computation, directs it to make a quantified response, by demanding a territory that will provide a sufficient and sustainable yield of food for winter use and the rearing of families nine months later. Change the bird's plane of nutrition experimentally beforehand and it will demand a territory of a different size (see p. 124). Conserving their food resources is not much less vital than reproduction itself to the long term survival of consumer species, and selective pressure to elicit the programming required will be correspondingly intense.

What emerges from this discussion is that, especially in mature ecological communities, particular food-producer species must often be preyed on by a number of vivivore species, which have managed to avoid making overt contact and thus involving themselves in interspecific conflict. The outcome of such conflict would normally be unequal and lead to the exclusion of the weaker. Avoidance of it can come from (1) specializing on different edible parts of the same 'host', such as the leaves, flowers and roots of a plant, which becomes progressively more feasible for animals of diminishing size; (2) feeding on segregated size-groups or life-stages of an animal or plant, such as the larva, pupa or imago of an insect; (3) feeding at different times of day or seasons; and (4) by deliberate avoidance, as when a weaker consumer species yields to a stronger one, so that it only attempts to feed when the stronger is absent, and gets out of the way immediately the stronger appears. This may develop especially within sets of polyphagous species, for example among tits, thrushes or crows, whose respective diets only partly overlap. Finally in (5), a few sets of closely related species seem able to form social unions in the non-breeding season, and feed under 'single species rules'.

All species of vivivores seem likely to be programmed to stop feeding in a particular area of habitat if the food resources show signs of being overtaxed. Those that are year-round residents on the same site must normally be able to forestall overexploitation by reducing their reproductive rate or expelling surplus residents; but many gregarious species (and some that are not) habitually move from area to area, moderating their demands on the resources in the short term by the rotational use of a large group territory. Rather a similar feeding pattern probably applies to a plankton-predator guild, such as the one centred on *Euphausia superba*, although being pelagic the habitat cannot be divided into permanent group territories. It may have taken thousands of years for the antarctic *Euphausia* complex to build up and become stabilized. If there have been consumer members in the past that broke the rules and overtaxed the resource, they will predictably have become extinct, at the expense of perhaps a large regional sector of the community that were carried to disaster with them.

For necrophages and plant-product feeders it is less easy to visualize the part that syntrophic adaptation plays. Individual flowering plants of many species are visited by rival kinds of consumers in search of pollen and nectar, and the same is true of many fruit-bearing plants and their frugivores. In some instances conspecific rivals are excluded, for example by territory-holding consumers; but virtually unopposable resource sharing between species must be extremely common. The same kinds of intra- and interspecific relationships also apply to necrophages, inasmuch as there are likely to be alternative routes by which most detritus is broken up and subsequently cycled by microorganisms. Many vertebrates and insects engaged in these trophic industries are niche specialists, showing signs of interspecific avoidance, for instance through the separation of day and night feeders on the same resource. Though adaptations for resource conservation are inessential and perhaps seldom evolved, most of the higher forms of detritivore clearly benefit by developing population homeostasis, rather than relying on simple opportunism; and if they share their resource with other consumer species, they must also share with them in its management.

It is a striking feature of the living world that the number of animal species greatly exceeds that of plants, whereas when their total biomasses are compared the relationship is reversed. For the biomasses the reasons are self-evident, in as much as the producers have to support the consumers, and a food conversion factor necessarily diminishes the biomass at each successive stage along the food chain. The great majority of plants are and must always have been primary producers, all using the same basic resources and performing the same basic metabolic routines in a far from infinite number of types of habitat. Excluding the fungi, only a minority have moved along the food chain themselves, and become wholly or partially heterotrophic as symbionts, parasites, saprophytes or carnivores. In species diversity the animals have come to exceed them by an order of magnitude, in part because for animals the potential subdivision of trophic niches is so enormously greater, and in part perhaps because of their greater capacity for organization, coordination and control.

Since animals are all ultimately dependent for resources on the primary production of plants it goes without saying that an average plant species will have several animal species to support; and although of those supported the different herbivores can segregate from each other by specializing and confining their consumption to different tissues or seasons in the manner just indicated, the fact remains that all herbivore species in the same syntrophic guild, dependent on the same plant population, must rely on one another not to overtax the resource or endanger its productivity. These herbivores have of course to raise their own demands and productivity sufficiently to pay the tax that is exacted from them by carnivores and parasites, and that expenditure is also met by the producer plants. The reliance of consumer species on each other's homeostasis within their trophic community, such as I postulated for the *Euphausia* guild (incidentaly at the second trophic stage), appears to be an absolutely general and necessary condition for the evolution of so enormous a diversity of animals.

Summary of Chapter 6

1 All animals stand to benefit in fitness by matching their numbers to the amount of food available, whether they are vivivores that must conserve their resources, or are free to demolish all the food in sight. Matching can only be accomplished by the self-regulation of numbers, and the general method used for achieving it is to have intraspecific competition diverted away from food, towards substitute conventional objects or rewards instead. The easiest of these to understand is the feeding territory because, as an object, it is obviously related to the food it contains; and it is readily apparent also that if individuals are programmed to demand sufficiently ample territories there will be no necessity to overtax the resources they contain. Feeding territories serve their purpose well as long as the food is more or less uniformly dispersed; but if it is clumped or patchy and productive enough to satisfy a number of individuals feeding at the same time, or is very sparsely scattered indeed, other more indirect or abstract kinds of substitute have necessarily evolved instead. These include e.g. the admission as breeders only of those individuals that can win one of the traditional nest sites, either in a traditional colony or in isolation, according to circumstances; or alternatively, admission to the right to reside, feed and/or breed in the habitat on qualifications of personal rank alone.

Vivivore populations will be most productive and secure where and when their food organisms are enjoying a similar prosperity themselves, and are making the fullest potential use of their own resources. Something approaching this optimal state probably exists in flourishing natural habitats. Cooperative restraints can always be exercised to advantage by consumer populations that live up to their carrying capacity, and food only becomes superabundant when they fail to reach that level. Self-regulating vivivore populations living in the optimal trophic state will surpass all rivals in productivity, and at the same time will maximize mean individual fitness. Uncooperative self-interest is inimical to such a state, and the effect of such antisocial genotypes, if they infiltrated a cooperative group, would be to push the density above the carrying capacity. The ensuing decline of resources would result in a frequency-dependent selection against the non-cooperators, by lowering the productivity of the infiltrated group.

2 Some organisms have defences that give them a measure of immunity against being used as food by vivivores. Others, more palatable or easily procurable, have to serve as food for several or even many kinds. The evidence suggests that if two vivivore species attempt to fill the same niche by exploiting the same staple resource simultaneously by the same methods, the more dominant of them will exclude the weaker; but if they can avoid confrontation by feeding one by day and the other by night, or at different times of year, or by using different hunting methods (e.g. fishing by plunging from the air or by underwater search), or confining their attentions to separate life-stages or organs or tissues of their food-provider, they can co-exist. There is nothing to prevent the accumulation of a syntrophic set of consumer species all exploiting the same food-yielding population, provided (1) that they do not collectively overtax its carrying capacity, and (2) that each can be independently relied on to control its own consumption, and conserve the resource as it finds it.

A striking illustration of syntrophism is found in the antarctic krill *Euphausia superba*, itself a planktonic herbivore which serves as food in summer to perhaps 50 consumer species at the same season, formerly including large numbers of six kinds of large whale, various penguins, tubinarine birds, fishes and a seal. Possibly it may constitute a superabundant resource, for consumers that are mostly prevented by pack-ice from reaching it in winter, and are limited by winter carrying-capacities in other waters. More likely however it is (or was) under full exploitation.

If the provisos of the last paragraph but one are fulfilled, collaborative management of such a food resource appears quite feasible. It is up to each individual consumer to see that the provisos are carried out; and with interspecific conflict eliminated by mutual avoidance, the individual only competes in its own social hierarchy, and homeostatic exclusion mechanisms will be applied as and when necessary within the hierarchy to which it belongs. Every individual consumer, regardless of species, will receive effectively the same feedbacks and instructions.

A much scarcer type of syntrophism is found in some vertebrate genera, e.g. among African weaverbirds, where instead of evolving a mutual segregation of closely related species, they come to resemble one another remarkably closely in the non-breeding season and feed in mixed flocks, which suggests they may all belong, on a common footing, to the same social hierarchies.

The subdivision and proliferation of food niches, as the general method of harmonizing syntrophism, has had an inevitable tendency to multiply the number of socially-independent animal species and compound the complexity of syntrophisms. Those consisting of vivivore species, most or all of which would be capable of injuring their common food-provider stock if their homeostatic mechanisms were to break down, seem necessarily to require a mutual and concurrent response by all the species in order to conserve the resource.

CHAPTER 7
THE RED GROUSE

7.1 A short history

A red grouse research project was started in 1956, based on Aberdeen University, with the principal scientific aim of testing the *Animal Dispersion* hypothesis, or as much of it as was amenable to test. The hypothesis was only then taking shape (Wynne-Edwards 1955). There was also a concurrent economic motive, to find out why the stocks of grouse in Britain had undergone a major decline over the preceding 30 years.

Lagopus lagopus scoticus is the commonest and most valuable gamebird in upland Britain and Ireland. It lives on the heather moors—tracts of unenclosed hill-land lying mostly between 150 and 600 m above sea-level, which are too rough, boggy or exposed for farming and are covered with a low shrub heath, more or less completely dominated by ling heather (*Calluna vulgaris*). The grouse are herbivorous and the heather is their main food plant. Grouse shooting emerged as a field sport in the middle of the 19th century, and the art of managing moors to produce large bags of grouse rapidly developed. The moors were, and still are, in private ownership, and estate gamekeepers are employed to control predators, prevent poaching and improve productivity by draining and heather-burning.

Most of the moors lie below what would be the tree-line if the land were to revert to its prehistoric state. The original forests largely disappeared long ago, through overexploitation for timber and fuel, and for making charcoal for smelting iron. In the maritime British climate, woodlands readily gave way to heaths suitable for summer grazing by cattle or sheep. Ling is an evergreen and where it is plentiful it can carry a large biomass of wild vertebrate herbivores all the year round, but much of the best grouse country is liable to snowfalls that preclude the out-wintering of farm livestock. The heather plant itself is a low woody shrub with minute leaves and small but often prolific purple flowers; in sheltered places it grows to 60 cm, but severe exposure to wind reduces it to a prostrate sward less than 10 cm deep. The plant lives about 30 years. It is not only very resistant to being burnt, but periodical burning increases its dominance over more vulnerable species, including trees, and leads in due course to the regeneration of younger, more nutritious heather plants. Heather grows best on well-drained soils, though on much of the poorer moorland it is rooted in peat. In the eastern Highlands of Scotland the Callunetum is replaced above about 1000 m by an arctic-alpine vegetation which provides a habitat for the ptarmigan (*Lagopus mutus*).

Before the turn of the century, good management of the best moors had succeeded in raising the game bags roughly tenfold. They were yielding annual crops as high as 500–600 grouse per km^2 (Leslie 1912: 333–42); but success was often interrupted. In 1872—73, for example, many of the moors suffered a particularly severe collapse, and attention became focused, not for the first time, on 'grouse disease', a reputed cause of heavy mortality among adult grouse, particularly in certain winters. T. S. Cobbold, a well-known helminthologist, expressed the opinion in 1873 that the killer organism was the threadworm *Trichostrongylus tenuis*, which inhabits the rectal caeca of the grouse, sometimes in thousands. Supposed outbreaks recurred, with the result that in 1905 a 'Committee of enquiry on grouse disease' was set up by the Board of Agriculture and Fisheries. Their commission lasted six years and led to a report being published in 1911; a better-known version of it, abridged for a wider public, appeared as *The Grouse in Health and Disease* (Leslie 1912). Briefly, the committee concluded that Cobbold had been right; and that, while there was no effective way of eradicating the trichostrongyle worm, the birds' resistance to this and other infections would be increased, and mortality greatly diminished, if certain principles of moor management which they prescribed were carefully followed.

The 'field observer' the committee had appointed, and the main author of the report, was a many-gifted man, Dr Edward Wilson of Antarctic fame, who soon after died with Captain Scott in a blizzard on their journey back from the South Pole in March 1912. It seems clear enough, reading between the lines, that he was never really convinced they had discovered the true cause of the grouse mortality; and the Aberdeen research was to show, 50 years later, that his misgivings were well justified.

The constructive message of the report was that the key to grouse production was good moor-management. This was acted upon by the moor owners, with the result that grouse shooting continued to flourish until the 1920's; but after that a gradual deterioration set in, which eventually led to the inception of the Aberdeen study as a sequel to the 1905 enquiry. Two game-bird ecologists were appointed to start it, Dr David Jenkins in 1956 and Dr Adam Watson in 1957. They were joined by a plant ecologist, Dr Gordon Miller, in 1959, and a biochemist, Dr Robert Moss, in 1963. Watson and Moss are still with the project, which was detached from the University in 1966 and is now in the Institute of Terrestrial Ecology of the Natural Environment Research Council.

I accepted the opportunity to direct the work because it could hardly have been better suited for testing my own ideas. The surest way to find a remedy for the grouse decline would be to discover how their population densities were controlled. I put this to Dr Jenkins when we arrived, and told him I expected the ultimate limiting factor would turn out to be the food supply, and that the proximate factor would most likely be the birds' territorial system (or its equivalent if the grouse turned out not to have territories); and by this and other domestic adaptations the birds would determine their own population density and would be able to vary it if their food resources changed in productivity from year to year. His reaction was sceptical. Red grouse are herbivores, reputed to subsist largely on heather in which they spend most of their lives hidden up to their necks. At a first guess, considering their present density and the deep and almost exclusive vegetation of heather as far as the eye could see, there might even be a ton of food perpetually available to every bird. Whatever the limiting factor was, it seemed unlikely it could be food.

Facilities for the work had been provided on a famous grouse moor in Glen Esk, Angus, by the Earl of Dalhousie. Jenkins started in October, and taking advantage of the fact that grouse were coming into the fields of a moorside farm to feed on fallen grain in the stubble, he constructed traps of wire netting, baited them with sheaves of oats, and caught over a hundred birds in the first six weeks. He tabbed them with easily read, boldly numbered plastic ribbons, held by a light harness and worn on their backs—the same as he had previously used on partridges; and thus he introduced the most valuable of all the research tools that have been employed, the labelling of birds as individuals so that their personal histories could be followed in the field. With only occasional help from estate staff, Jenkins was able to discover in the first six months that about half his tabbed birds disappeared during or soon after the trapping period, whereas the other half remained where they were, occupying their customary individual places on the moor, suffering very little mortality, and emerging in the spring as members of the 1957 breeding stock.

When Dr Watson arrived 12 months later, he quickly realized that the cock grouse were already establishing a territorial system in September–October; but it was then only functional in the early hours of the morning and was allowed to lapse for the rest of the day, like another extended lek system. His prolonged and illuminating studies of territorial patterns and behaviour to which this afterwards led are frequently quoted later.

The unit worked in Glen Esk for five years, and in 1962 moved nearer to Aberdeen, with offices and laboratories at Blackhall, Banchory. The 3000 ha moor of Kerloch nearby was leased for research, and the shooting of grouse for sport on it suspended. It remained in use till 1978, when a change of ownership led to another move, to Rickarton, 10 km east of Kerloch.

The grouse-moor owners have been given an answer. Grouse disease as a lethal epidemic was a misunderstanding; instead, as predicted at the outset, it is the birds themselves that, by mutual competition, determine their own population density; and, in the main, they predestine their own lives and deaths through a system of territorial ownership that is imposed afresh each autumn, and lasts until the young are hatched the following May or June. All reasonably good grouse moors (an exception is mentioned below) normally produce a surplus of young in the breeding season—that is to say, more young than the territorial system can accommodate in the ensuing October. Shooting, as traditionally practised, rarely takes as many birds as would become surplus anyway, through the natural process of social competition and elimination. The young males of the year can compete on fairly even terms with their fathers' generation in their first autumn, and the result is that in an average year less than half the available cocks succeed in establishing territories. The successful ones occupy the habitat (the Callunetum) in a virtually continuous mosaic. The surplus ones become outcasts, and expendable; all of them are normally dead from secondary causes before the spring when breeding starts. 'Grouse disease' was a mistaken diagnosis of the after-effects of social exclusion (see pp. 6–9).

Individual territory sizes vary, for the most part in response to the nutrient state of the heather. To the untrained eye, one heather plant can look very like another, yet contain a critically greater or lesser amount of nutrients, especially the nitrogen and phosphorus that are incorporated into protein and other essential molecules (see p. 111). Grouse are extremely selective feeders. Only the green-shoot tips, comprising one part in a thousand or so of the live heather tissues, contain sufficient protein to supply their needs. They are not really living submerged in a sea of food at all. The prescription for having more grouse on the moor is to provide a higher plane of nutrition, and that means burning; but especially in the winter and early spring months, when the birds are most strictly territorial, they require not only good feeding, that is to say sufficient heather plants of the preferred age and quality, but cover from predators and storms as well, which means old tall heather. Both need to be present on the one territory, which usually extends to one or two hectares. Ideally therefore, fires should burn only small patches and be carefully controlled, so that a wide range of ages and heights of heather come to be juxtaposed. Burning in such detail is expensive, and it is clear that the long-term cause of the grouse decline was the doubling or trebling of rural wages that occurred between 1920 and 1950. Fires had become larger and the rotation less systematic as the number of estate workers diminished. An experimental revival of the standards of the 1911 report, undertaken by the Unit on the then neglected moor of Kerloch, showed that burning alone could double the number of grouse in four years, compared with an unburned control area nearby (Picozzi 1968).

An interesting exception to the general picture in the Grampian Region of Scotland where most of the research has been done, occurs in a few localities where the grazing of hill sheep on the moors has resulted in heavy infestations of sheep-ticks (*Ixodes ricinus*). The ticks are transient ectoparasites, and need to engorge themselves with blood from a warm-blooded host three times in their lives, as larvae, nymphs and adults, although adults are not often found on small hosts like grouse. Very locally, larvae and nymphs are so abundant that by sheer numbers they can weaken and kill some grouse chicks less than a fortnight old. But *Ixodes* is also the vector of a virus that causes louping-ill in sheep. Some tick larvae pick up the virus from their host during their first meal and pass it on to the grouse at their second feed, when they are nymphs; and it is clear that the virus causes a heavy mortality among chicks (Duncan *et al.* 1978). The four worst-infested areas studied (near Grantown-on-Spey, Morayshire) over two consecutive years, taken together, raised less than enough full-grown grouse to sustain the population, assuming a normal rate of adult mortality (which may not be justified).

From the evolutionary standpoint this tick-borne disease is a short-lived event, due to the grouse being challenged by what is to them a new virus, to which they are susceptible; in the long term one would expect them to build up a resistance to it. The challenge has occurred in this case because of human interference, in introducing to upland Britain the combined 'package' of domestic sheep, sheep-ticks and louping-ill virus, none of them native to the microcosm of the red grouse.

7.2 Lagopus lagopus scoticus

The genus *Lagopus* was long considered to have four species. Three were alike in having colour adaptations for living in snowy habitats, namely white flight-feathers in the wing at all seasons and a change of body plumage from dark to white in winter. The fourth, our red grouse *L. scoticus* Latham, 1787, was different in having neither of these characteristics; its wing quills are always dark brown and it retains a ruddy body-colouring all the year round. Two of the white-winged species are more or less circumpolar, namely the willow grouse *L. lagopus* and the rock ptarmigan *L. mutus*; the third, the white-tailed ptarmigan *L. leucurus*, is confined to the cordilleran region of North America.

In the British Isles there are two species. One is an unquestionable subspecies of *L. mutus*, accordingly named *L. mutus millaisi*; it is confined to the Scottish Highlands and is known as the ptarmigan, an old rendition of the native Gaelic *tarmachan*. The other, the red grouse, is equally clearly an insular derivative of the willow grouse; but the striking differences in coloration were accepted, long after the general introduction of trinomial names, as justifying full specific rank for *L. scoticus*.

I got to know the willow grouse in Labrador, the District of Mackenzie and the Yukon Territory when I lived in Canada, and on later occasions when visiting Lapland. I found the resemblances much more impressive than the differences between *lagopus* and *scoticus*. The famous 'becking' threat call, *go-bak, go-bak, go-bak* (or as Watson & Jenkins 1964, put it, *kohway - kohway - kohway*) is closely the same, in regions up to 6000–7000 km apart. In summer the willow grouse, when standing in tundra vegetation, shows little or none of its white plumage and looks like a red grouse, though usually some shades paler. All the *Lagopus* species have dark remiges in their first or juvenile plumage, and the retention of this livery throughout life by *scoticus* does not seem to have entailed any very profound genetic change. Neoteny—the retention of juvenile traits into adult life—is a familiar evolutionary phenomenon; it affords an escape route by which organisms manage to withdraw from their former adult specializations and make a fresh start. There are functional parallels in various mammals such as the boreal and arctic species of stoats, weasels (*Mustela* spp.) and hares (*Lepus* spp.), which show similar differences in the presence or absence or the extent of their seasonal colour change, varying according to climatic conditions.

These reasons for treating the red grouse as a subspecies of the willow grouse, rather than a separate full species as formerly, were generally accepted (cf. Ibis **98:** 160, 1956; Voous 1960; Vaurie 1965).

7.3 The red grouse as a research animal

The suitability of the red grouse for population research was only vaguely foreseen when we committed ourselves to it. Some good points were obvious enough, for example, the birds live out in open country where they can easily be seen, and they have loud voices and use them; the sexes are easily distinguishable at close quarters; and if and when one needs to kill them for experimental purposes, much less public objection attaches to killing gamebirds than, say, robins or blue tits. These are valuable traits; but as the work progressed, one unforeseen bonus emerged after another, and after ten years it was clear that the red grouse is an exceptionally amenable animal for experimental study in the wild because it poses so few insurmountable difficulties.

One of the earliest advantages to be disclosed was the fact that a line of experienced men with trained gun-dogs can walk once across a moor, find all the birds present and count them. If the area chosen for counting is level or slightly dished in profile, the birds that rise in front of the counters either fly back over their heads or out to the sides, or settle again not far in front, where their position can be noted so that they are not counted twice; an absolute number with a high repeatability is thus obtained. In fact much of the population in the winter half-year consists of individuals more or less confined to their territories and thus committed to resume a fixed pattern after being disturbed. If the individuals are tabbed, not only can their presence and positions be checked, but it is quickly noticed if they are absent or if another bird has come in their place. A few rough tracks exist over the moors along which an observer can drive in a Land-Rover; the grouse pay little or no heed and the vehicle can be stationed day after day as an observation post. The birds' relatively large size makes it possible to use tabs that bear ciphers legible at a distance; for the same reason the birds can carry miniature radio transmitters without discomfort. Even their individual variability in plumage, especially on the head and underparts, is exceptionally wide and allows the recognition of many birds without having to mark them at all (Watson & Miller 1971).

Catching is unfortunately not easy. For adults the best method is the one already mentioned, of trapping them on arable land adjacent to the moor, during the weeks after the grain has been harvested. If it is necessary to catch a particular bird for experimental purposes, mistnets can be used, though the monofilament nylon easily gets tangled and torn in the heather. Chicks about a month old, before they can fly too far, can be caught by bringing a hoop-net, mounted on a long pole, down over them on the ground; their tarsi are then sufficiently grown for ringing, but for tabbing they have to be caught later, after they are full grown.

Adult grouse vary considerably in weight, both seasonally and between years; the average for cocks in August–September is close to 690 g and for hens close to 600 g (Jenkins et al. 1963: 348). Full body size is reached in about 4 months from hatching, and the recruits can then contend immediately for adult social status. There are no cohorts of full-grown adolescents waiting around for several years, such as one finds in some bird species of similar size, complicating the population structure. Their mean generation time is similar to that of small passerine birds: with an annual mortality rate for adults of 0.65 (Jenkins et al. 1967: 106) the mean expectation of further life for full-grown birds, calculated by Lack's formula, $s = (2-m)/2m$, where m is the annual mortality rate, works out at 1.0 year. The oldest marked birds recovered, a tiny minority, were 6 years old (Jenkins et al. 1967: 107). Shortness of natural life speeds up the succession of generations and eases the homeostatic adjustment of numbers; these are good features because they shorten the gap between cause and effect. It also largely suppresses one of the social complications common to many birds, namely the rise in social rank that goes with age.

Mortality is largely homeostatic and compensatory (see p. 102). Consequently shooting grouse in moderation or not shooting them at all on a moor make almost no difference to mean life expectancy in the population, nor do they affect the subsequent population density. If some birds are shot, the result is that there are fewer to be eliminated afterwards by self-regulation, in the late autumn and winter; and if none are shot, more have to yield to fate as outcasts.

Most cohorts of red grouse are highly sedentary. They are completely non-migratory in the ordinary ornithological sense. In a sample of over

900 ringed as chicks and subsequently shot, 90% were killed within a mile (or 1.5 km) of the place of ringing and 95% within 2 miles; only one bird was found as far as 20 miles away (Jenkins *et al.* 1963: 323). Another experiment in a later year reported the recovery of birds that had been caught and tabbed in September–December when they were 4–7 months old, and were subsequently shot or found dead; the results summarized in Table 7.1 reveal a striking difference between the sexes (see p. 238). Their extreme sedentariness means it is normally possible to keep almost any cock and most of the hens under surveillance, if necessary for the whole of their lives, and to study the dispersion of offspring in relation to that of their parents. Although the usual autumnal dispersal phase is thus for most of the birds a very local affair, there is also another quite distinct dispersal phenomenon that occurs at much longer intervals and is not represented in these figures; it is discussed in Chapter 12.

Table 7.1. *Distances from point of marking of the subsequent recoveries of tabbed grouse.*

		Percentages recovered at various distances (km)					
	N	0–1.5	1.5–5	5–10	10–20	32	42
Males	147	98.0	2.0	0	0	0	0
Females	198	79.3	11.6	4.5	3.5	0.5*	0.5*

Data from Jenkins *et al.* (1967: 150).
*One female at each distance.

A different group of characteristics that help to simplify the solving of the homeostatic question belong to the habitat and its least constant resource, food. The openness of the habitat and the consequent observability of its inhabitants I have already mentioned. There are parts of Scotland where heather moors extend almost continuously for scores and less often hundreds of square kilometres, within which the dominant Callunetum is broken by boggy ground, water courses, rock outcrops and loftier 'islands' of arctic-alpine vegetation, but on a scale that does not offer serious obstacles to the grouse. More or less continuous populations of thousands of birds can therefore exist. Microclimates, soils, drainage, burning regimes, and rates of predation and disease, differ locally, sometimes with observable effects; but the overriding common feature is the

Callunetum. *Calluna* provides some nine-tenths of their food intake; for practical purposes they are monophagous, and this enormously facilitates the estimation of the food resource in quantity and quality, and in its seasonal, local and annual variations.

It has been suggested by others that moor management creates an artificial habitat, on which the grouse maintain populations so unnaturally dense that it may invalidate the results obtained in population research. There are two independent sources of evidence that refute this. The first is that the spectrum of moors available for study extends from the best managed, that is to say the most artificial, to the completely natural, neither burnt, nor grazed by domestic livestock. Significant differences are found between these extremes in the mean densities of grouse they carry; but on every moor, managed or not, grouse densities vary in the course of years, commonly three to tenfold; and none of the management differences either between moors or between years have been found to impair or defeat the homeostatic control system. The second source of evidence comes from the congeneric ptarmigan, living in the arctic-alpine life-zone which has never been managed or significantly altered by man. Ptarmigan breeding densities in the most favourable habitats and years reach 1 pair per 2 ha (Watson 1965: 146), which is within the range of red grouse densities found on the poorer managed moors in the same region (see Jenkins *et al.* 1967). The homeostatic adaptations of the two species are evidently similar. Moor management can be regarded as an illuminating experimental technique, the results of which show that the homeostatic adaptations of the grouse remain competent at all observable density levels.

It is an advantage to work with a herbivore when one attempts to analyse its trophic adaptations, and especially a herbivore that depends on a single food plant all the year round. Part of what the grouse eats and swallows, especially as evening approaches, passes into its crop, where some of it may remain for hours, moistened by a small amount of saliva but not undergoing any visible digestive change. Identification of food items and assessing the quantities of each can thus be done accurately, on large samples taken from the crops of replete birds shot for the purpose or obtained from hunters. On matters of nutrition the fact that the red grouse and the domestic fowl belong to the same avian order (though to different families) has proved helpful, because the nutritional physiology of the domestic fowl is much better known than that of any other bird.

Wild grouse eggs can be hatched in incubators intended for poultry eggs, or set under bantam hens that will brood and foster them. Herbivorous birds are in general the easiest to keep and breed in captivity, and poultry pellet-food manufacturers can make up special prescriptions for grouse. The Grouse Research Unit have kept captive flocks of all four British species of grouse for more than 10 years, and have used them for studies of metabolism, behaviour and genetics, and perhaps most significantly for revealing the natural variability of egg-quality, and consequently of chick survival, as a function of maternal nutrition.

The sum of these advantages makes the red grouse a particularly favourable vertebrate for population research. The pages which follow show that it has been uniquely possible to make observations and experiments, which demonstrate beyond question that mean territory size is the immediate factor that normally determines population density; and that the ultimate factor which accounts for most of the differences in density that arise in different years and places is the varying amount and quality of food available. They show moreover that in a normal thriving population, most of the mortality of full-grown adults is socially induced. Similar social mechanisms obviously occur in other animal species that are less amenable to close analysis, but where there is no reason to doubt that the same mechanisms work in broadly the same way.

CHAPTER 8

TERRITORY AND FOOD

8.1 Territoriality and the spring sex-ratio determine the breeding density

The cocks stop defending their old territories when their broods hatch. The chicks, always accompanied by their mother and usually by both parents, often move off the territory on their first day out of the nest (Lance 1978b). Aggression between cocks almost ceases in June and July while they are attending their young and are at the same time heavily in moult. During August some cocks resume their displays in the early mornings on their old territories, and a few weeks later in the second half of September the young males of the year also begin prospecting for available ground.

Acquiring a territory is a gradual process. Lance caught three young cocks that were brothers and fitted them with radio transmitters in September–October 1969. He called them *A*, *B* and *C*. *A* was much the heaviest (680 g) at the time of marking; *B* and *C* weighed 600 and 615 g, and *C* was in fact killed by a predator before he had fully established himself on a territory. They moved about independently during the day, but frequented the vicinity of their natal territory in the early mornings.

Having been pullets together in a family party, a dominance hierarchy had probably grown up between them from the first, in which *A* ranked highest. The goal of at least some male recruits seems to be to establish themselves as near as they can to their father's territory; and in this particular family *A* actually challenged his father (whom I shall call *P*) for its possession. *A* was found to have displaced *P* temporarily on 3rd November, and permanently the following day. Thereafter *P*, a tabbed bird without a transmitter, was never seen again and presumably became an outcast. *A*'s brother *B* secured a contiguous territory which also included a small fringe of *P*'s former ground (Figure 8.1). Both *A* and *B* survived the winter and bred in their territories in 1970, and again in 1971. It is normal to hold the same territorial focus for life, though the boundaries may have to be modified in succeeding years.

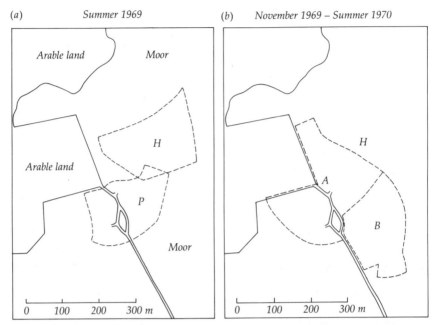

(a)   *Summer 1969*   (b)   *November 1969 – Summer 1970*

Arable land   *Moor*   *H*

H

A

Arable land   *B*

P

Moor

0   100   200   300 m   0   100   200   300 m

Fig. 8.1 (a) *Territories held by the adult male grouse* P *and* H *in the 1969 breeding season, near the edge of Kerloch Moor.* (b) *Territories taken by P's telemetered sons* A *and* B *in November 1969 and held in the 1970 breeding season. Redrawn from Lance 1978b.*

Next year (1970) A had two sons, *d* and *e*, who were also captured in the autumn and fitted with transmitters before release. Neither of them displaced their father A nor uncle B, but, at a time of increasing population density and diminishing mean territory size, they were able to carve out two territories for themselves just to the east of those of A and B. There was then a block of four contiguous territories all held by male descendants of P.

Harking back to P once more, he had an immediate neighbour to the north-east in the breeding season of 1969 called H, who was then holding his territory for the second year at least and, after another successful season, re-occupied it for the third time and bred again in 1970. Two young cocks from his 1970 brood, *f* and *g*, were given transmitters that autumn, and succeeded in wedging themselves in, side by side, between H and A, taking over about half of H's former territory and pushing him and his territory 100 m further to the north-east (Figure 8.2). Unfortunately Lance had to depart before the contest was finally resolved, and the curtain came down on this most illuminating drama.

Autumn 1970 – Summer 1971

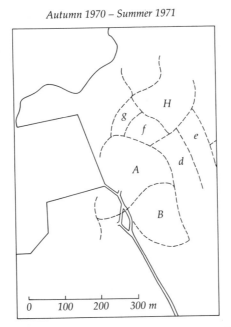

Fig. 8.2 *Territories held from autumn 1970 until the 1971 summer by A and B, A's sons d and e, and by H and his two sons f and g. Redrawn from Lance 1978b.*

Obtaining possession of a territory for the first time begins with frequenting the ground and presumably discovering empirically where the potential resistance is slack. The next step is to make an overt bid for the ground by joining in the dawn displays. These last for a few hours on clear calm mornings, and include vocal and visual rituals, a description of which has been given by Watson & Jenkins (1964). I quote or paraphrase part of it here. The red grouse has a pair of combs, one above each eye, much larger and redder in cocks than hens. When contracted they can be concealed by the small feathers of the crown; but so highly erectile are they that, when fully dilated, those of the cock are 2½ cm long and 1 cm high, bright scarlet in colour and deeply serrated along the edge. The hen's are smaller and paler. The combs are sensitive indicators of their owner's mood and determination. Asserting possession of one's territory begins with the 'becking' flight, in which the cock flies steeply up to about 10 m, sails poised on stiff wings for less than a second and then descends to the ground with rapidly beating wings, tail fanned out, head and neck extended and combs up. On the ascent he gives a loud barking *aa*, and on the descent a rattling *ka-ka-ka-ka* . . . of about 8 to 12 notes, gradually

slowing, and finishing with a gruffer, slowing *kohwa - kohwa - kohwa. . . .*
The cock also displays spontaneously from the ground, preferably from
an eminence, swelling his neck as he calls a vibratory *kok - kok - kok - kok -
kr-r-r-r*.

Cock displays draw attacks from adjoining claimants and from unestablished males, which the displayer must rebut with a counter-attack
to drive the intruder off. Claimants seek of course to acquire a holding of
adequate size, and each tends to become dominant to all its neighbours
on its own ground; but its assurance diminishes near the periphery, and
mutual boundaries come to be fixed along the line where the contestants
are aggressively balanced. Sometimes they meet there and walk with
threatening postures on parallel courses a few metres apart, each keeping
to his own side of the invisible line. Not infrequently the boundary
follows the crest of a knoll or a ridge which hides each from the other as
long as they stay on the ground and neither appears provocatively at the
top.

The eventual outcome of the dawn contests is to produce a territorial
mosaic with only occasional small gaps left unclaimed. As long as the
cocks are prepared to display, intruders on their ground receive short
shrift. But the early morning threat parade is soon over and for most of the
autumn day, birds of every social rank can meet, feed and move freely
over the moor, singly or in coveys and packs. The fortnightly Glen Esk
counts nevertheless showed that already in October the population
density underwent a sharp drop; some of the stressed subordinates were
leaving the moor, and their increased vulnerability led to a correlated rise
in recorded deaths from predation. On other occasions in October, for
example on the High Area in Glen Esk in 1959, the local population went
up as a result of displaced birds coming in, presumably from more
over-populated ground nearby, and staying till January when a winter
rise in social pressure forced them out. In most years and study areas the
largest decline occurred, usually rapidly, sometime between December
and February, marking the time when the territorial mosaic became
stabilized, and the owners were spending most of their time on their
claims, and would attack intruders at all hours of the day.

The population by that time is sharply divided in status into two
categories, an establishment, and a remaining surplus of outcasts. The
latter are subject to harassment by the former and are largely excluded
from the Callunetum. Mortality from starvation and predation among the
outcasts is high, although a remnant usually survive (as 'resident out-

casts') right through until April. One of the dramatic results of the e
tabbing of grouse was to show that, among birds living as outcasts in
November, the subsequent winter mortality from predation was almost
nine times as heavy as it was on established birds with territorial rights;
the difference is highly significant (figures recalculated from Jenkins *et al.*
1963, Table 37). Another striking result has been the discovery that,
in 20 years, on more than 10 study areas (i.e. 200 instances or more), there
has always been a population surplus to get rid of in the winter. The only
likely exceptions may have been on the few moors infested with a tick-
borne disease, already mentioned (p. 88), which were not included in the
regular study areas, and on Kerloch moor in 1977–78, at the end of the
exodus years, just when the research team had to leave (see p. 153).

The two smallest territories mapped in Glen Esk were only 0.2 ha in
extent, one in 1957–58 and the other in 1961–62, both of which were
nutritionally good seasons; both territories were held by cocks that failed
to mate. The smallest territory that supported a pair was 0.3 ha, also in
1957–58 (Watson & Miller 1971). These three were respectively about a
sixth and about a quarter of the current mean territory size for those years,
and were equivalent to circular areas of about 50 and 60 m across. There is
no way of knowing whether males abandon their holding voluntarily if it
gets whittled down below that size, or whether they find it indefensible
and are ejected by their neighbours.

The process by which the females became ranked as established or
outcast is less clear. Female dominance is not readily detected in the field
but probably exists. Some hens are seen to associate with individual
territorial cocks from late autumn onwards, others are not. With a normal
sex-ratio there will be more than enough hens in autumn to match the
number of established cocks, so that a female surplus roughly equal to the
male surplus is to be expected. Studying tabbed or telemetered birds has
shown that the accepted hens do not form permanent pair-bonds from
the first; mate-switching and adjustment of numbers tend to continue
during the winter. Lance (1978b) recorded that cock *A* became permanently paired by 8th January 1970, and cock *H* by 3rd February and
cock *B* by 15th April; but they had all had transient female partners at
earlier dates since November or early December.

In Glen Esk on the Low area in 1960–61 the sex-ratio on an intensively
studied 120 ha plot changed from 0.78 males per female in December–
January to 1.03 males per female in February–March, because 10 hens
and only 2 cocks had disappeared in the interval (Jenkins *et al.* 1963: 336).

Another analysis showed that in 1958–61 two-thirds of the territorial cocks which had no consort at 1st January subsequently disappeared from the establishment before 1st May; but of the cocks that harboured one or two hens at 1st January, only 7% disappeared (Watson & Miller 1971: 377).

The breeding season sex-ratio varies significantly from one year to another, and in the best years, when food quality and population density are rising or high, it may swing far enough to give an excess of hens over cocks because of the number of bigamous matings that occur. More often it is the other way round, leaving many of the territorial cocks unmated. Thus in April–May in different years in Glen Esk, cock-to-hen ratios varying between 1.4:1 and 0.9:1 were recorded (Jenkins *et al.* 1963).

Most of the unmated territory holders were birds that had only managed to obtain an undersized territory; and it seems likely that if the territory is below a certain threshold size (which would vary with the nutrient state of the heather) the male can be inhibited from taking a mate, just as he is inhibited from sexual maturation if he has no territory at all. Otherwise it is difficult to explain why surplus hens continue to be squeezed out of the habitat while there are still established males lacking a mate. It might be expected, at least at first sight, that a borderline female would consort with any established male, even one with an undersized territory, if he would have her and if matrimony offered even a small possibility of surviving the rest of the winter beside him. Moreover accidents do occur to higher ranking territory holders during the winter, and gaps are thus opened in the establishment which some small-territoried individuals would have a chance to fill.

It will appear later (p. 129) that both sexes, and especially the females, make their heaviest demands on the food supply in spring, when the hen starts building up the reserves she needs to produce a clutch of eggs. Savory (1975) found her daily food intake had already increased by a half in March, and attained almost double its winter level by May, when her body-weight reached its yearly maximum. She also eats smaller particles at that season, suggesting that she is being specially selective of the most nutritive parts of the plant (Moss 1972b). She sticks more than usual to the preferred age-class of heather; so too does the cock. Savory calculated that, even of this preferred age-class, no more than 20% of the annual production of edible material is consumed in the year by grouse; but his figure applies to the moor as a whole, and it is possible that in early spring when the food stocks are at their lowest, just before the new season's growth begins, a small territory might be hard pressed to provide the needs of a female as well as a male. Thus the cock might be inhibited from taking a mate simply to save his own skin.

What nutrient concentration points there are still left in the heather foliage in April, will be mobilized by the plant as soon as new growth is possible, in developing the first shoots and starting the production cycle. The grouse will go on depleting what buds remain right up to that moment, and almost certainly therefore, this is the month when over-browsing could most easily injure the plants. The fact that it is also the time of highest individual food-intake makes it doubly urgent that grouse numbers be brought down to a prudential minimum level which the heather can stand without detriment to its coming production. This is presumably the level that it was the function of the previous autumn's contest to predict and match, when nutritional feedback was being trans-lated into territory size, and the surplus outcasts were disfranchised.

A prediction in October can only be of limited accuracy, seeing that heather is much subject to winter injury from variable causes. We shall see shortly that the nutritional plane available to females on their terri-tories in April and May does vary considerably from year to year, and their reproductive output goes up or down accordingly. Under their homeostatic regime, their task is to optimize the production of young, within the now more predictable summer carrying capacity of the heather.

In the critical month or two before the heather growing season starts the population clearly needs to be stripped of supernumeraries. The outcasts, male and female, are mostly dead. Possibly the survival of batchelor cocks, which in years of depression can mean a third or even more of the established male population, is justified by the fact that, as late as March or April, a gentle winter and the prospect of an early spring would vindicate them all in taking mates and breeding. The territorial system may thus incorporate a male reserve, in anticipation that some of the subordinate territory holders will be inhibited from mating by their nutritional state, unless the winter has proved exceptionally kind to the food supply; on the other hand there are years when, at that time, some of them are found to have resigned their territories and disappeared. This is consistent with the otherwise unexplained delay on the part of many established cocks in entering into pair-bonds until the winter is well advanced, and it suggests that the highest ranking cocks ought to be the first to mate.

The most powerful controllable factor influencing the size of the reproductive output in most bird species is, of course, the number of females that are admitted to breed per unit area; and it appears to be regulated in the red grouse, first by varying the mean territory size, and second by programming the territorial contest so that the lowest ranking of the successful males only get small territories, which later inhibit them and their potential mates from pairing unless nutritional conditions are sufficiently good to justify it. Some of the inhibited males do survive, and they then have a chance of rising socially and breeding the following year; but the same opportunity is denied to the spinsters because females can have no territorial rights except as mates.

The hypothesis predicts that in sexually dimorphic vertebrates like the red grouse, the males, being more ornate, will be the epideictic sex, that is to say they will carry the main responsibility for monitoring and regulating the population density. I suggested long ago that in polygamous species such as the black grouse and prairie chicken, the cocks would be found to fertilize a sufficient quota of females and decline to inseminate the rest. This has never been demonstrated, partly perhaps because it is difficult to prove a negative; but the red grouse may be exhibiting a parallel phenomenon, in the rejection of the opportunity to mate by the batchelor males, and the expulsion of female outcasts while there are still unmated cocks in the establishment.

It is obvious that the social competition on which the homeostatic process depends has vital consequences for every member of the population, particularly between autumn and spring. Some young recruits like cock C in the telemetry experiments may get picked off by a predator while their social life is still unfolding; but even C could have already been put through social tests and emerged the loser, in his youthful tussles with his sibs, and thus been less secure than they. Lance (1978b) did show that cock A, who ultimately took their father's territory, had been the ace from the outset, the first to show territorial behaviour, the most active and dominant in combat, and the heaviest bird of the three.

For the vast majority of grouse that survive the first few weeks of infancy, it is the autumn competition for territories and establishment that decides their fate. In an average year some 60% of all cocks, regardless of age, fail to win a territory; they and a similar proportion of hens become outcasts. Only a handful of them subsequently get the chance of reversing their lot. Deprived of property ownership and the right to a portion of the food resource, they become vulnerable to 'Darwin's checks' (see p. 3); but these checks—starvation, predation, climate and communicable disease—are, from their point of view, only acting as executioners of the dropouts which the social system has already rejected.

8.2 Individual territory size, aggression, and some endocrine experiments

Not all territories are the same size even in the same habitat and year, so that the overall density of territorial cocks is equal to the reciprocal of *mean* territory size. The density of the total breeding population is usually rather less than twice this figure because, as we have just seen, in an average year not all the territorial cocks are mated.

Studies in Glen Esk quickly confirmed the variability of the breeding density from year to year. Annual surveys of the actual territorial mosaic were compiled by Adam Watson on parts A and C of the Low area for the five years 1958–62 (Watson & Miller 1971), and his set of plans for the first four of these seasons on part A are shown in Figure 8.3. The plans include whole territories, and the total areas they cover vary somewhat from year to year. The numbers and mean sizes of territories given in Table 8.1 take in some extra territories not shown in the plans, especially in 1961.

Table 8.1. *Annual changes in number and mean size of territories on part A of low area, Glen Esk (from Watson & Miller 1971: Table 2).*

Spring of	Number of territories	Mean size (ha)
1958	44	1.18
1959	21	2.48
1960	18	3.92
1961	42	1.42
1962	42	1.30

Every year at least half the grouse on part A were tabbed. In 1960–61, about 90% of the birds in the territories shown in the figure were so marked; some others in all years could be recognized individually by their plumage or by leg-rings. Without this high proportion of known birds the long and difficult task of mapping so many territories would have been

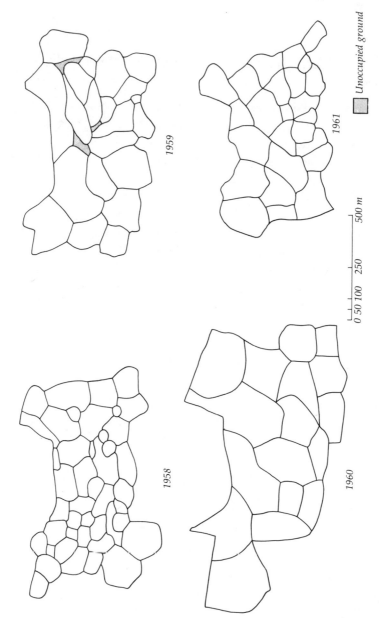

Fig. 8.3 Plans of the territorial mosaics on Part A of the Low area in four successive springs. The total area mapped in 1958 and 1959 covered about 50 ha; for 1960 it was rather larger (> 60 ha), and for 1961 rather smaller (~ 40 ha). From Watson & Miller 1971.

impossible. While making the survey Dr Watson sat day after day in a Land Rover, moving between different vantage points as required in order to watch every bird's activities at close range.

The maps reveal a between-year variation of 3.3 times in mean territory size, and a within-year spread of individual territory sizes that is even greater. The boundaries of individual territories depend to some extent on the lie of the land and what the owner can monitor from a central look-out. As mentioned already, boundaries are likely to run along the tops of ridges and hillocks, and the varying height of the vegetation can have a similar effect. But overwhelmingly the size depends on how much ground the claimant is programmed to demand and what vigour and dominance he can command in enforcing his purpose. Cocks assert and uphold their claims by their vocal and visible self-displays, and by attacking challengers and driving them away; and thus individuals can be scored for aggressiveness by counting the frequency of their threat calls or their wins in one-to-one encounters. The visibility factor within the territory can also be empirically measured; and it is no surprise to find that aggressiveness is a far more important factor than topography and cover in deciding individual territory size (Watson & Miller 1971).

Cock grouse that have bred and undergone the heaviest part of their moult resume their territorial dawn-displays in the late summer. The resumption was found to coincide with the reappearance of interstitial tissue in their previously inactive testes. The young males most precocious in showing territorial behaviour also have more advanced testes than their less forward contemporaries. This points to the androgen hormones being involved in arousing combative behaviour, and experiments were therefore carried out to test it (Watson 1970). A captive-reared cock was given, by a minor operation, a pellet of male hormone beneath the skin, of the kind administered to stimulate growth in poultry pullets. The pellet took about a week to act but after that the cock, who had previously been dispirited by the July moult, became very aggressive, pecking the fingers of anyone who came within reach and continually assuming postures of attack or making threatening calls. After a month he quickly subsided once more into a listless state. A control bird injected with corn oil instead (as a null substitute) showed no behavioural change.

Acting on this trial in the spring of 1965, two of the last surviving outcast cocks (named X and Y) were caught in the wild on 31st March and

3rd April. Each was given two 15-milligram pellets of male steroid hormone in the form of testosterone propionate. The presumption was that without the treatment both would soon have been dead. After a week they started challenging the local territory owners, and by 9th April both had pushed in and were gaining small new territories at the expense of established cocks (Figure 8.4). Neither obtained a mate that year (1965), but both were still alive in August. A third outcast caught at the same time and injected with corn-oil as a control, died as expected before April was out.

The implanted cock that survived longest after this experiment had an interesting further history. He obtained a much bigger territory in the same vicinity in the autumn of 1965, mated and bred in 1966, had a female companion during the winter 1966–67 but nevertheless spent the 1967 breeding season unmated. He finally became an outcast in 1967–68, his dead remains having been discovered in June 1967. He appears to have been consistently near the border-line between success and failure. An outcast in his first winter, he contrived to live through most of it before being experimentally revivified, but he did not take a mate; he bred in his second summer (aged 2) but reverted to unmated territorial status in his third year. He lived longer than most, into his fourth year, when he finished his life as an outcast.

A few weeks before this experiment began, a well-established territorial cock called A (not the same as the A in the telemetry experiment) had been caught close by (Figure 8.4) on 5th March 1965, and given a single testosterone pellet. The immediate effect of being caught and handled was to frighten him away, whereupon his neighbours lost no time expanding their holdings into his ground. He came back on the 8th and was at first unable to eject the neighbours; but within days he was feeling the euphoria of superdominance. He not only recovered his lost territory but was able to beat the neighbours even on their own ground. He exchanged his usual look-out post for one that had belonged to neighbour B, won back his own mate and, after two more weeks, acquired B's mate as well. He spent much more time than before in courtship; and his original moderate-sized territory of 1.7 ha grew to 3.0 ha. The effect of the pellet must have worn off soon afterwards; nevertheless, cock A retained all he had won, having evidently gained the acquiescence of his neighbours to his changed estate. As usual, a control bird given a dummy implant showed no alteration in his way of life.

Finally, a cock bird implanted with an oestrogen (female hormone)

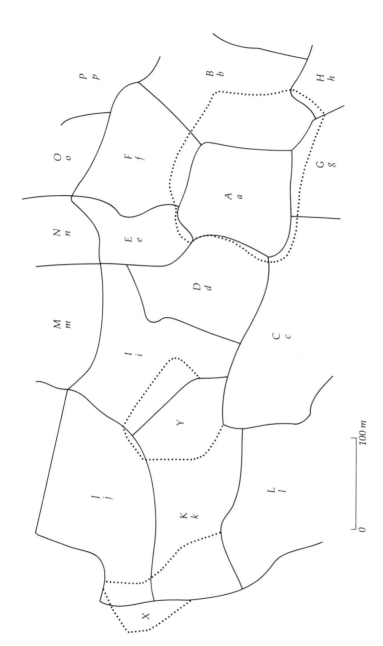

Fig. 8.4 Territory sizes and pairing attachments of red grouse at Kerloch in the spring of 1965, before and after three local cocks were given androgen implantations. X and Y had previously been resident outcasts. A was already an established territory owner. From Watson 1970.

pellet in the spring of 1968 lost first his mate, next his territory, and thereafter disappeared.

These experiments confirm the expectation that the individual's endocrine system will be involved, as a stimulant to success in social contests; and they give us some inkling as to why neighbouring cocks come to acquire territories of different sizes. Obviously there is a strong link between the individual's physiology and the dominance ranking he obtains; and since this is a classical setting for psychosomatic interplay, his social success or failure are likely to react upon his physiological state. One can predict that, in so far as social status has an underlying genetic component, it will not be a question of possessing a few crucial genes, but of a diffuse, broad-based effect. Success is likely to be attainable with any of a large variety of better-than-average genotypes, which contain many advantageous gene-combinations and few or no harmful ones.

The need for aggressiveness arises from the threat of displacement or encroachment by rivals. The nearer the rival comes the more dangerous he appears and the greater the stimulus to combat him. In any given year and habitat, the individuals with the largest territories automatically have the most neighbours to keep at bay, and they require to spend the most effort on aggression. Because mean territory size varies in different years and places the territorial mosaic keeps changing in scale. When the scale is small the neighbours are closer; but the *relatively* big territory owner still has the same number of neighbours as when the scale was big, and because they are closer he is spared nothing in his need for aggressive defence.

Since a bird's social status determines whether it is to live or die in the winter, it must follow that those which have been successful in establishing themselves in their first year will ultimately fail to repeat their success, and become outcasts in their last winter. (A minority get killed even while they are members of the establishment, almost always by predators or accidents.) Watson & Miller (1971) followed nine of the males that were territory holders in 1960 through to 1961. Between the two breeding seasons the number of territories on part A increased from 18 to 42, with the result that mean territory size was reduced to rather less than half. The figures are reproduced in Table 8.2 and show that the responses of the nine individuals varied widely. Seven out of the nine did reduce their territory sizes, but two increased them. The correlation coefficient between the territory sizes of the same individuals in successive years is 0.45, but the sample is small and it is significant only at $P < 0.1$ (t test).

The data draw attention to a second point, namely that when a male retains his established status in successive years there may be substantial changes in his territory size, with consequent adjustments in its exact location and boundaries. Thirdly, the table shows the strong correlation between the aggressiveness of the owner and his territory size. In 1960 the cocks were assigned the letters A to I in the rank order of their indexes of aggression (B is out of place because of a change in the method of plotting territories), and it turned out very much the same as the size order of their territories. And in 1961 when the aggression sequence had changed to B, G, I, A, E, C, D, H, F, it had become still closer to the territory size sequence (correlation coefficient = 0.851, P < 0.001).

Table 8.2. *Aggression and territory size of the same cock grouse in two consecutive years (from Watson & Miller 1971: 376).*

	1960		1961	
Bird	Territory size (ha)	Index of aggression	Territory size (ha)	Index of aggression
A	8.1	6.0	1.9	3.5
B	8.2	4.9	3.6	4.3
C	3.9	5.5	1.8	2.7
D	3.2	3.5	1.2	2.1
E	4.4	3.5	2.0	3.3
F	3.3	2.5	0.4	0.9
G	1.9	2.7	2.6	4.2
H	2.0	2.6	1.1	2.0
I	1.8	2.5	2.1	4.0

Even on the same moor in the same year, territories must vary among themselves in quality. Some will contain more 'good' heather than others, in terms both of nutrient availability and of the cover and shelter it affords, and some will give better visibility of approaching danger. It seems probable that the more dominant cocks will tend to select the best ground, and, because the food quality and quantity per unit area is greater they could thus be satisfied with smaller territories than would be necessary on poorer ground.

8.3 Food and nutrition as ultimate determinants of mean territory size

It has long been evident that the population densities of red grouse vary much from place to place, and that some moors consistently yield larger bags of game per unit area than others. Jenkins *et al.* (1967) gave a table showing the spring and August densities found on a set of six study areas which were spaced out in a belt some 80 km long, in the counties of Aberdeen, Kincardine and Banff, in the period 1957–65. The best moor among them, Corndavon, averaged 101 birds per km² in spring and 253 per km² in August over the nine years, which is two or three times the average for the poorest of the set. Improbable though it was thought to be in the 1950's, it was obviously necessary to check whether or not these differences reflected differences in the carrying capacity of the food resource, of which *Calluna* heather makes up some 90%.

There was already a straw in the wind. In the autumn of 1958–59, as previously mentioned, the number of territories on part A of the Low area in Glen Esk had dropped to less than half what it had been in the previous winter and spring, that is, from 44 to 21 (see Figure 8.3 and Table 8.1, p. 103); a dramatic die-back of the heather was associated with the drop. Back in March–April 1958 there had been a long spell of clear weather with hard frost at night and sunshine by day, and only patchy snow cover remaining on the south-facing slopes. The heather was exposed and it transpired in the sun, losing moisture it could not replenish because the roots were in frozen soil. The foliage dried up, and over many hectares the leaves turned reddish brown, ultimately withered, and fell off. So severe was the damage that it was not nearly made good by the following summer's growth. The grouse held on to the existing pattern of 44 territories, but the average breeding performance was poor. In the following territorial contest, in the autumn of 1958, a sharp increase in mean territory size occurred, as if in a delayed response to the poor state of the Callunetum.

In that winter there was a second incident, almost a repetition of the first. This time (1959) the foliar colour change was noticed at the end of January. A weather station had been set up in the interval, and an unusual duration of low temperature and low humidity had been recorded. When the new territorial pattern was formed in October–December 1959 the number of established cocks fell again to reach its lowest figure, 18 (Jenkins *et al.* 1963: 363).

These were the first events to suggest there might after all be some truth in the initial hypothesis, that the state of the heather was an ultimate factor in determining grouse numbers. They were quickly linked with the fact that the most productive moor under study in terms of grouse numbers, Corndavon, differed geologically from the majority of study areas in being underlain by basic bedrock, rather than acid bedrock or peat. Dr Miller therefore started a three-year survey of heather quantity and quality on 17 moors, with a view to comparing the measurements with the densities of grouse breeding on each.

Thirteen of the moors gave an almost linear correlation between the proportion of ground that was covered by *Calluna* and the number of grouse present. Two of the four exceptions could be accounted for by major differences in the average age of the heather found growing on them, since young heather is more nutritious than old: one had an exceptional amount of young growth due to active management, and held many grouse; the other had the oldest heather of any of the 17, and very few grouse. The remaining two were the ones on basic soils; both carried exceptionally high grouse stocks in spite of having a below-average cover of *Calluna*. Corndavon was one, and samples of heather shoots from there, when compared with others from a nearby acid moor, were found to contain 19% more potassium, 36% more phosphorus (P), 2.4 times as much cobalt and 21% more copper (Miller *et al.* 1966). These analyses were confirmed the following winter by Dr Moss, who showed that in addition the Corndavon heather contained 25% more nitrogen (N).

Moss (1967) produced two other illuminating pieces of evidence at the same time. He used captive-reared birds for feeding trials and established the fact that the grouse have no difficulty in eating and digesting, with bacterial aid, enough carbohydrate (cellulose and lignin) to provide for their energy needs, but their diet is relatively low in P, N, calcium and sodium. He watched some wild grouse feeding at dusk in November, and after they had plucked and swallowed a good quantity of heather he shot them. Immediately he cut samples from the plants where they had been feeding, and later hand-picked the most edible part, which is the current year's growth. When he compared the heather in the birds crops with his own samples he found that the grouse were very discerning in what they ate. On the average they had obtained about 30% more N and P than he had in his samples. By repeating the experiment on other occasions he showed they tended to be more selective when the heather was poorer in nutrients, and vice versa; moreover it was N and P they selected for, more than calcium or sodium (see also Moss 1972b).

This suggested that protein might be the limiting factor in grouse nutrition, at least in the winter half-year. Whether N or P was the more critical nutrient was uncertain at the time because they tended to vary in parallel, but subsequent experiments (see p. 120) pointed to nitrogen as being more often critical than phosphorus. Food of adequate quality was evidently fairly scarce, in the sense that by no means every heather plant would have shoots that could give a grouse a sufficiency of either; and the naive belief that grouse were buried up to their necks in food was evidently mistaken. It appeared that their territories needed to contain at least some substantial patches of heather that offered a high enough nutrient content; and since the birds are not seen to be feeding incessantly even on the shortest winter days, the limiting effect is most likely the rate at which they can extract N or P from the food in their guts by digestion.

Moss had three captive-reared grouse in the late autumn of 1967 which had become used to living in individual cages set out on the moor, where there was a good cover of 8-year-old heather (Moss & Parkinson 1972). The cages were moved to fresh heather each day and the occupants weighed; the heather was sampled, and the bird's droppings collected and weighed, prior to chemical analysis. The food intakes of the three birds were consistently different, averaging 62, 71 and 80 g (dry matter) a day; and analysis showed an almost 1:1 inverse ratio between the individual's food intake and the proportion of the cellulose content that it digested and utilized. Actually the middle bird was the one deriving most energy (in joules) from its food. We shall be led later to the conclusion that there are constraints on the bird's nutritional metabolism which apply particularly to its uptake of nutrients; so that if it is consuming the optimum quantity of food each day but is not succeeding in picking up enough nutrients to meet its physiological needs, there is no possibility of making up the deficit simply by eating more of the same kind: as hinted already, the only remedy is eating better food, of higher nutrient content.

These conclusions are reinforced by evidence of another kind. Grouse have the faculty of varying the size and length of their alimentary canal in accordance with the nutritional plane of their diet. Changing the size is a delayed process involving the growth of new tissues or the resorption of existing ones. The most labile parts are the small intestine and the two very long caeca, which open as a pair of blind branches near the point where the small intestine joins the large intestine. Moss (1972a) found that his captive male grouse, after being fed for several generations on high quality pelleted food with only a supplement of heather, had caeca and small intestines averaging only 54 and 72 % respectively of the length of those of wild birds.

The value of this adaptation presumably lies in saving weight—a matter of special advantage to a flying animal. Using figures that indicate the dry weight of their food, the voluntary intake of the captive birds was only 30 g of pellets a day, whereas wild grouse have to process on average some 60 g of heather (Savory 1978). The greater gut capacity of the wild birds increases the amount of food that can be held and processed at any given moment, and thus the total weight digested per hour or per day. Presumably 60 g dry weight is about the optimum intake for wild birds, where the optimum this time refers to the compromise that gives the highest net recovery of nutrients and potential energy, consistent with a gut that is not so long and heavy as to handicap the bird in flight. The fact that vegetable food entails having a heavy gut but can yield an abundant ration of metabolizable energy, may account for the fondness of game-birds for running when they need not fly, and for their expensive high-powered take-off when they do rise into the air.

In winter, when established red grouse are virtually confined to their territories, the heather is dormant, and the nutrients it contains therefore exist as a standing crop that is progressively diminished by consumption. The birds need enough high-quality heather patches to last them out until growth is resumed in the spring, and above all to cater for their peak food-demand of the year, which begins in March before the new growth starts. The scarce nutrients are vital not only to them but to the heather itself for the initiation of its own new growth; and the hypothesis predicts that the crop the grouse cream off will normally be conservative enough not to impair the production of similar crops in future years.

Bearing this in mind and also the relative scarcity of food of adequate quality, it is interesting to find that red grouse have been estimated to consume only 1 to 2 % of the total annual production of green shoots on an average heather moor in north-east Scotland, and less then 10 % even where good quality heather is most plentiful (Savory 1978; Miller & Watson 1978b). Nevertheless if there were two moors equally well covered with heather, their respective densities of breeding grouse would most likely differ in correlation with the amount of protein present per unit area of each. Nitrogen compounds, perhaps in the form of an essential amino acid or a vitamin, may prove to be the ultimate variable

with which grouse densities normally conform. And if the grouse consume only some 2% or so of the green heather produced, it is probably
safe to conclude that 90% or more of the green growth on an average
moor is substandard as food and would not sustain them.

The inference is that grouse can tell the difference between adequate
and inadequate heather. When Miller (1968) was making sets of fertilizer
trials on square-metre plots laid out geometrically on the moor, he
noticed that on the plots to which he added the most nitrogen the foliage
could be seen to turn a darker green within a month; and he found that
the grouse, mountain hares and rabbits could also pick out these plots in
an otherwise uniform stand of *Calluna*. The grouse may however have
some physiological indication of nutrient quality more reliable than can
be obtained by unaided vision and flavour. As a class, birds have surprisingly meagre organs of taste and smell compared with mammals. This
may conceivably arise from their different methods of masticating food,
which is typically done in the gizzard and not in the mouth as in
mammals; and the gustatory senses may have largely been supplanted
for them by deeper-seated perceptions. All the normal constituents of the
blood are continually being monitored by the kidney tubules, for
example, including the nitrogenous and phophorous compounds; and it
is not impossible that life-supporting information about diet could, as
with the deep-seated sensations of hunger, thirst, fatigue, breathlessness
and their opposites, be passed from some such receptors to the voluntary
centres of the brain. It seems doubtful if even mammals could appraise
the protein level in a sample of food just by olfaction and taste.

CHAPTER 9
MORE FIELD EXPERIMENTS

9.1 Experiments with 'shoot-outs'

The first four years' work in Glen Esk had shown that social competition in autumn divided the grouse population into two classes—established birds consisting of territorial cocks and their companion females, and rejected surplus birds that were to become outcasts and die in the forthcoming winter. It had been found also that the average territory size could change considerably from one year to the next, making a corresponding inverse difference in the population density of the breeding establishment the following spring. There were reasons for suspecting that the changes were connected with year-to-year differences in feeding conditions. The question arose, therefore, whether the changes were initiated by the male birds in response to feedbacks derived from the carrying capacity of the habitat, as the hypothesis predicted, or whether there was some other reason that rendered a variable proportion of the adults incapable of breeding, and allowed whatever cocks remained fit, to carve up the habitat among themselves.

Setting the stage to test this required that a sample population of individually tabbed birds should first be observed in detail, to ascertain the status of each resident and map the territories of the established cocks, before an experimental intervention was made. This was done in the first instance at Glen Esk, as a by-product of Dr Watson's territorial studies. Fuller details of the experiments described below, and others omitted for brevity, are given in Watson & Miller (1971) and Watson & Jenkins (1968).

Two tabbed first-year cocks, A and B, were caught on their territories 400 m apart, in late December 1960, and removed into captivity for 2½ days. While A was gone, an unmarked first-year cock took over his territory within 24 hours and consorted with his hen. B's territory was just as promptly occupied by a neighbour, N, who added it on to his own and was seen courting B's hen. When the owners were brought back and released they were little the worse for their confinement, and within 24 hours each had turned the tables on the invaders and completely restored the status quo. The unknown cock disappeared.

On 22nd February 1961 another three first-year territorial cocks were caught and held for a week. No unknown birds appeared this time, but all three territories and hens were added to the possessions of neighbouring territory holders. Once more it needed only 24 hours to restore the status quo after the owners were released.

These results were too ambiguous to throw much light on the question. On the one hand a number of neighbouring territory owners had been ready to take over the vacant spaces that appeared, and on the other, one strange cock who could have been an outcast had stepped in, and was obviously far from being unfit. Progress with the experiments was held up for a period by the removal of the research unit to Kerloch and the time required to set up baselines and appropriate conditions there (see p. 87). When they were finally resumed in 1963 it was clear in advance that larger and more drastic methods were going to be necessary.

In August 1963 on the Garrol Hill, a 32-hectare part of Kerloch moor, there were 14 old cocks, now resuming the occupation of their former territories, and 13 old hens who were their mates; also there were 16 young males and 23 young females just reaching full development. None of the young males had yet shown signs of territorial behaviour or had offered to challenge their elders.

Then, on 30th and 31st August, a marksman shot 13 of the old cocks one by one; the 14th managed to escape. Two days later, 14 of the young males, with another old cock that had appeared from elsewhere, had established positions on the vacant ground; the old cock who escaped shooting made a 16th, on his same territory; and the previous 13 old hens and one of the young ones were now consorting with the owners. Two of the cocks had small territories and no hens, but otherwise the size and disposition of the mosaic assumed much the same pattern as it had before the shooting. The most striking effect was that so many of the young cocks (all but four of the residents) were able to establish themselves, and do it so early, several weeks before they would normally have begun to challenge the older generation.

They were not given long to enjoy their success. In the second half of November all 15 of the new territory holders were shot; only the original old cock was spared. Recolonization at this date was much slower. Within the first fortnight there were only nine arrivals, seven young males and two old. Not all showed territorial behaviour immediately and it was a month before 10 territories were fully established. There were now 16 hens remaining. Months later, in April 1964, one of the old cocks disappeared, and in early May three new cocks came in and established themselves. The eventual breeding stock was 12 cocks and 14 surviving hens, all of which bred.

Here the inference was that there were fewer replacement males available than in September, and not all of them in a state to be instantly aggressive. Presumably the majority were being recruited from the outcast category, among whom the survivors were by this time declining in vigour. The experimental area may have remained understocked for the rest of the winter, judging by the later intrusion of three more cocks in May. The final population quota was not very different from the two that had been established the previous August and September.

These results were confirmed when the process was repeated, with slight variations, the following year on the same ground. In September 1964 11 of the breeding males were still there, with 10 young males, all the previous 14 hens and 10 young hens. None of the young cocks had yet engaged in territorial display. On 29th and 30th September, 10 of the old cocks and all 10 of the young ones were shot; so were six old and five young hens. Next day, 1st October, only the survivors were to be seen, that is, one old cock and 13 hens of varied age; but on the 2nd, two young cocks and two old ones, all four of which must have come from elsewhere, had appeared, and two old unknown hens as well. These six newcomers were forthwith shot also.

After that, recolonization was allowed to take its course. There were five new cocks after three days, seven after four days and nine after seven days, consisting of seven young and two old. With the surviving old cock, that resulted in the establishment of ten territories in place of the original 11. An equal number of hens stayed on, consorting with the cocks.

This time an additional 12 ha of moor adjoining Garrol Hill had been included in the first shoot-out. Its entire population of five cocks and four hens were killed at the end of September. Three days later the area had been completely reoccupied by four pairs of young birds and one pair of old. All the pairs on the surrounding ground outside the boundary were still in place and none had participated in the recolonization. Taking the two areas together, 16 territories had been replaced by 15.

Then on 16th December the 12-hectare annexe was cleared once more; all 10 of the birds living there were shot. Next day a young cock was there displaying; a second appeared after seven days and a third after 12–14 days. By that date four young hens had also come in and joined up with the three cocks. Four months later, in late April 1965, another young cock appeared; he established an extra territory and was joined by a hen, to bring the breeding stock up to four cocks and five hens, again close to the complement found there the previous September.

Conveniently, the 1963–65 experiments coincided with a period of static population affecting the 3000 ha Kerloch moor as a whole. There was a striking contrast between the movement of birds to recolonize the areas that had been experimentally cleared, and the virtual constancy that prevailed throughout on the control areas, which were being monitored for comparison. The repeated replenishment of the cleared areas up to approximately the same density as before, at least eventually, supported the theoretical expectation that the breeding establishment is related to the carrying capacity of the habitat, and consequently, in a given year and place, the majority of grouse should be programmed to seek the same norm of territory size.

There were of course other regular studies going on elsewhere on Kerloch at the same time. Some of the incoming birds had already been marked, and their immediate place of origin was known. They showed that a minority of the cocks moved in from territories already established elsewhere, up to 1 km away, but always leaving behind a smaller claim to take possession of a larger one. Many of the replacements came from nearer at hand, where their previous status was also known. The large majority were unestablished birds, either uninitiated recruits in August and September before the new territorial regime had been established, or, from October to May, non-territorial outcasts. The latter included birds of the year that had failed to win territories (or win acceptance if female) and, in smaller numbers, older birds that had since lost the territories or status they had held the previous season.

Thus a clear answer had been found to half the original question. The established population density *is* primarily determined by the sizes of territory that individual cocks take; and many of those that fail to establish themselves (as a result of social competition) are perfectly capable of holding territories and breeding if suitable space becomes vacant. The other half of the question was to be answered next.

9.2 Mineral fertilizers, and the uptake of nutrients by grouse

There remained a more direct way of testing whether or not the density of red grouse depends ultimately on the food supply. That was to upgrade the birds' plane of nutrition by giving the moorland soil a top dressing of fertilizer which would increase the nutrient contents of the heather. A pilot trial was begun at Kerloch in 1962. Miller (1968) set up experimental plots to test the effect of nitrogen and phosphorus in different dosages and combinations on the growth and chemical composition of the heather. A stand of *Calluna* in the 'building' stage was chosen, 6–7 years old, and five similar blocks were marked out each containing nine plots of a metre square. The two fertilizers, ammonium nitrate and ground mineral phosphate, were applied at three dosage rates so as to give the same nine factorial permutations in each block. The rates used were equivalent to 0, 20 and 60 kg of nitrogen and 0, 20 and 70 kg of phosphorus per hectare. The compounds were spread in June 1962, and random samples of heather were taken from each plot for measuring growth and analysing nutrient contents, in October–November 1962, 6–7 months later in May 1963, and 2½ years from the start, in October 1964.

On the appropriate plots both the N and P contents of the green shoots rose in the first summer. The increase in phosphorus was proportionally much the greater, and it had the side-effect of facilitating the uptake of calcium by the plants, and possibly that of nitrogen. As to heather growth, the nitrate brought the quicker results; where the dosage was greatest, single shoots were produced averaging 35–40% longer by the end of the growing season than those on the no-N plots. This occurred both in 1962 and 1963. The phosphate had roughly half as much effect on shoot growth, and none until the second season, 1963. In 1964 shoot growth was similar on all plots, indicating that the effect of the fertilizers had passed.

Although the experimental blocks were fenced to keep out domestic sheep and cattle, they were accessible to the smaller native herbivores, namely red grouse, mountain hares and rabbits. As a measure of the feeding intensity, the faeces of all three species were collected from the plots every 2–3 months; they were oven-dried and weighed. The grouse deposited an average of 14 mg per m^2 per week in summer and 21 mg per m^2 a week in winter on the plots. Savory (1975) later found that grouse actually eat less in winter when the days are short than they do in summer; but here they seemed at first sight to be doing the opposite, and spending 1.5 times as many bird-hours on the plots in winter as they did in summer. The explanation appears to be that in summer they were feeding less selectively and showing no preference for the fertilized plots, whereas in winter they assembled in significant numbers to feed on them, particularly those with the high-nitrogen dose, which received three times as many droppings as the no-nitrogen ones.

To check on this seasonal change in selectivity, Miller designed a second more extensive experiment. In June 1963 he marked out plots in pairs, widely isolated pair from pair and scattered over 80 ha of moorland. Ammonium nitrate at a rate equivalent to 100 kg of N per ha was applied to one square, leaving the companion square as a control. Here again, in the summers of 1963 and 1964 there was no difference in the weight of excreta dropped on the fertilized and unfertilized plots, yet in the winter half-year a threefold difference reappeared.

As regards phosphate, the first experiment had indicated no selection for high-P plants at any time at Kerloch; and this was confirmed in 1964–66 in another plot test in which six different dosages of phosphate were used, equivalent to 0–170 kg of P per ha. Again there was no evidence for selectivity by grouse. The hares and rabbits on the other hand did select for phosphorus in the first winter of the original experiment, and for nitrogen in both the first summer and winter; but they made no selection for either nutrient thereafter. Mammal faeces were not monitored in the paired-plot experiments.

The conclusions from these trials are, first, that the grouse could not have been hard up for phosphorus at any time and were satisfied, incidentally, with heather of lower P-content than the rabbits and hares were. Second, that grouse can readily fill their N requirement in summer, whereas they need to be strongly selective of high-protein food in the winter half-year.

9.3 Why are grouse more selective in their winter feeding?

There was no ready explanation then of why the grouse should be hard pressed for protein only in the winter. Moss et al. (1975) and Miller (1979: 123–4) later showed that there is a seasonal variation in the N-content of the green shoots, which stands about 33 % higher in June when growth is active than it does in September to March, when the amount stays roughly constant (see p. 134). It is of course the evergreen character of the foliage that enables so much of the protein to be retained through the winter, and accounts for the high resident biomass of deer, hares and grouse that the Callunetum is able to support. But the midwinter deficiency felt at least by the grouse is no doubt considerably aggravated by the shortness of the days, or rather the length of the nights, in this latitude (57°N), for the following reason.

Grouse feed much of the time throughout the hours of daylight, as is shown by the observation that at whatever hour a bird is shot, fresh food is commonly found in its gizzard, being ground before entering the intestine. The large membranous crop on the other hand usually contains very little during the morning, and even until evening in summer, because, by matching its food intake more or less closely to the rate at which the gizzard can masticate it, the bird avoids loading itself up with extra weight. It is not until the last few hours of the day that the crop fills up. Savory (1978) observed that captive-bred birds voided their woody droppings with remarkable constancy from hour to hour; and the fact that digestion continues during the night (although the bird does not feed then) is shown by the heap of excrement left by the roosting bird in the morning. Rock ptarmigan at 74°N in East Greenland hardly use their crops at all in the summer when there is continuous daylight, whereas in the shortest days their crops get crammed so full during the midday hours of twilight that their breasts visibly bulge (Gelting 1937). In January in arctic Alaska at 67°N the willow grouse were found similarly engorged when the dusk gathered at 15.30 h (Irving *et al.* 1967). The four heaviest crops the latter authors recorded were all just over 100 g fresh weight. (The birds from which they came were probably all cocks; almost all the hens were wintering further south.)

One must conclude that in *Lagopus* the crop is mainly used to supply the gut with food at night.

Wilson's (Leslie 1912) data on the crop contents of red grouse at different hours of the day (Figure 9.1) show that the mean crop weight of eight birds shot in winter at dusk (16.00 h) was 11 g (his fresh-weight figures have been converted to approximate dry weights by multiplying them by 0.45). The heaviest crop he recorded was 17.5 g. From Savory (1978) again we learn that the yearly period of minimum daily food intake occurs in December–January, when it is not much over 50 g. These figures, put together, suggest that an average red grouse will go to roost in midwinter with 20–35 % of its daily ration contained in its crop (and the arctic willow-grouse probably with more).

The need to prolong digestion into the night arises from the slow rate at which leafy materials can be digested. Their proteins are largely contained in cells with cellulose walls, and only a small proportion of the latter are ruptured even by thorough grinding in the gizzard. Like most

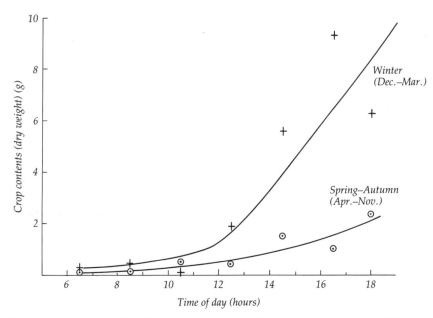

Fig. 9.1 *Mean dry weight of crop contents in small samples of grouse shot at different times of day, in winter and in spring/summer/autumn. Crop can hold more than 17 g. Drawn from E. A. Wilson's data in Leslie (1912:80).*

other herbivores, grouse cannot make cellulolytic enzymes for themselves and have to rely on intestinal bacteria for digesting the cell walls. Even so, a long residence time is necessary for the digesta in order to expose sufficient protein; and this suggests that the upper limit to gut weight which still allows the bird a sufficient agility when it flies, must also set a limit on the bird's daily digestive capacity. In the summer, with long days for feeding and a full crop at dusk, digestion can continue almost round the clock. In midwinter with a 16-hour night, there must be a considerable interval, beginning well before midnight, when the process is slowed or stopped. Because of this limitation less food can be digested, and a correspondingly higher quality of heather becomes imperative. There are, in addition, many days of bad weather and snow-covered food when the birds' intake falls even below their digestive capacity. Fortunately the volatile fatty acids the bacteria produce will normally provide them with ample energy.

In the arctic both willow grouse and rock ptarmigan eat quantities of dwarf-willow buds and twigs in winter (among other vegetation), and tend to increase the bit size of what they pluck in the reduced period of daylight (e.g. Gelting 1937). The red grouse, fortunate in being able to eat the same staple food all the year round, has little or no need to change the bit size; although females during their incubation period, when they take only enough time off the eggs to snatch some 20 g of food a day, do increase it by 50%, adding to the food bulk at some sacrifice to quality (Savory 1978). No doubt they also lose weight. Gasaway (1976a) actually found a large winter increase in caecum size in wild Alaskan rock ptarmigan; the caeca doubled in weight and grew 20% longer between August and October.

Being blind sacs the caeca necessarily work on a 'batch' process rather than a flow process like the intestine proper. They are the main organs of bacterial fermentation, and compared with those of most birds they are exceptionally large in the grouse family (cf. Leopold 1953). Only a liquid suspension of digesta is allowed to enter through their sphincter-controlled orifices, leaving behind an almost dry, coarse, fibrous residue which continues onwards through the large intestine and rectum, where it is articulated into faecal pellets. The caeca absorb, among other materials, much of the contained water. Besides breaking down celluloses and lignins the bateria synthesize useful substances that may include vitamins (McBee & West 1969; Gasaway 1976a, b; Gasaway et al. 1976). Little is known about the caecum physiology in grouse or of how the released proteins and the bacteria themselves are digested. Volatile fatty acids have been found in the red-grouse small intestine at concentrations similar to those in the caeca (Moss & Parkinson 1972); and Gasaway (1976c), who fed his rock ptarmigan on meals containing cellulose marked with radioactive carbon in order to make sure that they were absorbing and using the volatile fatty acids for energy production, found the marker beginning to appear in the CO_2 they exhaled from their lungs within 15 to 20 minutes, too quickly it seemed for any of the cellulose to have passed down the lengthy intestine and entered the caeca, to be digested there.

The caeca discharge their faeces back into the intestine and thence out through the vent, usually once a day soon after the bird awakens. One large, smooth, semifluid and coherent dollop is produced, completely different in character from the cylindrical packs of fibre in the dry pellets. A third to a half of the caecal contents are retained, and provide an inoculum for the following day.

9.4 Broadcast fertilizer trials

The fertilizer plot experiments prepared the way for a field-scale test of whether it was possible, by broadcasting nitrate over a moor, to elicit a response in the population density of the resident grouse. A uniform area had to be chosen, carrying enough grouse to show whether or not a significant change in density had taken place, and one sufficiently large to allow half the site to be left unfertilized as a control. It was also desirable to apply a dosage that would come somewhere near maximizing the possible uptake of nitrogen by the heather, so that its effect on the grouse, if any, would be large. That would mean using wheeled transport to distribute the tonnage required, and would place still more conditions on the possible sites, especially the need for accessible, firm, dry, nearly level ground without many rocks on the surface. There were only two such places on Kerloch, one on Garrol Hill (32 ha, half of which would be fertilized) and the other near Pitreadie farm (with 28 ha to be fertilized). The Garrol Hill ground was enclosed by a stock fence before the fertilizer was applied, but for lack of funds to perfect what was a costly experiment, the Pitreadie site was not enclosed, and the fertilized heather was prematurely demolished by cattle and sheep. The original plan to run duplicate experiments was thus abortive.

On the northern 16 ha of Garrol Hill a proprietary ammonium-nitrate fertilizer was spread as evenly as conditions allowed in May 1965, at a rate equivalent to 105 kg of N per hectare; no subsequent dressings were given. The southern 16 ha were left as they were (Miller *et al*. 1970). All the grouse present on the 32 ha were shot in July in order to leave the habitat in a virgin state, with no birds remaining that could be influenced by site-tradition and past experience.

The site was recolonized by equal numbers of birds on each half, namely seven; the previous establishment on each half before the breeding season had been eight. In other words, there was little change and no difference between the fertilized and unfertilized halves at that stage, notwithstanding that differences were already noticeable in the growth and colour of the heather. Analyses and measurements of the fertilized heather made 11 months after the application, in April 1966, showed 1.24 times as much biomass of green shoots, containing on the average 1.18 times as much nitrogen per gram, as were found in the unfertilized samples. Multiplying these together gives an estimated difference factor of about 1½ times in the nitrogen available per hectare.

The grouse on the fertilized part bred well in 1966, producing 1.8 young alive in August per breeding adult, compared with 0.6 young per adult on the control area, which gives a factor difference of 3 times. The territorial population established in autumn of 1966, rose to 17 birds on the enriched half, compared to a re-established quota of 7 on the control half—a difference of 2.4 times. These figures are all reasonably close, considering the difficulty of estimating how much nitrogen per unit area is actually accessible to the grouse, and the fact that there was only one pair of samples to compare, both of them small. By the following autumn (1967) the differences in grouse density and N-density had both disappeared. Most of the nitrogen is soon lost in the form of plant detritus, which under moorland conditions tends to accumulate, with little decomposition and recycling.

The results of the experiment agreed with predictions except for one anomaly, that at the first repopulation in the summer of 1965 the grouse showed no bias towards the higher nitrogen level already becoming apparent on the northern half. Instead they colonized both halves at the same density. The former stock had bred in April–June, and they and their progeny had all been removed in July, as already related. An influx of prospecting grouse began even while the shooting progressed; but these instant arrivals were also shot so as to leave the whole area briefly clear of grouse at the end of the month. Once shooting stopped, colonization was immediate (Dr A. Watson, verbally). *Calluna* is still in vigorous growth in August (the time when flowering occurs), and the build-up of crude protein presumably does not reach its peak until the end of September. Thus with two months growth still to come, the nutritional difference between the two halves of the site may then have been less than what was ultimately measured, in samples taken at the end of the following winter. There was no indication in August 1965 of the spectacular difference in flower production that was to strike the eye in 1966.

The settlers appear to have adopted a provisional mean territory size similar to that prevailing elsewhere in the neighbourhood at the time, and it is possible that in their hurry to establish claims they had not time or inclination to become conditioned to the higher fertility that was developing on the northern half. One of the shoot-out experiments already described (p. 116) showed that at the end of August, young cocks could be induced to seek territories prematurely, and it seems probable that much the same may have happened here. August is anyway the season of plenty, when the days are still longer than the nights and it is scarcely necessary to choose between plants on account of their nitrogen contents when feeding; the birds' responses to protein levels may thus have been less acute. Territories are not normally finalized until October or later, when the trophic situation is quite different, and there are standing food-crops, fixed in quality and quantity, in readiness for winter.

Two repeat experiments were made later on the Garrol Hill at intervals of five years, in 1970 and 1975. The first one gave positive results, the second negative. Both took place during a strong population fluctuation, and are discussed in Chapter 12. Seven others were done on various moors between 1964 and 1970 to gain experience of the economic possibilities of using fertilizers in moor management as a means of increasing grouse stocks. The outcome of these was to show that in most places where it is practicable to apply fertilizers, sheep or cattle are present on the moors in summer, and they wreck the fertilized plots or strips of Callunetum so rapidly that the grouse gain little or no benefit. The only remedy would be to exclude large herbivores altogether, and that would be economically impracticable.

However, one of the moors selected differed from the rest in having a substrate of limestone, and carrying heather with a naturally high nitrogen content. A commerical nitrate fertilizer was spread in May 1968 over an area of 35 ha at 0.5 ton per ha; and in spite of moderate to heavy grazing by sheep, the grouse increased. In the springs of 1969 and 1970 they reached densities 1.62 and 1.74 times those on the control area. The two following springs the densities on the treated and untreated moorland were converging once more, although breeding success was again higher on the fertilized ground (Watson et al. 1977). The more basic soil may have allowed more decomposition and recycling of the nitrogen.

Another experiment that included fencing was started at Glenamoy in County Mayo, Ireland, in May 1968 (Watson & O'Hare 1973, 1979b). It was done on flat wet peat-bog, where red grouse densities are nowadays extremely low, seldom reaching 7 birds per km^2 in spring, as against 50–100 per km^2 on moors in eastern Scotland. The difference is largely explained by the sparsity and heavy grazing of the Callunetum; but it may be due in some part to a general phosphorus deficiency. The nitrogen content of what little heather there is, on the other hand, was found to be quite high. Wet bogs are a poor habitat for Calluna. Phosphatic fertilizers were heavily applied to eight separate 1 ha plots, each newly drained, distributed over 71 ha area, and the whole area fenced all round to keep domestic herbivores out. The mixture used also contained sulphates of potash, copper, manganese and cobalt.

Six months later the N per gram of heather shoots had increased by about 50% although there had been no nitrogen in the fertilizer. The response of the grouse was delayed, apparently awaiting the growth in bulk as well as the rise in quality of the heather; and this growth continued for at least 3½ years until (in the autumn of 1971) the proportion of heather-covered ground was twice as much as on the control area. In the first spring (1969) the grouse numbers had actually fallen on both areas; but from that low point the spring numbers on the treated area rose to 1½ times in 1970 and 2½ times in 1971 what they were on the control ground.

Because of the difficulties that stand in the way of applying a uniform coverage of fertilizer to areas of moorland large enough to induce a significant rise in grouse numbers, it makes sense to try fertilizing separate plots over a wide area, as in the Glenamoy experiment in which only 11% of the experimental ground was actually treated. The results nevertheless affirmed that when the nutritional plane of a sparse population was artificially raised even to this extent, the grouse would respond by increasing their density. The Irish experiments emphasize that what is good for the heather is good for the grouse, and that it was phosphorus rather than nitrogen that was inhibiting heather growth at least on the bogs. On a hillside where the natural drainage was comparatively good and *Calluna* was already a main component of the vegetation, simple fencing, without any fertilizing, was successful in bringing grouse into a previously vacant site. Failure to obtain a response in heather-improvement experiments can obviously result from a wrong diagnosis and failure to supply the remedy that is critically required.

However, *Calluna* generally does make a strong response to nitrogenous fertilizers and speedily removes nitrogen from the soil; and this suggests that a deficiency of nitrogen must normally restrict its growth (Gimingham, 1972: 221–2). The natural input on moorlands comes largely from rainfall, and in eastern Scotland it averages only about 6 kg per ha per year, less then 6% of the amount applied in the Garrol Hill experiments. The quantity deposited varies with time and place and has positive correlation with local variations in annual precipitation (Robertson & Davies 1965). As yet few data are available for the Aberdeen region and they give no hint of any correlation with changes in grouse density.

The input of phosphorus to the Callunetum from any source is much less than that of nitrogen, but so also is the amount present in the heather plants. There is some indication that the considerable fund of phosphorus present in virgin peat soils can be depleted by long-continued grazing, with recurrent removal of the livestock for slaughter (Gimingham, l.c.).

A retrospective appraisal of the fertilizer experiments has been given in Watson *et al.* (1984a).

CHAPTER 10
FOOD AND BREEDING SUCCESS

10.1 Female nutrition, eggs, chicks and their survival

Established cock grouse are subject to stress for much of the year, starting in the autumn when their status is contested and won, continuing through the winter when territories are being held and defended, until the spring when they have to court mates as well. For the hens on the other hand a shorter and relatively more severe peak comes in the spring. The ovary and oviducts enlarge, a site is chosen for the nest, some 6–10 eggs are successively matured in the ovary with amazing speed and laid, and then incubated by the hen alone for 20 days or more. An average clutch of 8 eggs weighs about 175 g, and is 27% of the weight of a 650 g female.

She starts to prepare herself in late March, 5–6 weeks before laying begins, by progressively increasing her food intake until by laying time it may reach double the winter quantity. Both members of the pair become more selective in their diet than at any other time, picking heather particles only two-thirds the normal size (even smaller in hens than cocks), at rates that can average nearly 20000 pecks a day by the cocks at their highest in March, and more than 30000 by the hens at their still higher peak in April–May (Moss 1972b; Savory 1978).

Attention was originally drawn to this feature of the nutrition of female tetraonids by Siivonen (1957), who guessed correctly that the hen's success in building up her reserves is closely linked with her subsequent achievement in breeding. The Grouse Research Unit were soon on the same track. They began by bringing back clutches of eggs laid in the wild in order to hatch and rear the young in captivity, and after a few years discovered that not only were there great differences between years and moors in breeding success in the wild, but these were reflected in the results from artificial rearing. Between 1957 and 1963, 84 clutches placed in incubators or under bantams yielded a total of 515 newly hatched chicks, from six different moors; and the proportions of reared chicks that survived to the age of independence in August were compared, year by year and moor by moor, with the corresponding production of young in the wild. The latter had to be measured in an alternative way by the ratio of young birds to parents in August, which has all along been used as the standard yardstick of breeding success in the field (Jenkins *et al.* 1965).

The results are shown in Figure 10.1. Notwithstanding the small size of the individual samples and the correspondingly large variance, the correlation is highly significant, at P < 0.001. What the graph does not bring out is that this was only part of a long chain of correlated events that resulted in making some breeding seasons good and others poor. The

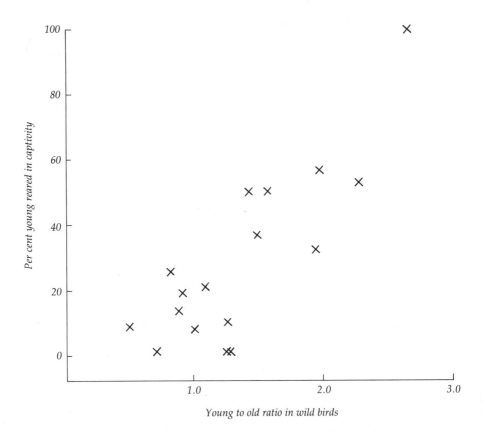

Fig. 10.1 *Correlation between the survival of young in the wild (measured by young-to-old ratios in August) and the percentage of success with young reared in captivity from eggs taken in the same localities and years (1957–63). The correlation is highly significant and indicates that much of the chicks' viability is predetermined in the egg. From Jenkins, Watson and Picozzi (1965).*

sequence began with the ratio of hens to cocks in the breeding stock being higher or lower, and laying starting earlier or later. Clutch size varied similarly, then hatchability (as a measure of infertility and embryonic failure), and the number of eggs lost in the wild through faulty incubation or desertion by the mother. After that there was variation in the diligence of the parents in caring for the young after hatching: in good years both of them generally accompanied their brood for two months, and they were bold in distracting predators including people, whereas in poor years often only one was found in attendance, distraction displays were seldom seen, and the family might be abandoned altogether within six weeks. Most of the chick mortality occurs in the 15 days after hatching, and from mid-June until August the young-to-parent ratios stay fairly steady. The

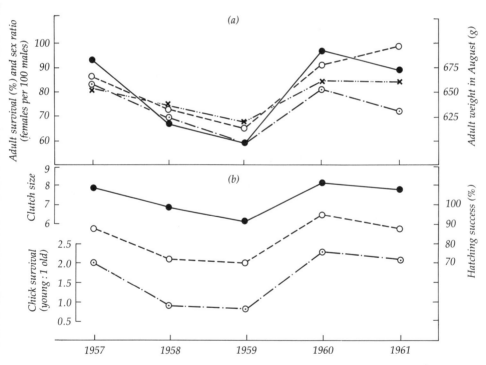

Fig. 10.2 *The Glen Esk correlations, 1957–61. (a) Adult weights, sex ratios and survival* ⊙, *adult survival in winter (1 January–31 March).* ●, *adult survival in summer (1 April–11 August).* O, *adult sex ratio in spring (number of females per 100 males).* ×, *adult weight in August (½ mean weight of cocks plus half that of hens). (b) Breeding performance.* ●, *clutch size.* O, *hatching success.* ⊙, *chick survival. From Jenkins et al* (1963:370).

Table 10.1. *Five years' data on relative breeding success in Glen Esk on the Low and High areas.*

Year	Mean hatching date		Clutch size		Young-to-old ratio in August	
	Low	High	Low	High	Low	High
1957	24 May	27 May	7.9	7.2	2.0	1.7
1958	4 June	12 June	6.9	6.2	0.9	0.8
1959	31 May	12 June	6.1	5.3	0.8*	0.6
1960	29 May	2 June	8.1	7.3	2.3	2.4
1961	24 May	30 May	7.8	7.1	2.0	1.9

Data from Jenkins, Watson & Miller (1963, Tables 19 and 23).
*July figure, as adopted by these authors, p. 346.

August young-to-old ratios, the August body-weights of adults, and the population densities established later in the autumns of the same years all followed the same course (see Table 10.1, Figure 10.2 and pp. 136–7).

It appears that nutritional factors were modulating the whole reproductive and homeostatic sequence from start to finish, through the responses evoked in the individuals concerned. It must be the mother's nutrition that affects the quality of the eggs and thus the viability of embryos and tender chicks. The same or other correlated factors must also be governing the quality of her breeding behaviour. Clutch-size depends almost entirely on how much nutriment she devotes to egg-production; and an essential dietary requirement she needs to incorporate in the eggs, possibly a vitamin, could occur in varying but often short supply and account for the difference between vigour and weakness in the developing chicks on a sufficiency/deficiency scale, as several vitamins are known to do in poultry. By the time the annual cycle has advanced that far, summer feeding conditions prevail; nights are very short, and individuals are not confined to territories but can roam at will in search of plentiful food. It may again be a particular nutrient deficiency rather than a general shortfall of food that tends to depress parental diligence even in June and July in poor years, and parental weight in August, at a time they are also burdened with summer moulting.

The maternal nutrition question was followed up by means of experiments with captive-reared adults (Moss *et al.* 1971), and by seeking additional data from birds in the wild (Moss *et al.* 1975). Captive grouse on their special pelleted diet have a higher nutritional plane than can be obtained from heather alone; but they also receive a small daily supplement of living heather because it has been found good for their health. The feeding experiment compared the breeding success of two sets of captive hens, one set being given a supplement of heather that had actually begun to grow, and the other set, dormant winter heather out of the deep-freeze.

The captives may do well, but they are far from being domesticated, and experiments run into difficulties, for instance with cock-hen relationships which can be hostile and disrupting. Nevertherless the females receiving growing heather laid nearly twice as many eggs, and in quicker succession, than those receiving dormant heather. (The eggs were removed as laid and set in clutches under bantams, and the experimental birds therefore continued laying until they could produce no more; two of them laid as many as 37.) The growing-heather birds also averaged 4.4 days earlier than the others in starting to lay, in each of two years although this difference was not significant. Possible differences in hatchability and chick survival were obscured by small sample sizes and large variances, so there was no clear indication whether or not the growing-heather factor had influenced the quality of the eggs. But proof had been obtained that growing heather does contain a factor of the kind postulated, which is deficient or absent in dormant heather, and leads to responses in the hen grouse that are known to be correlated with breeding success.

In the field study, the timing and relative success of breeding were compared with the earliness of heather growth on different moors, over 4–6 different years. The first few weeks of growth were found to produce a rapid rise in the N and P contents of the foliage (Figure 10.3). On some moors the factor that turned out to be most closely correlated with breeding success was how many days interval there had been between the start of heather growth and the mean date of the start of laying. This appears at first sight to conflict with the original finding in Glen Esk, that earliness *per se* was correlated with breeding success. But, as in many seasonally timed phenomena of living organisms, there is probably a photoperiodic pacemaker involved as well, prescribing a fixed target date for breeding, appropriate to the latitude and altitude; and if this were labile only within rather narrow limits, say a week either way in concession to current conditions, the laying date would vary less from year to

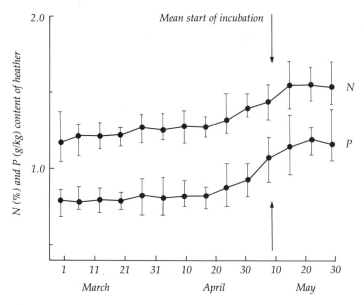

Fig. 10.3 *The N and P content of heather in spring at Kerloch in 1967. The closed circles are means, the bars ranges. From Moss, Watson and Parr (1975).*

year than the actual date of onset of heather growth, which is more dependent on temperature than on daylength. As a result, early laying *and* more days of growth before laying could both be expected to occur in the same year. The converse would hold in a late spring: that is, later laying, but sometimes before growth was visible at all.

Quite apart from the timing mechanism, the denseness of growth of the heather to which the hens had access on their territories in spring was also measured and found to be correlated with their breeding success. Density was averaged from samples of the biomass of last year's green shoots per set of 0.25 m² quadrats. As noted earlier in the chapter, it is not surprising that hens should benefit from having close-grown heather at the time they need to average 30–40 pecks a minute for 12–16 hours a day. This is also the time when both sexes eat substantial quantities of *Erica cinerea* and *E. tetralix*, amounting to nearly 10% of their diet (Moss 1972b); and though these heaths contain less N and P than *Calluna* they start growing earlier, which suggests they are being selected as a source of the unidentified laying factor.

A hen nests in her mate's territory, usually in heather tall enough to give shelter and concealment while she is sitting. The nest is a scrape lined with dry and fresh plant materials, some of which the hen can pull over the eggs when she wants to leave the nest. Red grouse lay eggs at intervals that vary with the individual, on the average about 1.5 days (Watson 1972: 8). The mean clutch in 112 nests over five years in Glen Esk was 7.1 eggs, with individual extremes of 2 and 15 (Jenkins *et al.* 1963, from whom most of the data that immediately follow are taken). As an index of the relative timing of the breeding cycle the easiest standard date to observe in the field is the hatching date, which occurs roughly a month after the first egg was laid. In the same five-year period (1957–61) in Glen Esk the mean hatching dates varied between 24th May and 4th June on the Low area, and between 27th May and 12th June on the High area (Table 10.1).

August had been chosen as the month for measuring breeding success because the young were then at nearly or quite full size, and the shooting season began on the 12th, providing large samples of dead birds for ageing, sexing and weighing; it also conveniently preceded the autumn territorial contest. The table shows that breeding success varied between 0.6 young to one old and 2.4 to 1, giving a fourfold range over five years, and also that the changes on the Low and High areas ran in close parallel. The general correlation between annual breeding success and the other population parameters is readily apparent in Figure 10.2, and in the supporting data of Tables 8.1 (p. 103) and 10.1.

10.2 Is the mean reproductive output per pair subject to homeostasis?

It is difficult to know whether the midsummer food supply is normally superabundant or not. The density of breeding females, on which population productivity and summer increase in numbers primarily depends, is regulated jointly by the density of territorial males and by the sex ratio. The latter showed a lower proportion of mated hens to cocks in years when the plane of nutrition and the territorial density were low, and a higher proportion when they were high (Jenkins *et al.* 1963: 336–7). In Glen Esk in the five seasons between 1957 and 1961 their combined effect was to vary the density of breeding females by a factor of roughly 2, between the poorest and best seasons, on both the Low and High areas. Population productivity, measured as recruits per unit area surviving in August, varied by a factor of 6.6 between the same poorest and best years. There can be little doubt that this larger variance reflects large differences in the summer food supply. The years of depressed recruit production showed lower average body-weights among the parent birds in August as well, suggesting that under poor conditions at least, homeostatic controls were operating on the reproductive output. In addition to this, recruit production, whether higher or lower, was correlated with the new territorial density established in the autumn of the same years. Later, at Kerloch, experiments were to demonstrate that territory size is inversely correlated with carrying capacity in September–October, and this makes it likely enough that recruit survival was being similarly adjusted to the carrying capacity already in evidence when their adolescence began in June.

Homeostasis may therefore be a continuous process, adjusting the population density to changes in food stocks and food production at each opportunity during the year, whether the chance arises through the autumn re-establishment of territories, or from winter adjustment of the establishment's sex-ratio, or from the effects of the spring nutrition of females on their clutch-size and the viability of the eggs and chicks. Watson et al. (1984a: 668) have noted that the effects of applying chemical fertilizers to the ground in spring and thereby inducing a significant increase in the quantity and quality of new heather growth after the young have hatched, did not result in increasing chick survival and brood size, which had presumably already been determined by that date.

The Glen Esk correlations could be described as the predictabilities of the reproductive process. Later years of the grouse project have revealed various unpredictabilities, and one of these is emigration, which redistributes the population on a larger than normal scale. In certain years it is indirectly detectable shortly after the young hatch, as soon, that is, as the establishment is released from its strictly territorial regime. Parents lead their chicks away, even before they can fly (which young grouse do very precociously while most of the body is still covered with down); and they travel sufficiently far to disappear and become lost to observation. The parents, especially the males, may soon come back, without their offspring. Except when it happens on a large scale it is difficult to detect, and this has made the whole subject elusive. The conditions that trigger it are still uncertain and further discussion is deferred till Chapter 12.

The perennial heather, making incremental growth each summer, renews the food crop for up to a year at a time, since grouse feed largely on parts of the plant that are less than one year old. Some residual effects on the plant and its productivity probably last longer than a year. Damage to heather from external causes can have similarly lasting effects. As just stated, this means that the same persisting nutritional plane can affect both grouse reproduction in a particular spring *and* the mean territory size taken up the following autumn; or alternatively it can run on from the autumn, when it governs territory size, right through the winter so as to predestine the summer's reproductive performance the following spring. Significant correlations can consequently occur between breeding success and the population density five months later, or between population density in winter and production ratios the following August (see Jenkins *et al.* 1967: 117–8). It needs re-emphasizing therefore that the population density in these grouse is independently established *de novo* by the territorial system, as the performance of naive young cocks and the shoot-out experiments (p. 115) clearly show; and when it happens that a good breeding season is followed by an increase in the winter establishment, to a level higher than in the previous year, the breeding success is not the cause of the increase that follows it. Each has been separately determined in relation to the same ultimate factor, nutrition.

Their independence is evident in various ways. The most striking is the fact that there never seems to be a shortage of recruits, but always a surplus instead, at the time the territorial system is being established. In other words, it is the territorial contest which cuts the recruitment down to size, not the recruitment which forces the establishment up.

There is also the fact that different moors are notably different in their capacity for producing recruits, reckoned either in young per adult or in young per unit area. When Jenkins *et al.* (1967, Table 2) compared the productions on six moors over periods of 4 to 9 years ending in 1965, the most productive averaged 1.74 young to 1 old, over 9 years, and the least productive only 1.06 to 1, over 4 years. The differences are largely correlated with the local productivity of *Calluna*, which depends on the exposure, drainage and type of soil. All the moors would probably be capable of sustaining a self-perpetuating grouse stock. Assuming a mean annual mortality of 60%, much of which is socially induced and homeostatically avoidable, an establishment of 100 birds would break even anyway on an August young-to-old ratio of 0.6, and on less than that if the rate of socially-induced mortality were relaxed. It is equally true that, over a decade or two, populations on the better moors show no persistent upward trend, nor do the poorer ones decline. Each is fluctuating up and down about its own mean carrying capacity, notwithstanding the substantial differences in their mean recruitment rates.

The best moors are celebrated for the harvests of grouse they yield; but under private ownership the shooting pressure has generally been sustainable, large crops only being taken in good years. The shooting season coincides with the run-up to the territorial contest, when the young cocks are looking for vacant ground. The result is that shooting mortality tends to be made good almost as fast as it occurs, from neighbouring areas where, because of their relative inaccessibility or for other reasons, grouse drives have not been organized and their natural surpluses are still intact. Jenkins *et al.* (1963, Table 31; 1967: 109) could find little or no relation between shooting mortality figures and population changes on their study areas between August and November. The grouse establishment seems to arrive at its homeostatically determined target-density, simply through territorial pressure between the cocks, which makes for give and take over a considerable area, with enough local redistribution to match density with capacity, whether the shooting is moderate, light or none at all.

If some moors have consistently higher productivities and young-to-old ratios than others, although the population of each moor remains steady in the long term and merely fluctuates up and down about its characteristic mean density, it must follow that high production is matched on average by high losses and low production by low losses. The losses occur through emigration as well as domestic mortality, and regrettably there are insufficient data to show whether the proportions in which these occur are more or less the same everywhere, or differ density-dependently. Socially-induced mortality is discussed in Chapter 13, emigration in Chapter 12 and annual dispersal in Chapter 14.

CHAPTER 11
GENERAL FACTORS THAT
REACT ON TERRITORY SIZE

11.1 The social hierarchy and plane of nutrition are independent factors

Dominance hierarchies develop in groups of mutually acquainted grouse in the manner typical for socially organized species; they express the two most characteristic and contradictory elements of a society, namely brotherhood and rivalry. Once formed, a hierarchy establishes the differences in rank that exist between its members and thus reduces the level of energy the members have to expend in further competition. It largely resolves the allocation of resources such as space, food and shelter. High rank entitles the bearer to a share in any new or temporary resource with little or no argument, while individuals of lower rank, according to precedence, accept whatever opportunities are left, in a stable situation in which quarrelling is unlikely to change anything and is therefore a waste of effort.

Many animals depending on foods that are immobile and occur in large patches, live and feed gregariously (the red grouse does so for part of the year); and as the individuals progress over the ground (or through the air or water) they avoid encroaching unnecessarily on each other's feeding space, to the mutual benefit of all. The hierarchy can still affect the organization of the flock, and if the feeding-patches are too small, or few and far between, subordinates can find themselves being crowded out of them. The group nevertheless holds the food crop as common property, often within a defined group-territory over which their rights extend.

There is some intergradation between this form of dispersion and the holding of fixed territories by individuals (and their dependants if any), even though the extremes are very different. Individual territories, whether for feeding, resting or breeding, require outright possession of property and dominance by the owner over any rival, at least at the focal centre of his site. Order, stability, and particularly the achievement of successful reproduction, under this kind of dispersionary regime, depend on the owners being granted a security of tenure for a sufficient period of time. The system is a good example of a mutual-benefit adaptation. Once the contest for possession is over, it obliges owners to content themselves with what they have got and respect their neighbours' rights. The hormone experiments (p. 105) give us some insight into how it works. The implanted cocks in their unseasonal exuberance were prompted to break the rules and infringe the established rights of others; but when, soon afterwards, they had sobered down and lost the power to resist, their neighbours calmly continued to accept the imposed situation without attempting to change it back again. Presumably during a natural autumn contest, when the boundaries are still fluid, cocks yield to their superiors and let them have what ground they want, or enough of it to satisfy them. Strong candidates finish up with holdings that contain ample resources for a pair; others have to be content with what is only marginally enough; and below them there are a residue of cocks that are lucky to have found even a batchelor corner, when the alternative is expulsion and almost certain death. Pressure within the establishment must thus ease off, until only the outcasts are left unsatisfied; and it is not long before most of them begin to suffer and deteriorate as a result of their defeat. The shoot-out experiments (p. 115) emphasize particularly how entirely personal these ownership rights to property are, and how quickly a death in the male establishment can lead to the takeover of his ground. Early in the winter especially, the owners may need to be present to claim their possessions every day.

When the territories have finally been stabilized and the outcasts relegated, the results will have been decided almost entirely on the basis of opinions, formed by each cock (and perhaps each hen also) of their rivals' chances of winning if it came to a fight. Scuffles between cocks do occur, but they are generally over in a matter of seconds, with no physical injury sustained bar the occasional loss of small feathers (Watson & Jenkins 1964: 167). Bearing in mind the vigour one can expect in any candidate that manages to win a territory, and in any consort that gets accepted by the male establishment, and even in some of those that fail to win such places, it is remarkable that the birds are satisfied to trust so much to appearances, and dispense with physical trials of strength. With the stakes so high, the deterrent is no doubt the great risk of exhaustion and irreparable bodily injury to closely-matched rivals, were they to fight it out. Selection would be particularly strong against losers that refused to concede defeat in time, before they got hurt, and perhaps not much less strong against optimists that resorted to lethal weapons in trying to force a decision. It will appear again later (p. 167) that the phenomenon of social selection, or self-adjudication between individual winners and losers, must be uniquely shrewd and discriminating. It depends on intuitions that are of great importance to the welfare and survival of the stock and must be as sharp and reliable as long-term adaptation can make them. The decision the opposing individuals reach therefore leaves little to chance, compared with the arbitrariness and 'luck of the draw' that sometimes figure so largely when Darwin's checks are the selectors.

Dominance and aggression are clearly phenomena of different kinds. Dominance is an abstract relationship established between individuals, whereas aggression is a concrete act or series of acts, used in the process of winning and keeping the dominant upper hand. The two are not always closely associated, especially in time, because the individuals that are acting most aggressively may just be the ones whose status is temporarily most in doubt, whereas tacit dominance can reign unchallenged.

Rank being a matter of relativity, each individual has to be programmed to act alternatively, as a dominant towards some conspecifics (or in some situations) and as a subordinate towards others (or in different surroundings). Rank is connected with the individual's physiological state and it will therefore have a partly genetic basis; but dominance is not a 'genetic trait'. Its genetic component must result from an integration of the effects of many genes, and individuals matched in rank can, like prize fighters or civil servants, be genetically very different. There are species in which the individual's age can be the most powerful single status factor of all. Aggressiveness as a tool in competition, on the other hand, might have a much less composite basis than dominance.

Social competition, apart from its role in deciding the precedence of individuals and so bringing harmony and order into the life of social groups, appears to have a second function which I term epideictic (see p. 23). This is to provide the competitors with an estimate of the current population density, which can be fed back into their homeostatic control centres. Many animals conduct what appear to be *ad hoc* communal demonstrations or epideictic displays, which are mildly or strongly competitive in character and thus density-dependent in their intensity; and to these displays I have attributed the function of attuning the participants to the density index, which their brains can then integrate with a corresponding food-supply index derived from their state of nutrition. The response elicited from the cock grouse in the autumn brings the first index into line with the second, by making the size of the territory appropriate to the current crop of food, and by ensuring that supernumerary cocks and hens are expelled; and later, in the early spring, by accepting or declining the formation of one or more pair bonds. In the fertilizer experiments when the nutrient quality and quantity of the heather was increased, some such integration must have taken place, otherwise no compensatory change would have been evoked in the population density and reproductive output of the grouse, such as Watson and his colleagues observed.

To provide an accurate population estimate, the epideictic demonstration should ideally secure the participation and/or the presence of all the potential rivals in a local group; and since quite a brief ceremony is often all that is needed, it tends to be synchronized at the appropriate season, and especially at the hours of the day when cooperation is convenient and when it is easy to tell the time, namely dawn and dusk (*Animal Dispersion*, Chapter 15). Sometimes, though not necessarily, the ceremony also entails assembling the participants at a rendezvous. The red grouse cocks do not foregather, but they commence their aggression at a set time, for an hour or two at dawn on each fine morning (except in high summer when the territories have lapsed). Under the best conditions their becking calls carry a kilometre or more, so that any particular cock may hear perhaps 25 or 50 others, and, in the course of time, pick up an accurate impression of the surrounding population density; individual voices will become known to him, and any sound gap caused by a casualty among the neighbours will soon be detected. The males of the lekking species of grouse such as blackcocks (*Lyrurus*) and prairie-chickens (*Tympanuchus*), on the other hand, do foregather at local arenas for their displays, held similarly at dawn. With them, dominance and aggressiveness are very conspicuously associated: the dominant cock or cocks hold their stances in or near the centre of the lek all the time, and sustain extravagant postures, with repeated war-cries and skirmishes, for an hour or even two or three without relaxing, while the lowliest cocks in poorer plumage stand inhibited on the sidelines, unable to perform any self-promoting act other than being present as spectators.

Watson (1964a) found there was a highly significant correlation in the red grouse, in any given year and place, between the 'index of aggression' shown by the owners, and the size of their territories. Watson's index took into account the frequency of encounters between owner and neighbours, and the outcome of the bouts, according to whether the owner won (3 points), drew (2 points) or lost (1 point). Failure either to challenge or retaliate at all scored 0 points. The most aggressive males tended to have the largest territories, and there was a linear relationship between the index of aggression and the territory size, with a minor residual variance due to secondary factors such as the amount of cover available, the presence or absence of visual barriers in the shape of ridges or hillocks, and patch-sized variations in heather density or quality. Accepting that the chief determinant of territory size is the male's mental drive, and his image of the area of ground he would like to obtain in a particular autumn and place, the extent to which he realizes his ideal presumably depends on his ability to compete with sufficient aggression to dominate his neighbours and exclude them from his chosen ground.

Watson's data were based on the scores he obtained from 19 individual cocks tallied in 1960. Nine of them again became territory owners in 1961 (see Table 8.2, p. 109). In the interval the population density more than doubled, and the mean territory size for these nine birds fell accordingly from 4.01 to 1.86 ha. Their average index of aggression, however, dropped only from 3.74 to 3.00; and considering that the sample is small and the calculations pertain to a rather subjective kind of index, this suggests that much the same mean individual effort was required to defend territories each year, notwithstanding the large difference between them in mean territory size. In fact, Watson and Parr (1969), making a generalized comment, stated that 'grouse interact more frequently at high than at low densities. . . . In the fertilizer experiment, cocks at higher density on the fertilized area interacted together more than on the control area alongside'. Similarly Watson and Moss (1971) say 'there was much strife on areas consistently at highest density and almost none at lowest density'.

In subsequent studies made in Ireland (Watson & O'Hare 1979b) where the grouse were dispersed at very low densities indeed, aggressive behaviour was virtually confined to an hour or less at dawn and dusk; nevertheless the duration of the calling period averaged longer on those study areas where the territorial density was rather higher. The daily number of song-flights per cock also tended to be greater on more populous areas, especially in spring. This again indicates a direct rather than an inverse correlation between aggressiveness and density, and the same conclusion receives strong confirmation when we compare the universally lower densities and aggressiveness of grouse on the Irish bogs with the far higher densities and levels of aggression found in eastern

Scotland, as the authors were well aware. In Scotland, moreover, aggressive behaviour in spring lasts longer at dawn and recurs repeatedly during the day.

The conclusion emerging from these facts is already clear. We have two independent main factors that influence territory size, one being the status of the individual cock and the other the plane of nutrition that is currently available. The first accounts for the individual differences in territory size in any given year and place, and the second for the mean size around which these differences are ranged. Confusion has arisen in the past because individual status is positively correlated with relative aggressiveness, and *being more aggressive than average procures the owner a bigger territory*. But contrarily, average aggressiveness is linked with the plane of nutrition: both rise and fall together, and both are negatively correlated with territory size; for that reason *average aggressiveness rises as average territory size gets smaller*.

This may be confusing at first sight but it is easily understood. The social hierarchy on the one hand, and the nutritional/epideictic computations in the brains of individual birds on the other, perform two completely different functions in the homeostatic machine, each affecting territory sizes. The function of the hierarchy is to arrange the competitors in an order of rank; and the function of the index computations is to determine what the target population density ought to be. These two are simply superimposed, with the result that individuals sufficiently near the head of the hierarchy succeed in appropriating enough territories (of various sizes) to fill the habitat at the right target density per km^2, whereafter the surplus tail of the hierarchy are left with no territories and no room, and become, *ipso facto*, outcasts.

11.2 Year-cohorts of males and mean territory size

Watson and Miller (1971) analysed some other interesting data from the Low area of Glen Esk, showing how the size of territories belonging to the same individual males changed in successive years during the 1957–61 population depression (see Chapter 8.2, p. 103). In the late winter, each year's male establishment was further classified into two age groups—young cocks barely a year old, and older cocks that had bred before, the great majority of them being in their second year. Four successive breeding establishments were analysed, and for three of them the authors found there were significant differences in mean territory size

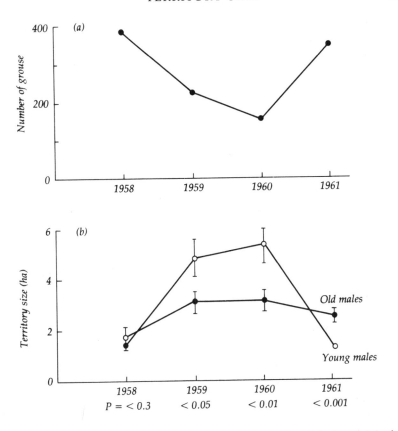

Fig. 11.1 (a) Breeding populations on the Low Area in Glen Esk (460 ha) in four successive years. Data from Jenkins et al. 1963:329). (b) differences in mean territory size of O, first-year males, and ●, older males in same four years (see Table 11.1). Standard errors and P, significance levels of differences are shown. Redrawn from Watson & Miller (1965).

between the one- and two-year-olds. The remaining year showed a similar but non-significant difference (Table 11.1). Populations were high in the first and fourth years of the series, and in these years the young cocks had on average smaller territories than their seniors; in the two intervening low-population years the juniors had much the bigger ones. Figure 11.1 shows the four different, successive sets of young cocks taking territories that comprise a strong fluctuation in mean size, while the four sets of older cocks reveal a weaker series varying in the same direction. The authors put the difference in a nutshell when they say that

Table 11.1. *Mean territory sizes of young (Y) and older (O) males on the Low area, Glen Esk, in March each year.*

Year	1958		1959		1960		1961	
Age group	Y	O	Y	O	Y	O	Y	O
No. of territories mapped	30	10	14	15	19	10	61	21
Mean size (ha)	1.36	1.74	4.85	3.10	5.32	3.12	1.43	2.53
Standard error of mean	0.12	0.37	0.74	0.41	0.66	0.39	0.11	0.27
Significance of size difference = P	< 0.3		< 0.05		< 0.01		< 0.001	

Data from Watson & Miller (1971, Table 9).

the changes in breeding density over the four years 'were mainly due to changes in the size of territories taken in autumn by the young cocks'.

Each year the researchers had mapped some 30 to 80 territories in advance of the breeding season; they also caught and tabbed their owners and classed them individually as young or older. To make it clear what happened it is necessary to take each of the year-cohorts in turn and follow it through its essentially two-year membership of the breeding stock (Figure 11.2). The first cohort were hatched in 1957, a good year as far as the state of the heather was concerned. The young cocks entered the establishment that autumn, taking on average small territories on which to breed in 1958, smaller in fact than those of their older companions in the establishment at the time. In their second year the survivors, themselves now classed as older, responded to the much impoverished food resource by more than doubling their territory size, and, after passing a second winter, they bred again in 1959. At the same time the new 1958 cohort were making their debut as the young breeding cocks of 1959, on territories even larger than those of their elders. Moving ahead another year, we find the breeding stock still further diminished to its lowest point in 1959–60. The young 1959 cohort had now entered and taken the largest mean territories of all; but the now senior 1958 cohort, on the contrary, had moved against the trend and actually contracted their territories to only two-thirds of what they had been the year before. The breeding season and summer of 1960, however, saw an upturn in grouse fortunes as the heather recovered from its two previous years of winter damage, and that autumn the 1959 survivors reverted right down to the small

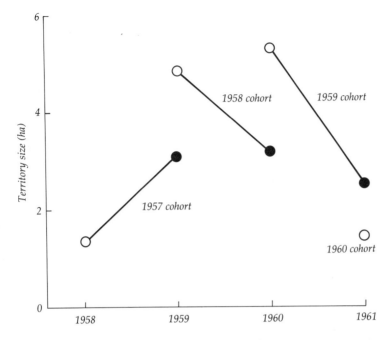

Fig. 11.2 *The data of Fig. 11.1b are re-plotted to compare the mean territory sizes taken by males of the same year cohort, when 1-year-olds (O) and when 2-year-olds (●).*

territories seen when the series began. But, as the graph shows, the newest cohort of 1960 once again appeared to outdo their elders in responding to the improved conditions, and took territories that were even smaller than the fifty-niners'.

The conclusion reached in Section 11.1 was that, in any given place and year, individual differences in territory size are mainly attributable to differences in the dominance rank of their owners. If so, in the first and last years of this cycle when feeding conditions were good, the older cocks must, on the average, have ranked higher than the young ones, whereas in the two middle years when feeding was poor the older cocks ranked a good deal lower than the youngsters. What can be the explanation of the switch-over?

In Chapter 10.1 (p. 129) we saw there were a notable series of population parameters in Glen Esk that all fluctuated in parallel and stemmed from the nutritional state of the heather in early spring (Figure 10.2); and it will be recalled that the last items in the series were adult survival rates in summer, and adult body-weights in August. Like the population density, these both fell to a low level in 1958 and a still lower one in 1959, during the months immediately preceding the autumn territorial contests of those years; on the other hand, in the years before and after the depression, i.e. in 1957 and 1960, they were both high. The mean weights obtained from large samples of cocks shot each summer were 679 and 661 g respectively in the two low years, and 695 and 707 g in the two high years, the differences being significant. A table of adult survival rates (Jenkins *et al.* 1963: 338) for the period 1st April–11th August each year, which refers to established birds between the beginning of spring and the end of the close season, puts their survival at 93, 67, 59 and 97% in the four summers of 1957–1960 respectively. It seems safe to conclude therefore that the cocks were in much poorer condition in the two 'low' summers than they were in the two 'high' ones. In contrast we are told that 'there was no significant variation from year to year in the mean weights of the young birds' in August–September.

Summer survival rates and August body-weights are the most relevant of the Glen Esk parameters to the present context; the others relate to the sex-ratio, to various components of breeding success, and to winter survival. When one recalls the hormone experiments and the deficiency of androgens evidently experienced by poor-conditioned subordinate males (p. 105–6), there seems at least a reasonable likelihood that the switches in relative dominance and mean territory size which occurred between the young and old sets of cocks were due (i) to the old cocks having a slight advantage over the young in autumn contests when they were in good condition, and (ii) to the young cocks, varying little in their health in autumn from year to year, being easily able to dominate the old cocks when the latter were down in weight and condition in the poor years, which is what Figure 11.1 shows.

Watson and Miller mention other possible explanations, the most important of which is that there might have been an unusually strong selection of dominant, aggressive types in the depression cohorts of 1958 and 1959, which resulted in the large mean size of the territories they initially took up. This is in line with Chitty's (1967) theory of population fluctuations; but the facts seem not to support it. There is a strong selection of dominant individuals for places in the establishment every year in the red grouse. When they compete for territories in autumn and the cocks are trying to repel each other, the pressure they encounter depends on the number of candidates there are per unit area and the mean size of territory that they will settle for. In any given year and place, much the same pressure must be felt by all the competitors, old or young, dominants or subordinates; it will gradually relax as the surplus birds give up the attempt and stability is attained along all the territorial boundaries. The proportion of candidates that fail is the measure of the selective intensity.

Jenkins *et al.* (1967: 116) show that, on this criterion, the selection pressure was not significantly different when stable and decreasing populations are compared together, though it was significantly less when increasing ones are compared with either of these. The size of the autumn surpluses expelled from the population in Glen Esk are approximately known for the years 1957–60; they were 61, 71, 56 and 37% respectively. In 1960, the only year when the selection rate was significantly different and fell to 37%, the young cocks actually took territories similar to and if anything larger than those held four years earlier by the young cocks of 1957, when the exclusion rate had been 61% (compare young cocks breeding in 1958 and 1961 in Figure 11.2). Conversely, the cohort of 1959, subjected to a very similar selection rate to that of 1957, took territories over three times as big as the 1957 cohort had taken. These results indicate that on each occasion the birds' response was conditioned by the food situation, and had little or nothing to do with the severity of selection for dominant qualities they had undergone. The behaviour of the 1958 cohort, which took very large territories on their first appearance and reduced them the following year, is more equivocal. On either hypothesis they could be presumed to have been dominated, when in poor condition, by their successors the fifty-niners. The fifty-niners however, which had also taken very large territories in the poor feeding conditions that prevailed at their first appearance, lowered their demands even more sharply in 1960, their second autumn. They were then back in excellent health (average August weight 707 g), which suggests they were not an innately aggressive lot but were again responding predictably to altered feeding conditions.

Thus the conclusion reached in the previous section is again borne out, namely that the mean territory size in a given locality and year is adjusted in response to the prevailing nutritional plane, whereas individual territories vary in size in much the same way in every year and local group, principally in correlation with the vigour and rank of their claimants. The hierarchy has to be involved in the territorial system because it is the instrument for producing social selection, and without it the population surplus could not be identified and removed.

CHAPTER 12

DO INERTIAL CYCLES PLAY A PART IN POPULATION CHANGE?

12.1 The Kerloch fluctuation, 1970–77

The preceding chapter did not find evidence to challenge the conclusions of fifteen years' research, that the red grouse continually adjust their population densities so as to match their changeable food supply, and that the food supply acts as an ultimate independent variable. No indication has been shown of other external factors influencing the birds' brains which result in determining mean territory size. The evidence collected from field, laboratory and controlled experiment has all been pointing the same way, to food as the main factor that induces homeostatic change; and by 1970 it appeared in fact that the *Animal Dispersion* hypothesis, of consumer populations moderating their food demands so as to insure a sustained supply, had been completely confirmed as far as the red grouse was concerned.

In 1969 a population fluctuation of unusual size began at Kerloch, which was then the team's home study-area. For three (and on the lower ground four) years in succession, grouse numbers built up and up, until in 1972 and '73 mean territory size had fallen to 1.1 ha over large areas, and more locally to 0.7 ha (e.g. on Garrol Hill, Watson *et al.* 1984a: 669–670). Concentrations as dense as these had never been seen there or anywhere else in the research team's experience. The population then declined just as dramatically for another 3–5 years, reaching a low in 1977–78.

The fluctuation was unexpected and its course unforseeable as it progressed through the 7–8 year period. Before it began, routine measurements of annual changes in the food-quality of heather had been discontinued because that aspect of the work was regarded as virtually complete; the research effort had turned to newer topics. Documentation of the event consists of standard spring counts, recording the yearly pre-breeding numbers on each of some 20 subdivisions of the moor; and of fuller and more varied demographic data for substantial parts of it, which included observations of tabbed birds, sex-ratios, egg-hatch-abilities and chick survivals, young-to-old ratios, and visible states of the heather. During the period two more fertilizer experiments were performed on Garrol Hill, starting in 1970 and 1975 respectively, and these did entail making regular analyses of nutrients in the heather on the experimental and control areas (Watson *et al.* 1984a).

By the time the fluctuation reached its low in 1977 there were convincing data to show that, especially during the decline, the population density changes were no longer tracking the current status of the food resources, and some rethinking and amendment of the previous hypothesis would be necessary in order to account for them. The fluctuation had taken on the characteristics of a population cycle of the type familiar in subarctic and continental tetraonids, including the willow grouse, and also in microtine lemmings and voles. It differed from the earlier fluctuation in Glen Esk (p. 110) in several respects, most conspicuously by starting with an upsurge that carried the density far above normal levels, and then swung over into a decline, to reach depths that were just as phenomenally low. At Glen Esk there had simply been a depression of density followed eventually by a return to the norm, which was sufficiently explained by a corresponding injury to and recovery of the state of the heather.

The Kerloch cycle moreover was not merely a local event. The newsletters circulated by the research unit at the time reported widespread simultaneous increases in other parts of the country in 1970–71, and lasting for another two years or more. Nevertheless it was not a universal occurrence, and many places were unaffected; but the initial synchrony suggests a climatic triggering factor. At Kerloch the upswing coincided with a run of mild winters causing little die-back of heather, and/or early commencement of growth in the spring. The Garrol Hill experiment data showed the control-area heather had better than average nitrogen contents in 1969, '70, '72 and '74. Young-to-old ratios in August were high or very high, reaching 3:1 locally on Kerloch in 1970, '71 and '72. In the four autumn contests of 1969–72 the cocks successively decreased their mean territory sizes, with the result that by the 1973 breeding season the density had risen by more than a factor of three since the cycle began.

On Garrol Hill, which is isolated by farmland from the rest of the moor, the number of territory-holders increased from 8 in 1970, the year the second fertilizer experiment began there, to 21 in 1973 on the untreated half, and from 7 to 33 on the fertilized half, multiplying their densities by 2.6 and 4.7 respectively. The effect of the fertilizer was thus to speed the population growth by 4.7/2.6 = 1.8 times, not much different from the outcome of the original experiment which produced a 2.4-fold increase. Incidentally the treatments of the north and south halves of the hill were reversed in each successive experiment (including the 1975 one) to counteract any residual effects of previous fertilizer applications.

As the population peak approached on the moor as a whole, an unusual proportion of the individually-marked parents were seen to lead their broods off, out of the study area and up to 2 km away, even before the young could fly. This type of emigration had been noted sporadically in former years on a small scale and had been attributed to the patchiness and changeability of the food supply at a time when the growing chicks were especially in need of high protein diets (Savory 1977). In and after the summer of 1973 it grew to such proportions that few young remained to join the local population, and valid young-to-old ratios were impossible to obtain. Nearly all the parent cocks and most of the hens that had left, afterwards returned home without their progeny. In 1973 and 1974 about 20% of broods left, but in 1975–76–77 the corresponding estimates rose to 57, 68 and 83% respectively (unpublished figures, and published data from Watson *et al.* 1984b).

During the decline mean territory size progressively increased again from 1974 onwards, and latterly the ratio of breeding females to males dropped to roughly 1:2, meaning that half the territorial males declined to mate, though not for want of available females (Watson & Moss 1980). There was no visible deterioration in the state of the heather; and estimates of heather production, based on Miller's (1979) finding that mean air temperature and total rainfall, taken together, accounted for 94% of its annual variation, suggested that in 1975 and 1976 the food supply should have reached its highest mass for a decade (Watson *et al.* 1984b: 655). The inference was clear that so severe a decline could not have been due to any commensurate fall in the plane of nutrition; and this was reinforced by the last Garrol Hill experiment, begun in 1975 with the express purpose of showing whether it could stop the slump: it failed to have any effect. As in the earlier experiments, the heather absorbed extra nitrogen from the fertilizer, although on this occasion its growth was not enhanced, being equally high on both the treated and untreated ground. Possibly this was caused by a delay in applying the fertilizer until June, instead of April or May, followed by dry weather which left some of the granules lying undissolved on the soil surface till July. The population cycle continued its unrelenting course on both halves of the hill until, in 1979, no grouse were breeding there at all.

The cycle could not be followed through to the return of normal numbers at Kerloch because a change of ownership of the moor brought the research group's lease to an end, and obliged them to move elsewhere once more. Before I conclude this narrative section, however, there are a few other details to mention and preliminary conclusions to be drawn. First, the cycle must be regarded as unique as far as this research project is concerned, and as forming an isolated episode quite different in character from the normal homeostatic fluctuations which reflect annual changes in the state of the food resource. Second, although there was a synchronized start affecting a very much wider region than Kerloch moor itself, the cycle did not take hold everywhere, and in many intervening habitats the status quo was not appreciably disturbed. Thirdly, where it did start, its further course appears to have been locally autonomous and self-fulfilling, as indicated by the fact that even on the different parts of Kerloch itself the synchrony of progress was soon lost. The 'low ground' took until 1973 to reach its peak, though some other sections had reached their turning point the previous year and were then starting to decline. On the higher east section the cycle ran so quickly that grouse numbers were down to a subnormal low by the spring of 1975, whereas on the low ground it took two more years to reach the corresponding level.

I agree fully with Watson and Moss that the phenomenon was new to their experience, and resembled one of the major cycles that willow grouse and rock ptarmigan more frequently undergo in other regions; but, as this and later chapters show, my view differs from theirs on the likely biological functions the cycles may perform.

12.2 A possible function

Tetraonid population cycles in their most emphatic form, for example in the rock ptarmigan in Iceland (see p. 176), are synchronized over large regions and have been known to follow one another at regular intervals of about 10 years for decades at a time before the rhythm is broken and a more stable regime supervenes. Variable climates are generally involved, at least in producing conditions that are propitious for rapid population growth, and in most places where the cycles occur they tend therefore to be less regular and often more localized in their incidence than is typical of Iceland. A single cycle isolated in time like the one at Kerloch is perhaps not uncommon as a weak expression of the phenomenon.

One of the remaining obscurities about group selection is the way in which group fitness is realized and takes effect. Presumably it is through the survival and expansion of the fitter groups, or through the dispersal of their genotypes to infiltrate other groups. In the red grouse, which I assume to be typical of many vertebrates, most of the annual surplus resulting from reproduction is promptly discarded in the form of outcasts, in the process of population regulation. Outcasts are normally produced in substantial numbers, and, being unable to gain admission to the establishment, their genetic value must generally be lower than that of the established birds. For that reason they would not provide an adequate body of phenotypes for export, to join forces with or supplant populations elsewhere. What seems ideally suited to give effect to group selection is the export of a fully representative cross-section from each strong and viable local gene pool.

In periods of environmental stability the furtherance of fitter local groups may normally be a low-key process. Local stocks of animals appear to be more or less isolated, with only a little, if any, gene-flow proceeding between them; and in the long term it is in fact considered to be of survival value to the species to harbour some degree of local genetic diversity (see Chapter 15, p. 202). Perhaps during the calmer epochs of evolutionary time, local gene pools whose members and organization are fitter than those of their neighbour groups just gradually supplant them along their common boundaries; and as the successful gene pools spread by this means, they also begin to acquire their own internal local variances. Thus the sum of diversity held within the wider population need not diminish in the long term as a result of local expansions and contractions.

Another on-going demographic requirement must commonly be the discovery and colonization of habitats that have newly developed, or fallen vacant by the loss of their previous owners; and if they are isolated from conspecific neighbours, that could necessitate a foreign-going type of dispersal. It would no doubt require an alternative brain program to the normal one that governs the homeostasis and structuring of philopatric stocks; and this might then be allocated to a proportion of the members of every generation genetically, or perhaps be possessed by all members but only switched on by the occurrence of an appropriate set of circumstances. Either mechanism could be made to result in the dispatch of a complete cross-section of the gene pool; but the second one could be closely identified with an intermittent phenomenon of mass exodus.

These are the tentative ideas that come to mind, in considering the possible function of the population cycle at Kerloch, which resulted in the wholesale emigration of phenotypes, most of them still very young and not yet thinned down in their variety by the local agents of selection. The ideas are developed further in Chapter 14.

CHAPTER 13

A REVIEW AND COMMENTARY ON THE RED GROUSE RESEARCH

13.1 Population density is self-regulated

This chapter concludes the middle part of the book which began with Chapter 7 and is devoted to the red grouse project. Two of its sections provide an overview and selective summary and their substance is mainly repetitive; the third introduces a new topic. Chapter 12 extended still further our conception of the complexity in the coded programming the birds inherit for promoting population management and population survival, in as much as 15 years of research had to pass, and almost as many generations of grouse, before the mechanism the chapter describes even started to reveal itself. Nevertheless the species has been so favourable a subject for study, even compared with other grouse species, that two of the most elusive questions about population control have been answered, and the answers verified by experiment, namely that the birds can control their own density, and normally vary it according to the changing quality and carrying capacity of the food supply. The research has laid much stress on nutrition, in attempting to expose as clearly as possible the homeostatic balance that exists between the birds and their food resources, and the reactions that are instrumental in preserving it.

The results emerging have implications that extend far beyond the red grouse, to the obligatory status that all animals possess as consumers of organic food, and the necessity for population regulation and structuring for its maintenance. One need not expect that any two species will be identical in the details of their homeostatic programs, but it is obvious that a vast array of other animal types have to cope with much the same consumer-versus-resource problems as the red grouse; and that because they show similar conventions in competing for property and rank they can be inferred to possess similar programming systems for population control. These widespread conventional mechanisms are indicators of a common sociality, and they have presumably had long evolutionary histories going back to paleozoic times when the first 'higher' animals emerged; and though sociality may have originated more than once in different phyla, their wide occurrence today must owe very much to descent from common social ancestors. If the sociality that species can be seen to share is an ancient attribute enabling individuals to collaborate for the common good, it strengthens the likelihood that the basic social mechanisms, used to bring about population regulation in the red grouse, have been inherited for the same purpose by very many other species as well.

The first requisite for effective regulation is that the members of the population should be able to operate the process for themselves. Possessing the capability of adjusting the density, virtually at any time, so as to admit only a determinate quota of residents to occupy the habitat, which then allows them enough to eat without putting future food-production at risk, must clearly be an ace card in the survival game; and its high selective value no doubt accounts for the complexity of programs that have evolved to make it possible. In the red grouse such adaptations have been proved to exist, to work efficiently, and respond to experimental changes in the birds' plane of nutrition.

For clarity, I shall first recall the way the birds achieve their control, and come back afterwards to the food supply aspect. The primary regulator is the territorial system, which lapses in the summer and is reconstructed each autumn to last for another 10 months; and though the system has some singularities peculiar to the red grouse, affecting the type of property holding and the manner of its acquisition, these do not influence the general deduction that property possession has been widely and variously conventionalized in animal species to serve as a population regulator. The principle of control is that by their mutual interaction and responses the males determine among themselves how many territories are to be held in a given area and year; and by hypothesis the size of territory a cock is programmed to claim varies from year to year according to his current feeding conditions and state of nutrition, with the result that the population density is re-set at a level that is unlikely to overstep the carrying capacity of the resources in the year to come. At the same time the males admit a corresponding quota of females to share the habitat with them; and any surplus birds for which no room is left after these requirements are met, are excluded.

The established birds do not have to kill the outcasts in order to get rid of them. The latter are sufficiently disadvantaged by their loss of feeding rights to cause their body-weights, gonads, sex-hormones and self-assurance to regress. They have been made homeless and exposed, and before the winter is over they succumb to predators or starvation or disease.

The exclusion of one individual by another from rights and privileges depends on their respective adoption of submissive and dominant roles, which is often decided without combat simply by threat and concession. Once the successful cock has carved out a territory which satisfies his motivation, the territory becomes a social passport for survival in the ensuing year, and he defends it accordingly. Boundaries are defined by the line where the pressures that neighbours are prepared to exert against each other are equal and opposite, and before long they become accepted and fixed, and the system is stable. All that owners really need to do thereafter is to exhibit to their neighbours that they are alive and still prepared to defend the status quo.

Stability is also assured by the virtual invincibility of an established cock on his own ground. There is actually an individual hierarchy present among near neighbours, shown by the fact that some hold considerably larger territories than others in any given season. Each has presumably won on merit enough ground to inhibit any further ambitions, taking all circumstances into account, and including a tacit decision that acquiring more at the expense of neighbours will not repay the extra cost of patrolling it. In practice, therefore, serious challenges to an established owner's possession are probably few. Stability is important, because if it did not prevail from midwinter till June the presumptive breeders would be deflected by constant harassment from the tightly scheduled feeding regime to which they need to adhere, and that might result in denying successful reproduction to some or even all of them.

Territorial rights lapse again when the eggs hatch in May–June. Parents, broods and unmated cocks are then freed to feed anywhere for a couple of months. Prospecting for territories by the young cocks begins in August, and old cocks, then through the worst of their moult, start to resume their former holdings. At that stage, territorial displays by cocks and the defiance of rivals are largely confined to the conventional hour or two after first light, and then cease for the rest of the day.

The mosaic of territories gradually stabilizes in October–November, by which time the birds' staple food plant, heather (*Calluna vulgaris*), has virtually stopped growing. A standing crop of food is consequently left for their winter use, to last until growth begins again in April or early May. Heather is a low-grade diet by agricultural standards, but it has the great virtue of being an evergreen that does not die back much in a normal winter. It is relatively non-toxic and is dominant and abundant over enormous areas. The parts of the plant with enough nitrogen to support the grouse are confined to the youngest shoots, grown the previous summer; and the highest concentrations of nitrogen are present in the dormant buds that are waiting for the coming of spring, and in the small seeds which stay on the plant but are gradually spilled by the wind.

Assuming that predators do not prevent it, any species of browser like the grouse should theoretically be able to raise its numbers and food-consumption rate above the limit the food-plant can withstand, to a level where not enough leaves and nutrients are being left in the plants to make good their recovery next season. Moorland soils are mostly poor, and the heather takes up all the nitrogen as fast as it is replaced; there is no reserve fund in the soil. In our region we frequently see the heather sward mangled and sometimes eventually destroyed by browsing ruminants, usually domestic sheep and cattle, and less often excessive stocks of red deer (*Cervus elaphus*); but neither red grouse nor mountain hares (*Lepus timidus*) are seen to inflict any gross damage on the heather in the wild. Since the birds themselves and not the predators have been shown to limit grouse numbers, there can be no doubt that their density programs are adjusted to maintain the Callunetum in perpetuity, and allow it a sufficient margin of safety to ensure, as far as lies within their power, that its productive capacity is not impaired by over-browsing.

The parts of the plant that contain enough nitrogen to nourish a grouse through the winter make up only a small fraction of its biomass, and some depletion must occur as the months go by. The hens in par-ticular will need about six weeks of high-nitrogen feeding between March and May to build up reserves for egg-laying, and at its peak their intake will reach twice its winter level. The cocks have a simultaneous but lesser surge of appetite. The heather is still dormant when the period starts; and in fact when the cocks originally take up their territories in the autumn, and mean territory size is determined in relation to their state of nourish-ment and feeding prospects at the time, they are in effect making forward provision for the resources a pair of birds are going to need in the critical months of spring. In practice the cocks' autumn forecasts may either prove over-optimistic six months later, if there happens to have been more than the average of winter die-back because of adverse weather; or they may be over-conservative if the winter has turned out unusually mild.

These deductions conform with the birds' actual mating habits. The established cocks and hens do not form permanent pair-bonds at least until January, and more often not till February or March. The most dominant males are probably the first to commit themselves to a particular female; and the tardiest males in the establishment often fail to mate at all, perhaps because they have smaller than average territories, and events now show that the food resources remaining are too small to support themselves and a hungry female as well, without being overtaxed. Some of these borderline cocks go forward to the breeding season as batchelors; others disappear, abandoning their territories altogether, and presumably becoming outcasts along with any remaining unmated hens.

On the other hand, when the early spring condition of the heather is considerably better than average and there has been little winter dieback, the most dominant 10% or more of the cocks, which hold the largest territories, mate bigamously, and in that event few or no hens may have to be off-loaded in April.

Thus the actual density of breeding females is not decided until the last moment, when the final, most accurate prediction can be made of the food that will be available for the breeding season. To avoid any misunderstanding I must make it clear that these responses and predictions are presumably subconscious and automatic, and chiefly reflect each individual's state of nutrition, of which his body-weight is a useful objective indicator. It would be a futile waste of effort, and of food production, to attempt to rear more chicks than the Callunetum is able to support without depletion. If the breeders were able to do just that, they and their progeny would be faced with reduced provisions when the next winter came because the summer feeding had been overtaxed, and as a result an unusually large quota of them would have to be sacrificed in the autumn as outcasts.

The indications are therefore that updating the female breeding quota is an ancillary homeostatic response, serving to forestall the likelihood of a general depression of the birds' individual body-weights in August–October, at the time of year when their collective biomass is swelled on average to more than double its April size by the members of the newest generation, now full grown. Adults are known to have fallen on occasion to subnormal body-weights in August after summers poor for heather growth, although at the time, the new recruits appeared to be cushioned against sharing their disadvantage and to suffer less privation than their parents. The difference is thought to have been responsible for the young cocks being able to achieve a larger mean territory size than the old ones in autumns that followed poor summer feeding, whereas normally the balance tends to swing slightly the other way.

These two mechanisms for adjusting the density of the population each year, the prime one depending on the number of territories set up per unit area, and the second on how many pair-bonds are actually formed, are separated by an interval of six months because of the red grouse's habit of taking up territories in autumn rather than spring. Most species of birds wait until nearer the breeding time before competing for the property that qualifies them for social status as breeders; and thus the number of pairs per unit area tends to be decided on up-to-date information anyway, and there is no point in or opportunity for subsequently varying the mating ratio. The peculiarity of the red grouse has exposed for the first time the fact that some socially and territorially qualified males and females, already in position and on the eve of the breeding season, decline the opportunity to mate; and this is done notwithstanding that grouse are such short-lived birds that more than half the male defectors and virtually all the female ones will never get another chance. The facts have been confirmed many times over the years. They appear difficult to interpret in terms of maximizing individual advantage. On the contrary, the shedding of supernumeraries to bring the breeding density into line with the carrying capacity of the food resource could obviously increase the chances of prosperity for the group as a whole. A couple of months later the territorial system fades out, when the young chicks appear and the summer feeding is thrown open to all the survivors and their progeny; and thus the benefits of individual cooperation can be seen to be shared for the common good.

A series of experiments in 1963–65 gave the clearest proof that male grouse regulate the territorial density each autumn. The individual territory holders on the experimental ground were shot, one by one, with a small rifle, so as to render the area vacant for reoccupation. When this was done in early autumn while there were plenty of adolescent cocks or surviving outcast males about, recolonization was rapid. In winter it took longer, but the net result was the same, that the new claimants took territories of almost exactly the same average size as their predecessors, and the establishment was thus replaced at the same density as before. In some experiments the ground was cleared a second time, giving three successive colonizations in one season with the same results.

This work brought out two critical facts, the first that there actually were many males in the neighbourhood that were being kept out of the establishment, although they were perfectly capable of holding territories and subsequently breeding if places became vacant; and it was the established cocks who possessed all the suitable ground that were keeping them out, and nothing else. The second crucial fact relates to mean territory size, the factor that governs the density of the established stock; and it is that, in a given year and place, all the males must be programmed in a similar manner as regards the size of territory they should seek, although in actual fact the system evens out the pressures by allowing rather more space than average to the more dominant birds and rather less to the more subordinate ones. There is no reason to doubt that the programmed size is about optimal for an average pair, in the prevailing state of the food supply.

Territory size here shows its complexities. Not only does it change in relation to the nutrient status and abundance of the heather, but at the same time and place some cocks have bigger territories than others, occasionally by a factor of 5 or 10 times between the largest and the smallest. It takes more effort to defend a larger territory than a smaller one. Experiments in which individual wild cocks received an implant of the male hormone testosterone showed that within a week they became hyper-aggressive. If the cock so treated already owned a territory he would enlarge it, even months after the mosaic had stabilized; if he were an outcast, even at the end of the winter, he would thrust his way into the establishment and win a place, thus escaping from his previous fate of death although the property he held was only small. In other words, social dominance has a strong connection with an individual's physiological state and in particular with the development and activity of his sex-hormone glands. Relative territory size is a tangible indicator of the owner's rank in the social hierarchy, although there are subsidiary factors such as topography and the amount of cover available, which affect his ability to supervise his property, and impose secondary modifications on its shape and size.

The adaptive value of the hierarchy lies in its being a low-energy instrument of exclusion within the social group. Where the number of interacting individuals is small enough it tends to take the ideal form, giving a different rank to each so that they stand in linear order. Then whatever the 'goods' may be that the individuals seek, they will be shared down the line in sufficient quantities as far as they will go, leaving the remaining tail of candidates to do without. If goods in short supply were divided into equal shares for every candidate, none of them would receive enough; and if there were no social conventions, they would expend energy and inflict injuries on one another, fighting to possess the goods.

This is why the hierarchy becomes incorporated into the allocation of grouse territories. A mechanism is required to sort out the candidates into two categories, the haves and have-nots, and the job is done most economically and advantageously to the group, by first establishing linear rankings that run from top to bottom of the small interacting groups concerned. After using them to find out which, and how many, of the members will obtain territories, and which will be excluded, the personal rankings still persist, and when the successful cocks proceed to settle among themselves where the mutual boundaries of their territories are to lie, the more dominant of them finish up by commanding more space than the subdominant ones.

Thus we find a system in which mean territory size is determined by the state of the food supply, and this basic response is overlaid by a second condition that provides a greater margin of safety to the higher ranking birds, and a greater risk of failure to those of lower standing. Because the basic size depends on nutrition and not on aggression, there is no absolute relationship between aggressiveness and territory size, such that x units of aggressiveness per hour will always secure or defend y ha. The contrary turns out to be true, and in fact when the food supply dictates a small mean territory size, the owners are brought closer together; the threat they actually perceive from neighbours is increased by their nearness, and in total they expend more energy in keeping one another at bay than they do at times and places when they are more spread out and further apart.

Another indication of the relativity of hierarchical differences in territory size is that, in the experiments where the first-rate cocks were shot and the second-raters took their places, both lots showed the same mean territory size, and within both lots the birds reacted according to their own relative status differences to produce a range of individual territory sizes. At first glance one might have expected the second-raters, being presumably less dominant and successful than their predecessors, to have all taken smaller territories; but that is not what the experiment showed. Instead, an individual's aggressiveness and territory size depended on his relative standing in the group he happened to be in, and could change when the membership of the group changed; and the mean territory size remained the same for both teams. This principle must of course apply generally to social hierarchies.

13.2 The plane of nutrition as the ultimate factor

The hypothesis that led to the red grouse research being undertaken, in the hope that it could be tested, predicted (1) that the population densities of vertebrates (at least) would generally turn out to be self-regulated, the self-regulation being achieved by social mechanisms; and (2) that the densities would be linked to the carrying capacity of the habitat, as determined by the productivity of its limiting resource, generally food. The second proposition has been confirmed as fully as the first by the research, although at the outset it had seemed a much more improbable one to the research team. Grouse live at much lower densities than one is accustomed to see with farm livestock, and yet the heather on which the grouse depends is usually present on the moors in greater biomass than the pasture grass on farms. It took a while to realize that heather productivity is small compared with that of grass, and its nitrogen content so critically low that for much of the year the grouse are obliged to pick out just the nutrient-richer tips of the green shoots, consisting mainly of buds and young growth.

Later these dietary circumstances were to prove especially opportune for the investigators. The fact that red grouse are virtually monophagous herbivores, feeding on a plant that makes a small and readily measured annual increment of growth, has made it easier to quantify, analyse and experiment with their food supply than has generally, or perhaps ever, been possible to previous workers when studying the feeding ecology of a wild bird.

Within the first ten years experiments were begun on a field scale, by broadcasting over the vegetation a proprietary nitrate fertilizer which previous trials had shown capable of upgrading the growth and nutrient quality of heather. There were practical difficulties to be overcome, one of them being the exclusion of cattle and sheep from the fertilized ground; but where this was done the most successful experiments showed the grouse would respond to the treatment by roughly doubling their population density, compared with that on a contiguous untreated area. The change in density was similar in magnitude to the change in the amount of dietary nitrogen per unit of heather.

Better nutrition prompts a cock grouse to seek a smaller territory and thus accept a higher population density, and vice versa. Quite apart from experiments, heather varies naturally in its nutrient quality from year to year and place to place. The differences that can be demonstrated by chemical analysis are not necessarily apparent to the eye of a human observer, but they are undoubtedly recognizable to the grouse. Consequently variations in grouse numbers also occur in the course of time, and some moors, usually those underlain by neutral or slightly basic soils as opposed to acid peat, have consistently higher stocks and higher productivities of young than others.

Experiments of a different kind showed that the nutrition of laying hen grouse can have great effects on the viability of their eggs and chicks. Poor nutrition in the run-up and laying period was found to lower egg-hatchability and increase postnatal mortality. Not surprisingly, a lowered nutritional plane (perhaps including shortfalls of accessory food-factors such as vitamins) has major effects on the parents, and in such seasons the average ratio of adult hens to cocks is lower, the hen's constancy in brooding the eggs falls, and so does the standard of care the parents expend on their young. As a general result the number of recruits per parent that survive to obtain their independence in August is low as well. When spring nutrition is better than average the opposite effects are produced. Nevertheless it is almost always true that more than enough recruits are produced to counterbalance the year's mortality that has occurred in the establishment, and this usually necessitates the shedding of a large surplus of full-grown young and adults in the autumn.

It is important to realize that when the population density is fixed each autumn by renewing the territorial system, the density is dictated by the apparent carrying capacity of the food resource; and whether the previous summer's breeding success has been good or poor is of no real consequence to the result. Breeding success and the subsequent new territorial density do in fact quite often turn out to be correlated, because they have both been independently determined by the state of the heather, and this has remained much the same in the autumn as it had been in the previous spring. But autumn density is not directly dependent on the summer's breeding success.

Interesting conclusions have followed from biochemical studies of grouse nutrition. Vertebrates cannot manufacture enzymes for digesting the most abundant carbohydrates that plants produce, namely cellulose, the hemi-celluloses and lignins, and herbivorous vertebrates generally culture symbiotic microbes in their digestive tracts to do the job instead. This is indispensable to the effective use of a plant diet, because the proteins and easily assimilated carbohydrates the plants produce are mostly boxed up inside cells with cellulosic walls; and though herbivores almost all have specialized teeth or gizzards for grinding the food in order to break the cells open, many if not most cells actually escape and are still intact after mastication. The microbes therefore not only provide the herbivore with digestible sources of energy derived from cellulose etc., but they unlock the nutrient-rich protoplasm inside the uncrushed cells. Large numbers of the micro-organisms themselves may eventually be digested in the host's intestine.

Cellulolytic bacteria work slowly, and enormous numbers are therefore required, preferably held in a sac-like organ or organs where the fermentation they produce can take its course. Leafy foods tend to contain far more carbohydrates than an animal consumer needs, in proportion to the available protein, and herbivores consequently have to process a large bulk and discard much waste, entailing a long residence time for the food and a correspondingly long, heavy gut. This presents special difficulties to winged herbivores whose flight depends on lightness. The insects mostly get round it by doing their main feeding and growth as larvae, before the wings develop. Some birds (and bats) confine their vegetable diet to fruit or nectar, as mentioned in Chapter 5, which are adapted to the consumer's dietary needs and readily digested; or they eat seeds which, once the integument is cracked open, can be an even better balanced food. But these do not bulk large in the diets of grouse; and so critical is the allocation of weight to their digestive tract that the research unit's captive red grouse, after being raised and fed for several generations on high-grade poultry-type pellets, were found to have reduced their intestines and caeca to a fraction of the bulk and mass they attain in wild grouse. In rock ptarmigan other workers have shown that even wild birds change the size of their digestive organs twice a year, growing them longer in autumn and shrinking them in spring, in response to dietary changes.

All species of grouse have exceptionally large crops and caeca. The caeca are a pair of sacs, but otherwise similar in anatomical position to the single blind caecum possessed by pigs, horses, rodents and hares; and as in these other animals they are the principal habitat of the bacterial flora. The crop is a raw-food reservoir placed anteriorly to the gizzard. It is completely filled by the grouse only in the last hours of daylight, and used in prolonging the input and digestion of food well into the night. Grouse are visual selectors of what they eat and cannot feed in the dark. In daylight their feeding can easily keep pace with their rate of digestion; but the recovery of protein from the food is often sufficiently slow to require that the digestive system be kept going for as much of the 24 hours as possible. At midsummer when nights are short, the full crop holds enough to keep the gizzard and intestine stoked practically all night, and under this regime winning a maintenance ration of protein from a diet of heather is usually no problem. The new growth present on almost any and every heather plant then actually contains more protein and becomes nutritionally acceptable. In midwinter's short days, on the other hand, the bird has no longer enough digestion time to process a summer-sized food intake. Though it can still easily feed fast enough to get the crop filled on the side, at the same time as it is also stoking the intestine, within 8 hours it has to stop feeding because darkness falls. The digestive tract is then full, but the bird has not now enough raw food in the packed crop to keep digestion supplied at full pace for another 16 hours, until dawn. Possibly a third of its daily digestive potential has consequently to go unused. If the digestive tract were made any heavier to accommodate extra food, presumably the bird's mobility, especially at take-off or when dodging an airborne attack, would be slowed and the nutritional benefit cancelled by the graver predation risk.

Red grouse consequently have to be more selective in their winter feeding in order to obtain a subsistence ration of protein. It becomes important to have access on one's territory to heather plants with a better than average nitrogen content, preferably growing in dense stands so that good food can be picked fast, with a minimum of moving about.

In brief, the nutritional studies and experiments have demonstrated beyond doubt that food is the ultimate factor to which the grouse's population homeostasis responds. The results indicate that in winter the birds are closely dependent on the dormant buds of the heather plants where nutrients for future growth are stored; and thus how easy it would be for an unregulated increase in grouse density to lead to their taking too much, and injuring the plants' regenerative powers.

13.3 Social selection

The maturation of young grouse is so rapid that the territorial competition in autumn usually occurs in a population consisting wholly of adults. The male recruits among them are normally able to compete on level terms with the older males. The general result is not only to readjust the population density to the carrying capacity of the food resources in the coming year, so far as this can be predicted, but also to draw the vital distinction between the males that become, and fail to become, established.

Old cocks normally resume their previous locations. Experiments with young cocks carrying radio transmitters and entering the fray for the first time—unfortunately rather few in number—showed them to have a strong inclination to settle very close to their natal site, and thus to create patrilineal enclaves. Details of this are presented in the following chapter, in the context of the spatial structuring of red grouse populations.

By early October most habitable parts of the moor have would-be claimants. They may circulate quite widely during the day, but return at night to their focal point, to be on location for tomorrow's dawn. Their bids for possession consist of noisy flight displays, in which they launch into the air from their lookout post, and are briefly airborne before fluttering to the ground again. They also make provocative sorties towards individual neighbours which result in chases and other tense but ritualized engagements, sometimes including minor buffets. During hostilities their scarlet combs are held erect (dilated comb size being correlated with sex-hormone titre). Each cock is conditioned (i.e. pro-grammed) to know how much ground he should claim, and presumably how much compression of it is acceptable. The feebler cocks, assailed from several sides, may fail to hold enough to make a viable territory, and have to resign. The more spirited manage to command an adequate space with boundaries where both parties' acquisitiveness is satisfied and neither is inclined or able to push the other back.

Those that do not secure a territory presumably give up because they sense they could not win a bruising fight against the individuals that are crowding them out. At this date they may have little chance of gaining a foothold elsewhere, and would only risk injury if they tried. None the less, some losers must have failed only by a hair's breadth whereas others never really stood any chance. Possibly one in ten of the losers may get a second chance to take over a territory made vacant by the unexpected death of its owner, or, for the outcast hens, to take the place of a deceased female consort.

In the autumn months active territorial defence only lasts for the early part of the day, after which the owners flock together and the outcasts are not harassed. Through the winter however the proprietors grow more intolerant of trespassers at any time of day, and since between them they control the habitat in an almost continuous mosaic, a heavier mortality of outcasts results.

For the females the process of social discrimination is more inscrutable, but it appears again likely to depend entirely on the established males. Although red grouse do not form permanent pair-bonds for several months after the territories are established, the males accept the presence of a sufficient number of females from the first to provide them with mates when the time comes. Unaccepted females are treated as intruders and become outcasts. To gain entry to the establishment a hen probably needs to be acceptable to one or more adjacent cocks with whom she is already acquainted, who grant her asylum in their territories. Sooner or later she may perhaps have to widen her acquaintance and acceptability. Such a fluid arrangement could give the more nubile cocks and hens a chance to court and exercise a choice before actually mating; and it would help to account for the observed early-season instability of cock and hen partnerships and for the residual expulsion of some of the established hens that sometimes occurs in April.

For both sexes it seems likely that admission to or rejection from the establishment largely depends on the scrutiny and appraisal of one another. Even though they may all be coded to respond for the good of the group rather than their own, a male seems unlikely to give up the attempt to win a territory and thus save his own life without being satisfied that he could not defeat his opponent by escalating the fight, the weapons as always being psychological as well as physical. Similarly an established cock acquiescing to the presence of a consort must be judicious about admitting any female with whom he may eventually have to pool his genes in reproduction, taking into account the probability that he will survive only for one breeding season and sire one clutch of eggs.

These decisions are probably automated more or less unconsciously in the central nervous system; but competence in making the appraisals on which they are based is crucial to individual fitness and, more importantly in the long run, to the survival of the group. It is therefore reasonable to expect that natural selection has elicited in such animals as these a high degree of skill in judging the phenotypic quality of their companions. The process is common to all social hierarchies and is itself a domestic kind of natural selection, not involving extrinsic agents. It results from a formal, compulsory contention for dominance, or just acceptance, between individuals, one to one or one against others, male to male, male to female, female to male, and less often in our grouse female to female; and the contention is pursued until a largely abstract but normally binding decision is reached. Manifestly it is a contrivance, a group adaptation, which seems appropriately described as social selection (Wynne-Edwards 1962: 246).

It is the most discerning type of selection to which red grouse are subject in as much as it contains only the irreducible minimum of chance, and no individual is exempt. When necessary it condemns the loser to die, by forfeiting the means of survival but without specifying the lethal agent. In doing so it normally precipitates more adult mortality than all the 'uncontrollable mortality' put together. It is therefore a very powerful force.

Social selection has proved rather unusually conspicuous in the red grouse, for two reasons. Firstly because its homeostatic function of dividing the population into an establishment and a surplus of outcasts (in average proportions of 40:60%) is so clean-cut, leaving no remainder in the shape of non-combatant adolescents that must be permitted to share the habitat as an extra junior social caste; and secondly because it takes place at a different time from mating and sexual selection, with which social selection has long been vaguely confused.

The social hierarchy itself is an integration of these one-to-one dominance rankings, and it is the primary instrument of social selection. Many vertebrates and some arthropods carry overt status symbols that improve in quality with individual condition and age, in the form of weapons, adornments, colour markings or body size; and these normally permit the instant recognition of relative rank even between strangers. Actual contests may then be necessary only for deciding between apparent equals. In the red grouse the combs in some degree fulfil this function. Status symbols are better developed in the males wherever they are the epideictic sex on whom the main function of density regulation falls. The involvement of the hierarchy in the nomination of winners and losers in the grouse's territorial competition is apparent from the range of individual territory sizes claimed by the various cocks that form a neighbourhood group. Their relative rankings show up as plus and minus differences distributed about the nutritionally determined mean size; and since these area differences are more or less randomly mixed together in the habitat they do not distort the overall male density.

CHAPTER 14

CYCLES IN POPULATION STRUCTURE

14.1 Structuring in red grouse populations

The upland grouse moors of eastern Scotland, with their shrubby carpet of *Calluna*, extend in some places for hundreds of square kilometres without major breaks, and red grouse populations are thus often continuous over areas 10 to 100 times as large as the home ranges of individual birds. Any localized genetic structuring their populations acquire must depend in large part on the attachment that individuals show to their native locality, measured by the distances between their birthplaces and the sites where they mate and breed. Early in the Aberdeen project a brief collaborative experiment was made in Scotland to catch and ring young grouse before they could fly, as mentioned in Chapter 7. Of the 900 marked birds that were shot and reported after they were full grown (including 700 the first autumn), all but 3% were found within 5 km of their place of marking. The research team also trapped adults each autumn in Glen Esk, most of which were eventually found dead. Ninety-five per cent of the recoveries were still within 5 km (see Table 7.1, p. 92). These birds can easily fly 5 km in 5 minutes, and one suspects therefore that their striking sedentariness or philopatry must be programmed; and this is confirmed by the discovery that the hens are several times as mobile as the cocks. When all the data from the ringed chicks that had been sexed on recovery are combined with those from young birds tabbed when just full grown, they provide a sample of more than 400 recoveries about equally divided as to sex. Only 3% of the cocks had moved more than 1.5 km, and none more than 8 km; whereas 20% of the hens had moved more than 1.5 km, 6% more than 8 km, and one had turned up 42 km away.

Even more striking results were obtained from the telemetry experiments (see p. 95) in which two young grouse brothers were observed while they prospected for and obtained their first territories, and afterwards lived on to breed for two successive seasons. It will be recalled that they (distinguished as *A* and *B*) took over their father *P*'s territory and established their own claims side by side so as to occupy the whole of it plus some adjacent ground. *A*, the more dominant of the two, who actually ousted their father from his estate, produced two sons, *d* and *e*, in the second year of study; they were also telemetered, and in turn established themselves in the final year in territories side by side and contiguous with those of their father and uncle, making a completely patrilocal block of four.

The original *P* had a neighbour *H* who continued to re-occupy the same site for two more years, with *A* and *B* as neighbours. There is a fair probability in the light of the events just described that *P* and *H* were also closely related, possibly as brothers or as the father of an earlier son. *H* had two sons belonging to the same year-cohort as *d* and *e*, and they were also telemetered; they too obtained contiguous territories sandwiched between those of *H* and *A*. That made a second patrilocal block of three, or perhaps extended the united consanguineous block to seven, in the final year. In terms of Hamilton's (1964) theory of kin selection, *A* is therefore known to have had an 'inclusive fitness' (see p. 330) of not less than $2\frac{1}{2}$ units before breeding started in the final spring, comprising his own genes (1 unit) plus half each of those of one brother and two sons. *B*'s inclusive fitness works out at 2 units and those of *d* and *e* at $2\frac{1}{4}$ each. If *H* is assumed to have been *P*'s brother or son, all these values would be augmented; *A* for example would gain an uncle ($\frac{1}{4}$) and two cousins ($2\times\frac{1}{8}$), or alternatively another brother ($\frac{1}{2}$) and two nephews ($2\times\frac{1}{4}$). All these values are minimal estimates in the absence of pedigrees and territorial data for other male neighbours, or of pedigrees of the mothers of the telemetered sons.

These shreds of information suffice to prove that red grouse have strong adaptations for the genetical structuring of their populations; and they reveal at the same time the dramatic effect that close philopatry in one sex can produce. The hens on the other hand showed much more inclination to scatter. Sexual differences in dispersal programs are known to be common in animals, and are discussed in a broader context in Chapter 16 (p. 238). It would be desirable to add to the red grouse data by repeating the telemetry experiments if the opportunity arose, because it offers much the most practicable method of following pullets of known parentage to their subsequent breeding sites. The recoveries of ringed birds have already made it clear that the dispersals of individuals of both sexes are distributed in a manner that is strongly skewed towards the shorter-distance end.

Having one sex more philopatric than the other allows the population to reach a desirable balance between conflicting benefits. It averts incestuous matings at one extreme (see next paragraph) and can export a small quota of emigrants at the other, while the great majority of the population observe a sufficient degree of philopatry. Philopatry is necessary because it is the trait that enables descendants to benefit from the habitat conservation of their ancestors; and it also creates a population infrastructure of local, largely isolated, hereditary-habitat groups or demes. The demes can be shown to differ from one another in their productivities or group fitnesses, as we have already seen (p. 137), and this is the basis of group selection.

One of the potential merits of having sexually different roles in dispersal is that a very close philopatry can be tolerated among individuals all belonging to the same sex because they cannot mate with one another. Instead they are bound to take their mates from a suitably controlled but wider provenance. It is a method of increasing the average degree of philopatry (i.e. the philopatric kurtosis) without incurring a corresponding rise in the gene fixation rate.

There are no detailed data about the lower end of the range of dispersal distances for the females, between birthplace and breeding site, but only the lumped information that 80% of the marked hens were eventually shot or found within 1.5 km of where they had first been caught as adolescents. This compares with a figure of 97% for the cocks. It appears therefore that a circle of 1.5 km radius would cover the lifelong range of the great majority of individuals of either sex. It would enclose about 7 km^2 and normally hold 200–700 residents in the spring, or two or three times as many in August. For genetical purposes it would be very desirable to know the approximate size of their effective interbreeding units, and for social purposes that of their individual circles of associates—the birds with which they are in daily or frequent contact as companions or rivals in a common hierarchy, which I am going to term their 'in-group'. We know that when a predator, especially a golden eagle, is hunting in their vicinity, grouse often pack into flocks for safety and these occasionally contain a few hundred individuals; but there seems little likelihood that any one bird could recognize personally as many individuals as that. More likely the in-group is confined to 50–100 individuals at most, perhaps somewhat biassed towards the noisier and more demonstrative male sex.

In an isolated population of 50 or less any new mutant allele would proceed to fixation within relatively few generations (cf. Baker & Marler 1980: 66). However the effective interbreeding population for a red grouse is probably larger than its in-group. In-groups of prospective breeders are formed as a sequel to the autumn dispersal, and may occasionally contain a female that has come several or many kilometres from her natal site. This must widen the effective interbreeding population size perhaps even by an order of magnitude compared with the size of the in-group.

Chapter 12 showed that there is another additional type of dispersal that occurs at intervals of one to several decades, during the peak and declining years of a population cycle. Only one such episode occurred in the first 25 years of the research project. It took 7–10 years to run its course, and after carrying the local population densities up to unprecedented heights in a run of years with ample food production, it ended by leaving an area, measured in tens if not hundreds of square kilometres, depleted to densities that were just as exceptionally low. The fluctuation shared many of the features that accompany the better known episodic emigrations of other species such as the European lemmings (*Lemmus lemmus*) and migratory locusts (*Locusta migratoria* etc.), which occupy inconstant habitats in arctic and arid biomes. Large emigration years are characteristic of arctic and subarctic populations of willow grouse and rock ptarmigan also; but most of the populations that have been adequately studied have cycled at rather more frequent intervals than the red grouse, averaging between 3 and 10 years, and in some instances rising and falling almost continually, without steadier intervals in between.

Fluctuations in the red grouse stocks of Great Britain and Ireland have been noted since 1800 or before (Leslie 1912: 152), though they have not often been identified with emigration. None have previously been accurately quantified by repeated counts of birds on a study area, nor has any been observed by scientists of comparable experience. Emigration is not easily detected by casual observers, and it is possible that evidence of it as a cause of falling numbers has been dismissed in the past, on the mistaken assumption that all declines were caused by disease. During the 1970–77 cycle at Kerloch few birds were being routinely ringed, and there is no evidence on how far away the emigrants travelled. Just as in the normal annual dispersal, however, more females and young birds than adult males were known to have left the area.

It is tantalizing not to have more information about such an infrequent and potentially important event. My tentative suggestion as to its function, mentioned in the previous chapter, is that populations of red grouse and other cyclic species are structurally programmed in time as well as space. For say 5–20 generations the demes, made up of numerous in-groups, preserve a distinctly viscous texture because the rate of gene flow is low. It can be assumed for the moment that demes which are almost isolated will inevitably come to differ genetically in their efficiency as conservators of their habitats and resources. After a sufficient number of generations have elapsed to allow these differences to materialize, and at a time when climatic conditions are favourable, the population will switch to a second regime, during which the texture is more fluid. This results in emigrations that vary in size between demes, in proportion to their differences in productivity. The more productive ones will be able to export more of their genotypes into other demes, and the not so productive ones less or none; and this differential will make a major contribution to the process of group selection. The same exodus carries 'new blood' into the localities where successful emigrants arrive and establish themselves, and provides an antidote to the gene fixations accumulated during the first regime. Clearly it is not necessary for the exodus phenomenon to occur simultaneously over the whole of a large region. Any single deme could produce its emigrant crop in isolation, receiving small and irregular inputs of immigrants unpredictably from other self-promoting demes, as and when they occurred.

If the Kerloch cycle can be taken as typical for the red grouse, there may be a genetic significance in the final collapse of the population which left some parts of the habitat completely depopulated. These gaps are eventually occupied by colonists, and that may be another method of assembling new in-groups and even demes (we shall see later that neither of these conceptually useful units need be sharply defined in real populations) with incomers of relatively varied origin. In that case a set of in-groups, if not the deme itself, would become a transient entity, repeatedly bursting like a bubble and as often being re-formed on the same site as the decades go by.

14.2 Population cycles in grouse

Episodic fluctuations similar to the Kerloch cycle occur in many tetraonids. They presumably account for the occasional arrival of emigrant willow and rock ptarmigan, and similarly of sharp-tailed grouse (*Tympanuchus phasianellus*), far south of their normal ranges, for example in southern Ontario and Quebec. It has long been noted that North American grouse species have the same tendency towards periodic cycles as the varying hare (cf. Criddle 1930; Leopold & Ball 1931); but the best documented example of regular periodicity in the tetraonids is probably that of the rock ptarmigan in Iceland (Gudmundsson 1960). Between 1864 and 1939 large numbers of this species shot by hunters were sold for export as a table delicacy, the numbers shipped abroad being annually recorded. As a population index the figures leave something to be desired. It has been learnt since that ptarmigan ringed as chicks may subsequently emigrate and be shot in winter up to 275 km from the place of marking. The data take no account of the varying locations and availability of large winter concentrations, or of the travel conditions affecting their availability to the hunters.

Nevertheless the amplitude was most of the time so great that the cycles stand out very clearly. An exception occurred in the years between the peak of 1902 and a spectacular crash in the winter of 1919, when the population suddenly dropped to the brink of extermination. Though the totals of exported birds varied irregularly through those 17 years they were always relatively high. Even in the poorest season of 1917–18, 20 000 were shipped. For comparison, the highest total in the series was 250 000 in 1927; and in the low phase of the typical cycles it generally fell well below 10 000. The exports ceased after 1939, but a comparable index relating to local trade in ptarmigan, at Husavik in northeast Iceland, extended from 1943 to 1957 and showed two typical cycles. Finally Gudmundsson (1972) set up a population study of the bird in the offshore island of Hrisey, which recorded a further peak in 1966. Incidentally Hrisey contained no ground predators and the ptarmigans' only effective consumer was the gyrfalcon (*Falco rusticolus*). Its numbers rose and fell with the ptarmigan cycle but on only a fraction of the scale. At the peak the prey saturated the predators' demand to such a degree that 'population changes in the ptarmigan could in no way be attributed to predator–prey oscillations'.

Over the whole series there were distinct peaks (separated by lows) in 1871, 1880, 1891 and 1902, the last initiating the 'non-periodic' regime; then again in 1923–27, 1934–35, 1945, 1954 and 1966. Omitting the non-cyclic interlude, that gives seven nearly equal inter-peak intervals in 72 years. Information is lacking to show whether, when the trading peaks continued for two or more winters as in the 1871, 1890 and 1923–27 cycles, the cause was merely a lack of exact synchrony between different parts of the country; nor whether, during the 1902–19 break, there were cycles still taking place regionally, but out of phase.

14.3 Population cycles of microtine rodents

Grouse fluctuations resemble those of various small rodents in several respects, for example in the unsustainably high numbers reached at the peak and vanishingly low numbers sometimes remaining in the trough, in the fact that not all the local populations in a district necessarily respond in synchrony, and in the tendency for isolated episodic cycles to be typical of less harsh and variable climates and for more regular periodic ones in climates that are more severe. Rodent cycles have been studied intently by ecologists for some 60 years, and for just as long the possibility has been recognized that they have evolved in parallel with those of grouse (and various other birds and mammals), to perform the same functions (see Krebs & Myers 1974, and Finerty 1980, for reviews). The Microtinae, in which the cycles are most widely developed, include the lemmings and voles. They are incomparably more difficult animals than grouse to study in the wild. Their numbers can rarely be counted directly because they hide underground, come out mostly at night, and cannot be heard except at close quarters; and they are usually seen only when caught in a trap. Live-trapping does not yield random samples, and at best gives an uncertain index of the true numbers. Instead of being territorial, during adult life the males usually occupy quite large home-ranges, overlapping those of other males; and the adults feed on mixed, chiefly plant, diets. Consequently the study of a single red-grouse cycle, though it had many shortcomings, has provided insights that have never been fully appreciated in the innumerable investigations of microtine cycles, which have involved more than 20 species.

In the red grouse we can readily see that year-to-year demographic changes can be of two types, which have distinct causes and functions. On the one hand, the quality and productivity of heather as food is much affected by climatic events, and the cock grouse have been proved to vary their average territory size year by year, in seeking to match the resulting changes in carrying capacity. The fluctuations this produces may reach an amplitude in the region of 50% above and below the running average density. The Kerloch cyclic fluctuation, on the other hand, took the peak density to 3 or 4 times the preceding norm, and at its succeeding low phase there were small blocks of habitat here and there that lay completely vacant (see p. 152). The cycle appeared to supervene or take charge, as a course of action overriding the normal homeostatic regime; and apart from the fact that its increasing phase coincided with good feeding conditions, the rest of it showed no evident correlation with changes in carrying capacity and nutrition. The fertilizer experiment carried out during the decline did nothing to check the fall of numbers.

Genetic considerations have been given prominence by many previous authors on the theory of population cycling, though, with the exception of Sewall Wright (1949: 381), not in relation to the alternation of inbreeding and outbreeding but in explaining how the cycles themselves might be driven. A well-known hypothesis originating with Chitty (1967, 1970) assumes that cycling is a method evolved, in some taxonomic groups, for the self-regulation of animal populations, and that it works by keeping their density continually oscillating about a mean. It postulates that, under suitable habitat conditions, the number of voles (on which Chitty's studies have centred) rise, and this leads to mounting interference between individuals. Aggressive phenotypes are then at an advantage, and their frequency therefore increases at the expense of unaggressive ones. The result is to thin the population out, because the aggressors demand greater individual space. Eventually they reduce the density to a level at which interference ceases to be acute, and because aggressive types are assumed to be more vulnerable to the external agents of mortality, selection switches back once more in favour of unaggressive phenotypes, and the cycle repeats. The key to the population regulation in the system is assumed to be the density-dependence of the aggressors' advantage.

Chitty's hypothesis has been strongly supported, especially by C. J. Krebs (e.g. 1964, 1978); but because its assumptions are virtually untestable, neither author has ever claimed to have proved or established it, though they have amassed some encouraging circumstantial support. Originally it paid no particular attention to dispersal, though of course emigration has long been the best-known feature of the Scandinavian lemming cycles. Local dispersal is not a readily detectable phenomenon in voles, and its general importance went unrealized until experiments were undertaken which involved enclosing small populations, in 2-acre fenced plots of natural habitat. These populations unexpectedly built up to twice or more times the peak densities recorded in unenclosed control plots nearby; and eventually they actually destroyed their food supplies (Gentry 1968). A so-called 'fence effect' was thus discovered, preventing the normal eviction and dispersal of surplus population. Myers and Krebs (1971) further demonstrated the reality of this by a reciprocal experiment, in which an unfenced plot had all its residents and subsequent invaders trapped out at regular intervals. At the same time the authors monitored adjacent control areas, and marked individually many of the voles residing on them. Some of these marked animals turned up as immigrants on the vole-free plot. From the survival data of marked residents on the control areas and the proportions of them that disappeared, and from the numbers of the latter that were caught on the cleared plots, it was possible to show that emigration occurred throughout a 9-month population cycle in 1969–70. It was especially heavy in the increasing and peak phases, during which 56 and 33% respectively of the known losses from the control areas were attributable to emigration.

Adopting a more direct technique, Beacham (1980b) confirmed these results by reverting to the use of two duplicate fenced populations; but he provided each with adjoining 'dispersal sinks' (pitfall traps) to receive would-be emigrants. He found emigration occurring right through the spring and summer of the high-density phase, during a 1976–78 cycle, at rates accounting for 56% of the male and 67% of the female losses. The remaining losses were due to mortality within the colony. In the following autumn and winter, emigration still continued and accounted for over 30% of the losses; but in the terminal decline in the spring of 1978, emigration fell to less than 5% of total 'mortality'.

Support for Chitty's hypothesis was at first thought to have been found from a study of allozymes as genic markers in the voles' blood plasma, which enabled an individual's genotype to be characterized by electrophoresis from small blood samples. When a vole population in central Scotland was monitored through a population cycle, the supposed allele frequencies were found to undergo significant change, though it was not actually known whether or how this might be physiologically connected with aggressive or migratory behaviour (Semeonoff & Robertson 1968). Myers and Krebs (1971, & cf. Krebs et al. 1973) later found significant differences in the frequencies of homo- and heterozygotes of two such (transferrin) 'alleles', between samples of apparently emigrating and apparently stay-at-home voles. Eventually, however, it was demonstrated by McGovern and Tracy (1981) that the supposed allelic markers were not stable: they could be altered in the same individual vole in response to changes in its physiological needs; and thus what had been assumed to be evidence for selective changes affecting allele frequencies in the gene pool were probably due to the individuals' normal responses to seasonal or other environmental change.

Remembering that the hypothesis was advanced to explain how vole numbers were regulated, one is surprised to find concurrent research which shows, on the contrary, that various small rodents including species of voles, do in fact regulate their densities in the normal way, by social hierarchies and spacing mechanisms identical in their effect to territories. The populations themselves are apparently no less structured than those of the red grouse. Each individual vole acquires the right to live and feed within a certain home range, which commonly overlaps the home ranges of several other individuals; and if so, they share the space and its resources together on a hierarchical basis. During the breeding season the mature females of many species become more exclusively territorial. Densities also go up and down according to changes in the plane of nutrition. Thus Cole and Batzli (1979) showed that the prairie voles (*Microtus ochrogaster*) living in three adjacent but mutually different types of habitat in central Illinois, went through typical fluctuations between 1972 and 1976. Their peak densities reached 244 per ha in an abandoned alfalfa field, 128 per ha in ungrazed bluegrass pasture, but only 38 in native tall-grass prairie. This reflected the differences in nutritional plane the three habitats provided. The alfalfa was shown to have a much higher digestibility and to contain much more protein than the bluegrass; and though the prairie forage was not similarly analysed, it was found to produce the slowest growth rate of the three when captive voles were subjected to feeding trials. Taitt and Krebs (1981), in an unfenced plot experiment with *Microtus townsendii* near Vancouver, provided extra food resources at dispensers placed at regular intervals, but they varied the interval unit to give different food densities on different plots. A plot without supplementary food served as a control for a period of 18 months, during which two of the experimental plots with extra food in different amounts built up to about three times, and over four times, the control vole density. On another set of plots, and at a time when all populations were increasing, the number of voles on the control grew at an average rate of 7% a week, compared with 10% a week where extra food was given. These experiments suggest that vole populations can and do vary their density in relation to the carrying capacity of the food supply. (A general account of the role of nutrition in wild vole fluctuations was given by Kalela 1962.)

Frank (1957), after working for many years on the continental European vole *Microtus arvalis*, gave a dramatic picture of its population structure and dynamics. Its potential fecundity is very great. Granted good food and climatic conditions, females can bear litters that average 7 young at intervals of 20 days. Males fertilize at least some of the young mice before they are weaned, and first litters can be born to mothers only 33–40 days old. Older females are territorial over an area surrounding their burrows 10–20 m in diameter, and most of the time they exclude all voles from it other than their own dependants. Males from outside the immediate vicinity wander about in search of oestrous females, and the territorial female when she requires his services admits a mate to her nest. This might allow the mother to exercise some control over the coefficient of inbreeding, in that she could, for example, refuse to allow the father of her latest litter to return and mate with his daughters, as long as they remained in her territory. When independent the daughters tend to settle down as near their mother as they can, but the sons leave home when they mature. The last two or three litters in autumn are not expelled, however, but remain to form a 'great family' for the winter, living together in a thickly lined nest and mutually contributing to the common heat budget. The population thins down in winter, when home ranges increase to four or five times the summer size.

According to whether or not the matriarch herself lives through the winter, she is either reinforced or supplanted by 2–4 of her daughters, all ultimately reproducing and bringing up litters in the same nest. When there is an episodic cycle this pattern of events is especially typical of the increase phase, and it enables voles to crowd at high concentrations. On such occasions, recruitment of resident males is outstripped by that of females.

Kalela (1957) found in Lapland that a species of red-backed vole, *Clethrionomys rufocanus*, had a basically similar social structure, with territory-holding mature females and overlapping males. A powerful factor in their population changes was whether, and what proportion of, the early-born young matured and bred in the same summer. This he inferred to be socially controlled, and inversely density-dependent.

Frank (1957) observed that the cycles of *M. arvalis* tended to be 'released' by human land-use changes, and the same is known to be true of *M. agrestis*, which sometimes produces plagues after woodlands have been felled, or when new forestry plantations are beginning to put on growth. Frank comments that highly intensive farming with its too frequent cultivations and disturbance is in fact inimical to cycling, whereas extensive stock-ranching tends to encourage it. He thus illustrates the episodic character of the cycles, revealing their tendency to become irregular or be held in abeyance while the voles await the return of propitious conditions.

Even on the tundra of northern Alaska, after an uninterrupted run of six lemming cycles that peaked at 3–5 year intervals, Pitelka (1973) recorded a lull period of five years during which the population stayed low. When the next peak did come, in the sixth year (the summer of 1971), it was evident from the lemmings' sudden appearance in large numbers that they were invading Point Barrow from somewhere else, possibly even from a coastal area 95 km away where they had been reported abundant the previous year. The peak he had expected in 1968 had never come, possibly because feeding conditions were poor at the crucial time, or predation prohibitively heavy. As far as the Point Barrow lemming stocks are concerned, however, or those of any other locality, the regime of inbreeding could be broken as effectively by immigration as by emigration. The phenomenon is the same: only the viewpoint is reversed. An earlier observer at Point Barrow had reported the opposite event—a rapid disappearance of a peak population by dispersal, in June 1953 (Thompson 1955). The species principally involved there is *Lemmus trimucronatus*.

The parallels between grouse and microtine cycles are apparently close. In addition to those already mentioned they show similarly autonomous sequences of change in density and mobility, requiring a more or less set period of time for their fulfilment, and extending over more than an average lifetime. If cycling is functionally the same in both, Chitty's hypothesis ought to apply to them equally. The fact that red grouse are known to regulate their numbers in an entirely different way from the one he postulated suggests to me that it is probably untrue of either.

I disagree incidentally with Beacham (1980a: 462) that 'it seems difficult to accept a behavioural mechanism that causes virtual extinction of the population, because as density approaches zero, the rate of behavioural interactions must necessarily decrease'. The emigrating Norway lemming needs no driver or antagonist. It marches purposefully, intent on holding as best it can to the same chosen direction that has steered it radially outward from the native habitat whence it came, at speeds of up to 1 m per second. It moves largely in twilight and darkness, alone and aloof from other lemmings, though often making use of paths that predecessors have recently taken. If one emigrant meets another, most often they pass without acknowledgement or hostility. Contrary to popular belief, lemmings aggregate only because there is a 'traffic jam' and further progress is blocked by obstacles such as wide or swift rivers, or lakes (cf. Collett 1895: 42–43, Myllimäki *et al.* 1962, and de Kock & Robinson 1966).

14.4 Population cycles of hares

Cyclic fluctuations are also well developed in some northern populations of hares, particularly the snowshoe or varying hare (*Lepus americanus*) in Canada. A population of this species at about latitude 49°N in southern Manitoba was monitored for 15 years and showed two well-marked peaks 11 years apart. The first decline was all over in a few months, during which the density fell from over 500 per 100 acres to about 20. After that the population rose very slowly during the next eight years, and then more rapidly in the last two, to reach the second 2-year peak, again at over 500 hares per 100 acres (Criddle 1938). The numbers were only roughly estimated.

Ten or 11-year cycles like these do not recur regularly in all parts of southern Canada, and the phenomenon becomes still weaker and more sporadic south of the American border. The species inhabits the forest, extending to the tree limit across the continent, and it is in the northern woods that cycling appears to be strongest. The fur-trade statistics of the Hudson's Bay Company showed a succession of seven evenly-spaced high peaks with intervening lows, between 1845 and 1905, giving an average periodicity of 10 years (Seton 1912: 105; Elton 1924); but the figures are lumped from vast areas and there is nothing to show whether all the cycles occurred in all parts of the region. Some of the highs produced spectacular concentrations, perhaps augmented by large emigrations from one point to another. Seton (1912) and Preble (1908) both witnessed the great production and peak years of 1903–1905 in the District of Mackenzie, where in some places the density rose to 100 or 1000 times the low-year minimum level. It was the biggest increase that had been seen there for many years, and involved hare populations at many locations within the 2500 km length of the Athabasca–Mackenzie valley. Preble makes it clear that not all places were equally affected, and those that were did not all reach their peaks in the same year. The majority were at their height in 1903–1904, and a minority in 1901, 1902 and 1905.

Of the crash that followed Preble wrote: 'Throughout the upper

Mackenzie region during January (1904), and to a lesser extent during February, many thousands of (hares) perished from disease.' Some of them were just 'sprawled in their tracks' in the snow, with emaciated bodies and empty stomachs. Seton (1909: 646) also attributed the mortality to a plague, which at that time seemed the most likely cause of their mysterious disappearance. In the same decade the red grouse declines in Britain were also being ascribed to 'grouse disease'. Movement away from eruption centres was not suspected, and the way the hares could melt away into less populous areas at no great distance made it seem 'by magic they were gone' (Soper 1921).

Recent experiments with snowshoe hares demonstrate that some dispersal may take place in all years and seasons (Windberg & Keith 1976). These authors adopted the same methods described already (p. 179) in use by Myers and Krebs (1971) with *Microtus pennsylvanicus* in British Columbia. The experiments consisted of clearing the resident hares from a couple of study areas and then, at regular intervals of 3 or 4 weeks, going back for a few days of intensive trapping to remove the incomers. Windberg and Keith kept this going for 4 years, and found that when hare densities on the control areas were highest, at 8.4 hares per ha in 1970–71, they could catch 65 immigrants per trapping session on an 11 ha area; but 3 years later when the control density had fallen to a low of 0.4 hares per ha, the corresponding 3-weekly catch of incomers averaged less than 2. In other words there appears to have been a non-linear density-dependent relationship between hare numbers and dispersal, similar to that found in the grouse, lemmings and voles. The white hares that Preble saw dead in January 1904 may in fact have walked till they dropped from starvation. Windberg and Keith quote an unpublished record of an individually marked snowshoe hare in Alaska being recovered 20 km from its place of marking.

Rather surprisingly the arctic hare (*Lepus arcticus*) seems generally to be a much weaker and less regular cycler. It replaces *L. americanus* ecologically at about the tree limit in North America, and is a very close relative of *L. timidus*, the mountain and tundra hare of the palearctic. Both *arcticus* and *timidus* do undergo sizeable changes in abundance with time, but too patchily and sporadically in *arcticus* to have a noticeable impact on the fur-trade in fox skins (see below, p. 192). A population study of mountain hares in eastern Scotland in which many individuals were tabbed, showed a fourfold range in density over 14 years but not much indication of cycling (Hewson 1976a). Long-term studies have not apparently been made of arctic hares (Smith & Wang 1977), though Soper (1928) reported irregular fluctuations in their abundance.

14.5 The function of intrinsic cycles

The hares add a third parallel, similar in all the essential characteristics to the other two cyclic taxa. There seems no reason to doubt that emigration is the primary adaptive consequence of cycling, and the normal cause of disappearance of the population swarms produced. In all three taxa the cycles seem most often to build up in local centres rather than being uniformly disseminated through every deme; and that probably helps to ease the overcrowding by giving the new recruits, when old enough to travel, the opportunity of moving out successively into less populous surroundings. Even though that may be so, microtines when the last emigrants depart sometimes leave devastated pastures behind (cf. Elton 1942: 132, 190; Pitelka 1973: 199).

That this occurs confirms the expectation that producing large emigrant contingents must put a strain on the local food resources, and can best be undertaken in propitious years. Some habitats will be more capable than others of producing the threshold of reserves required; and the focal centres may just reflect the fact that some demes are better situated and more quickly mobilized than others for eruption. The amplitude of the cycles, meaning the relative size of the emigrant force produced, certainly varies very much with time and place. On more favourable occasions many cyclic foci could be set going in a given district, where on minimal ones single units might erupt in virtual isolation.

Because it may have to wait for suitable climatic conditions, the next occasion for emigrant production tends to be unpredictable in advance, but once started, the duration or time the cycle takes to run its course is surprisingly regular, and characteristic of the species concerned. The start and finish are marked by the suspension and eventual resumption of normal homeostasis, initially switching the homeostatic program off and replacing it with a completely different kind of population control. After the cycle has run out and the exodus is over the habitat repopulates, and food-related homeostasis returns and usually runs for several years at least before a new cycle again interrupts it. On the hypothesis I am postulating, this alternation is adaptive and its survival value derives from the genetic structuring of the cyclic population. While the cyclic phase appears to need a fairly predictable time-span or run of generations, the length of the homeostatic phase is not critical and can continue for a longer period without detriment. If the investigator were able to follow the demography of a single deme of one of the more regular cyclic species such as the snowshoe hare or Norway lemming for 50 or 100 years, it would not surprise me if he found it only participated in one out of three or four of the cyclic events that hit the region in that time; and on the other occasions it just marked time, keeping the homeostatic program running for much longer periods than the gross phenomena of cycling might suggest.

On the other hand, two examples have already been quoted that illustrate the remarkable phenomenon of regularly consecutive cycling, in the rock ptarmigan and snowshoe hare. This seems to require that at least some of the demes in the periodical populations concerned, do participate on consecutive occasions, and need a certain minimum of time to perform the two-phase function when the alternate programs are following each other at top speed. Being able to adhere to a repetitive 10-year periodicity for half a century is a very positive indication of internal timing, because of the lack of any external pacemaker that could give the proper cue. The timing must result from programs and feedbacks integrated in the brains of the participants. It is difficult to conceive how individuals born during the cycle are clocked into phase, let alone how they take the timing function over from the generation they outlive. During the homeostatic phase there must also be a live but inhibited cycling program, and perhaps a gradual cumulation of reserve resources, pending the perception of cues that induce another cycle.

It was discovered fairly long ago in eastern Europe that some small-mammal populations customarily overflow in the summer, or in the emigration stage of a cycle, into adjacent 'colonization habitats' which remain unoccupied at other times. In low years (or winter) the animals are confined to their permanent 'survival habitats'. The phenomenon has been briefly reviewed by P. K. Anderson (1970: 304), who states that outbreeding and gene recombination are maximal in the colonized habitats. He quotes O. Kalela's observations that lemmings in Finnish Lapland, in their emigration years, first spread into these secondary breeding areas; and, having found the conditions favourable for 'early occupancy, heavy and prolonged reproduction', they generate the expeditionary forces that create the eruption. Presumably when the colonized habitats are abandoned again there is another dispersal, and the original survival habitats receive outbred recruits. Perhaps the same may sometimes follow after a cyclic low when the habitat is being repopulated; and if not, repopulation will occur through the reproduction of residual survivors as 'founder' parents.

The genetic effect of producing a run of generations with a small, controlled dispersal, followed by one or more generations of wide dispersers, cannot be otherwise than to alternate between higher and lower inbreeding coefficients, that is, between inbreeding and outbreeding. My hypothesis is that, by the cyclic process, the species or its metapopulation is made to operate an endless series of fitness experiments or tests. Small local units within the structured deme are first left undisturbed to reproduce philopatrically for a number of generations, under conditions conducive to genetic drift and lasting long enough for small differences in their habitat management practices to show their effects on the state of the food resources. In due course, in a year with better than average food prospects, the program switch occurs and releases the fastest possible reproduction of the units' gene pools. The climax is dispersal, followed in time by a return to normal densities and homeostasis, with the local gene pools somewhat differently constituted than before. I see this as a method of realizing and trading on the differences in fitness that are inherent in nearly isolated local breeding units as a result of individual selection and drift, whereby groups that survive and succeed better in resource management will be favoured (see Chapter 15).

The proliferation stage of the cycle cannot last long without causing density-dependent damage to the food resources; and how long it can last will depend partly on the rate of increase of the food demand and partly on the continuing favourability of the climate. The highest fecundities among the cyclic taxa are found in the microtines. Their females can usually produce a succession of large litters in a single summer, and sometimes they have resources enough to breed in the previous or succeeding winter. Their cyclic peaks more often last one year than two. It may be genetically important that early-born young can often themselves reproduce in the year of their birth, so that two generations are initiated in the same year. Not unexpectedly their cycles tend to be violent and short, with a normal interpeak interval of 3 or 4 years in the most repetitive stocks. The integer for food production is the year; and because they may have to wait for a favourable year before cycling can begin again, even the shortest delays have a relatively large effect on cycle succession, which commonly varies between two and five years.

It is possibly no accident that the cyclic hares and grouse have minimum peak intervals of roughly the same length, and it may be governed by the same factors as in the microtines, namely their potential fecundity and the permissible limits to resource depletion. Both the varying hare (Banfield 1974: 83) and the red grouse start to breed in the next year after their birth, and the females of both can probably produce a maximum of between 5 and 10 viable recruits a year. The build-up stage lasts some 2–3 years, and the exodus stage was found to continue for as long or longer than that in the grouse at Kerloch. To judge from Gudmundsson's (1960) data for rock ptarmigan in Iceland and Seton's (1912) data for the snowshoe hare in Canada, 5 or 6 years (and new generations) are the normal duration of the intervening homeostatic stage of the cycle; and adding these all together at maximum compression gives a cycle repeat frequency of close to 10 years.

The minimum number of homeostatic generations required in the dual program sequence may possibly have a genetical explanation. For example, the genetic drift that is expected to occur in a set of neighbouring but mutually isolated gene pools in a structured deme, would be due to the random selection of a quota of individuals to be the parents of each succeeding generation. This expectation is acceptable as being consistent with the process of social selection found in the red grouse, in which the competition to become established breeders occurs between individuals as whole genotypes, and not between alleles as such. Even if the allele frequencies in every gene pool in the set were the same to start with, the random sampling in the choice of parents in units held more or less constant in size through several generations by the carrying capacities of their habitats, would lead to progressive divergencies in their allele frequencies at individual loci. It can be shown, and confirmed by experiment, that it only takes a few generations under these conditions for the maximum dispersion of relative frequencies between any two alleles to develop, and reach the point where some gene pools in the set have one allele fixed and the other lost (frequencies of 1 and 0), and others provide between them a complete range of intermediate frequency combinations. In other words the maximum diversity of allele frequencies that drift can produce in a set of gene pools is rather quickly reached; and if a production cycle and emigration is mounted soon after this has occurred, there will be a maximum variance of productivities between the units, as a basis

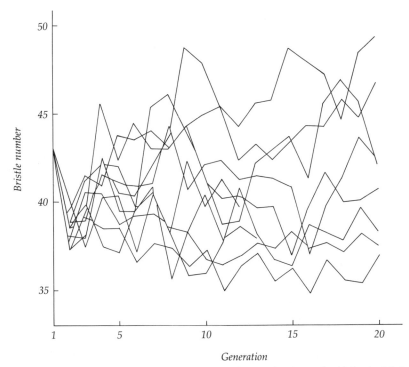

Fig. 14.1 *Differentiation between lines by random drift, shown by abdominal bristle number in* Drosophila melanogaster. *The graphs show the mean bristle number in each of 10 lines during full-sib inbreeding without artificial selection. From Falconer (1981), after Rasmuson (1952).*

for group selection (see Figure 14.1, and Rich *et al.* 1979; Falconer 1981: 45–69).

In the exodus stage that ensues, recessive genes that some of the gene pools have retained but others have lost, can presumably be redistributed through the deme and picked up by the reconstituted breeding units, though in somewhat different frequencies than they had before the cycle, as previously suggested; heterozygosity would thus be restored.

The arguments advanced in this section have rationalized most of the features of cycles that are paralleled in all three of the taxa principally concerned. All three are alike in being herbivores and being distributed mainly or wholly in the northern hemisphere. I have not found evidence of any comparable cycling south of the northern temperate and sub-tropical zones. The microtines do, however, share their high fecundity, and consequent propensity for becoming agricultural pests, with numerous other small rodents that inhabit most parts of the world; and damaging rodent outbreaks can occur almost anywhere that intermittent food bonanzas are provided by man. This is a related but functionally different phenomenon and is solely due to a homeostatic response to a transient steep rise in the plane of nutrition. Similar natural population surges can also occur, especially perhaps in arid climates, in response to exceptionally favourable climatic conditions.

The true cyclers, as understood here, carry out their cycles under natural conditions in order to induce production and dispersal at the culmination of a repetitive genetic process. The process seems likely to have originated, as a new adaptive development, from the normal homeostatic changes in population density that are unavoidable, if density is to be kept in balance with variable food supplies. Non-cyclic species, just as much as the cyclic ones, need to export emigrants: it is essential for gene flow, and is the means of substantiating and disseminating group advantage. Export is most likely to take place when a population is experiencing good feeding conditions and its density and breeding success are high. Cycling probably developed, making a virtue of necessity as it were, by accentuating the best opportunities for exporting genotypes, and thereby improving on the existing rudiments, to develop a prudential system of genetic structuring. (A different type of adaptation for genetic structuring is discussed in Chapter 16.) By the same token, it is likely to have originated in the northern tier of the northern continents, where incessant annual changes in food productivity provide the opportunity, where most of the best exponents of cycling still reside, and where regular periodic cycling is best developed. It is likely also to be especially necessary there because of the secondary effect of the exoduses, in repopulating habitats that had been rendered temporarily uninhabitable by harsh extremes of climate.

Especially in lower latitudes where the cycles are less regular and the homeostatic intervals longer, the signal that eventually starts up another cycle is presumably perceptible as a feedback from the environment. The gradual accumulation of nutritional reserves through programmed underexploitation (a suggestion for which there is yet no evidence) might produce such a delayed-action effect and indicate the readiness of the habitat to support a major proliferation-emigration event. Some kind of social facilitation would also be required to secure its concerted acceptance and support.

Other hypotheses to account for population cycles have of course been put forward, but in my view they are readily faulted. Chitty's (p. 178) refers primarily or solely to the microtines, and proposes that their cycles serve to prevent population densities from reaching excessive levels. It can be rejected on a number of grounds, the first being that the cycles of all three families show such close parallels that any acceptable explanation for one of them must be equally applicable to the other two. In the red grouse we *know* how populations are normally controlled, and the mechanisms involved do not include cycling, although grouse nevertheless exhibit cycling, as a different phenomenon. It has also been shown beyond reasonable doubt that the microtines and hares possess the same kind of social mechanisms for population homeostasis as the red grouse and put them to use, as the grouse does, in the intervals that elapse between one cycle and the next. In short, there is no evidence that cycling is connected with or necessary for homeostasis, and much evidence that it is not.

Another alternative hypothesis is that of Freeland (1974), who attributed vole cycles to dietary changes which the voles inflict on themselves. According to him, when vole numbers rise sufficiently high they overtax the supply of non-toxic vegetation, to an extent that allows species that are more toxic, and hence have been left uneaten, to spread and encroach. Eventually the voles are reduced to eating these obnoxious plants for want of other food, with the result that mortality from poisoning comes to exceed natality. From that stage the cycle is predicted to complete a circle. There is no evidence to support this from general observation, nor of the capacity of non-toxic plants to recover their supposedly lost ground as soon as the voles become scarce. In so far as the red grouse can and ought to provide a parallel, the hypothesis is known to be misconceived; and before it could possibly be credible of any cyclic species it would be necessary to dispose of the evidence assembled in this book that animals do not normally cause the extermination of their food species.

14.6 Dependent predator cycles

In regions where strong cycles occur in the microtine and hare populations, secondary cycles are frequently evoked in the birds and mammals that prey on them. The best known example is in the lynx (*Lynx lynx*) in Canada, whose fluctuating numbers were recorded in the fur-trade returns of the Hudon's Bay Company, already cited in earlier sections of this chapter. Its principal prey is, of course, the varying hare; and although the process of assembling the data on numbers of traded skins, of either species, was not free of imperfections from a demographic point of view, there is no mistaking the close synchrony in their striking fluctuations, over at least 65 years from 1845 to 1910. In that time the lynx, like the hare, produced seven peak years, with six intervals between them (Seton 1912: 105–6; Elton 1924: 135). Neither the lynx peaks nor the hare peaks occurred quite synchronously across the whole breadth of Canada, whence the annual data were summed, so that year-by-year comparison of the two species tends to be inconsistent; but Brand *et al.* (1976) showed in a more limited area of central Alberta that the lynx normally lags a year behind the peaks of the hares. No one seriously questions that the hares are the pacemakers in this relationship.

Similar dependent cycles occur between the coloured foxes (red, black and 'cross' morphs of *Vulpes vulpes*) in northern Canada, and their prey. Over a large part of the traditional Hudson Bay Company empire there was, and presumably still is, a 10-year periodicity in the trade in fox pelts running in parallel with that of lynx pelts, because the varying hare is also the staple prey of this species of fox. Fortunately we have another independent set of trade figures, collected by the Moravian missions on the Labrador coast, where no varying hares exist; and there the same species of fox followed a clear 4-year cycle, no doubt evoked by microtine prey. In 92 years of records there were 23 peaks, with extreme intervals of 2 and 6 years (Elton 1942: 264–9). The arctic fox (*Alopex lagopus*), also largely dependent on lemmings especially in their years of plenty, shows a similar 4-year type of cycle in the fur-trade figures. Other well-known predators on lemmings include the snowy owl (*Nyctea scandiaca*), rough-legged buzzard (*Buteo lagopus*), the skuas or jaegers (*Stercorarius* spp.), and various ermines and weasels (*Mustela* spp.), all of whose life patterns are also affected by the prey cycles.

Around the peak of a cycle the prey usually become superabundant, sometimes enormously so, with the result that the predators are swamped with food, and many of the prey consequently survive to undertake and no doubt complete their emigrant missions. The predators are far slower breeders than their prey, and much of their local increase in numbers is probably brought about by travelling and assembly at places where the prey abounds. Lynx, foxes, and snowy owls are especially notable as tramps and opportunists, wandering far and wide when their prey vanishes, and forgathering again whenever they manage to find a new herbivore cycle developing. Reduced fox and lynx populations can survive adequately while their prey are 'free-wheeling' at low densities between cycles, but they use the opportunities when they occur to step up their own reproductive rates, as do the owls and skuas. If my hypothesis is right, one might expect herbivore cycles to be generally patchy, weaker in some localities and stronger in others not necessarily very far apart. The heavy concentrations of predators where prey is superabundant occur while the cornucopia is briefly spilling its contents; but the last litters the predators manage to produce, as the prey production vanishes, must scatter as soon as they are weaned. Almost certainly it is the exodus of the prey that sets off the vagabond predators; and on occasion it drives snowy owls in autumn south into the agricultural and urban belt, for example into southern Quebec and Ontario and the States across the border. Even arctic foxes may forsake the tundra plains and travel up-country, as much as 300–400 km south into the forest in Canada, and 1300 km in Siberia (Banfield 1974). Arctic foxes also migrate laterally along the shore-ice through the arctic archipelago of Canada; and on one occasion a lynx, denizen of the forest as it is, turned up in southern Baffin Island, 400–500 km from the nearest forest tree. It must have crossed the Hudson Strait on the moving, milling pack-ice.

This picture must remain in part conjectural. It makes one suppose that the population structures of these boreal small-prey predators are very fluid, and especially those of species that depend on short-cycle microtines. The hares tend to have a more leisurely build-up to their peaks, as indicated on an earlier page (p. 183), and perhaps even give the predators a chance to adopt the prey cycles as their own, for genetic structuring purposes. The greatest problem in pursuing the nomadic way of life, from a structuring and group-selection point of view, is the lack of continuity, and not being able to keep hereditary stocks going on their own philopatric ground, where they can reap the rewards of homeostasis and habitat conservation; but the transient superabundant crops these predators seek and find are in fact neither vulnerable nor conservable.

As a general inference to be drawn from this chapter, it appears from the evidence presented that spontaneous population cycles, inherently capable of promoting frequency variances, and perpetuating allelic diversity, in structured demes with small, intermittently isolated gene pools, are largely confined to herbivorous species of rodent and lagomorph mammals, and galliform birds. They appear to be a rare, though because of their parallel evolution a very positive, phenomenon. It seems possible that the perennial, shrubby or arboreal vegetation on which the cyclic animals subsist, especially in winter, lends itself particularly well to an alternation of prolonged build-up and intermittent cropping, much in the way that man has traditionally adopted with coppiced woodland, and with slash-and-burn agriculture, though on rather longer rotations than the cyclic animals would practise. Some alternative mechanisms for demographic and genetic structuring, and thereby evolutionary change, are taken up in the next chapter.

Summary of Chapter 14

1   Red grouse are normally year-round residents in the same locality. There is a dispersal of young adults in autumn but for most of them it is unexpectedly restricted. Of those ringed as pre-flying chicks or undispersed adults and subsequently shot, 97% had moved less than 5 km. Among the males, few exceeded even 1.5 km, though 20% of the females did so. The six young males traced by telemetry (Chapter 8) all established themselves in or next to their own natal territories; and in contrast, one female had gone 42 km. The methods of catching birds for ringing did not pin-point their birth sites; consequently no short dispersals of females have been exactly measured, but it is known that on average they move further than males. Sexual differences in mean dispersal distance are common in birds and mammals (Chapter 16).

Philopatry is the foundation of sustainable population structuring; and having one very closely philopatric sex while the other is somewhat less so, permits a high average philopatry while avoiding matings that are harmfully close. Dispersal distance in conjunction with population density are the major factors also in determining the effective size of interbreeding units. The structuring itself is a group adaptation: it subdivides the population into small, almost isolated local stocks which become inheritors of their habitats; and if they conserve them it will benefit their posterity, just as the present holders have most likely benefitted from their prudent forbears. At the same time, isolation allows their gene pools to differentiate from one another; and genetic divergences so created may result in productivity differences between them, with consequent group selection (Chapter 15).

2   At long intervals the grouse exhibit a second type of dispersal, taking place *en masse* during the height and decline of a population cycle; the distances involved are unknown (Chapter 12). This cyclic phenomenon is shared by other tetraonids as well as by lemmings and other microtine rodents, and a few other vertebrates. For the red grouse population in any given district the cycle comes and goes as a sporadic episode, which runs its course in about 8–10 years, and is preceded and followed by a much longer interval of normal homeostasis. In the Icelandic stocks of rock ptarmigan, however, the cycles have shown strikingly regular recurrences of three or four in succession before the sequence broke down. The density amplitude of all such cycles, whether in birds or mammals, varies, but is generally greater than that of homeostatic fluctuations; and as well as rising to a high the density also typically falls to a deep low before returning to the norm.

3   The cyclic microtine rodents typically live—like the tetraonids—in arctic, boreal or arid habitats. Their dynamic changes have attracted much research; but, being secretive, largely nocturnal mammals they are much less amenable to study than grouse. They show the same phenomena both of intermittence and a proclivity for merging into sequences. Although independently evolved, cycling is remarkably similar in the grouse, microtines, and the one cyclic species of hare, and this points to an identity of function.

Chitty's hypothesis proposed that the rodent cycles serve to regulate the population density by keeping it alternating between growth and decline. As it grew, selection would increasingly favour aggressive genotypes who demanded more individual space, and so thinned the density down again. Selection would then turn against the aggressors and allow tolerant individuals to predominate and densities to rise, completing the cycle. Against this, in addition to a lack of corroborative evidence, is the fact that the microtines have repeatedly been shown to possess effective homeostatic density controls, which parallel those of the grouse and are similarly related to the state of their plant food supplies: no alternative controls should therefore be necessary. Moreover the red grouse, whose homeostatic regulation is both efficient and well understood, is itself an intermittent cycler, apparently for non-homeostatic purposes.

As in the red grouse, cycling in microtines is always, as far as can be ascertained, linked with the general dispersal of surplus populace which the upswing of the cycle creates—a sequel that has long been recognized in the lemmings of Scandinavia. In some circumstances the emigrant lemmings and voles do not migrate far but merely take over nearby 'colonist' habitats during the cyclic highs, which in other years are left vacant, once the population has shrunk back again into its 'survival' habitats.

4   Closely similar cycles occur in the snowshoe hare (*Lepus americanus*) in the northern forests of Canada. A striking feature here is the historical

evidence of cycles in the Hudson's Bay Company's fur-trade records, going back nearly 150 years. At their peaks hares in white pelage sometimes assemble locally in great numbers and then suddenly vanish without trace, no doubt by dispersal. On a study area regularly cleared of hares by trapping, a more than linear rise in the recolonization rate occurred as the surrounding hare density increased, indicating an enhanced dispersal. Though their cycles tend to be consecutive they are not usually synchronized over the whole transcontinental range of the species; and there is the same tendency to patchiness on a more local scale that characterizes other unrelated cyclic species. Thus although the long-term collective records suggest a regular, unbroken 10-year periodicity, not every deme appears to erupt during each regional peak. However, some must do so in order to supply the self-evident pacemakers.

5   In all the cyclers, increasing density puts pressure on the food resources especially while the impending emigrants are nurtured. In the variable climates they have to face, high numbers can be reached only in the more productive years. Climate consequently holds the trigger, and hence the well-known tendency for cycles to start in synchrony over wide areas. The period a cycle takes to run its course appears to depend on generation times and tends to stay the same, regardless of the height of the peak attained; hence its constant length; and between consecutive cycles it is only necessary to wait for the first suitable year before starting again. A delay of a year or two is short in relation to the 8–10 year cycles, and this gives them a greater semblance of regularity than the 4-year cycles of the rodents, which are liable to be kept waiting just as long.

The function suggested here is that cycles are necessary to fulfil the genetic structuring programs of the species involved. If numbers are plotted against time, the cyclic lows tend to be U-shaped and the peaks steep. In the former phase the units are isolated and inbred, and are also engaged in conserving or rebuilding their food resources. In the latter they make the best bid they can for productivity and then immediately disperse; and this leads to outbreeding for a generation or more. That redistributes diversity and brings back more heterozygosity to the local gene pools, which has been sacrificed through inbreeding and fixation. If it is true that many units let one or two successive cycles pass them by before they are ready to erupt again, the inbreeding phase will be prolonged; the unit's effective mating size will remain small, favouring the continuance of drift and thus between-group differentiation. When the program switch eventually comes, units with more efficient resource husbandry will be rewarded by better survival and preparedness and a higher productivity, and vice versa. That could provide a powerful opportunity for experimentation with small, relatively expendable and mutually differentiated gene pools, and in due course allowing each unit to contribute to the wider population in proportion to its merits, thus achieving group selection.

As hinted earlier, the long and short cycles pertain to animals that initiate, respectively, one or two new generations a year. Certainly in most and probably in all the cyclic microtines, the early-season young mature and breed later in their own birth year at least during the expansion phase; whereas grouse and the snowshoe hare are alike in breeding first at one year old. Four to six generations of isolation and drift measure the time it takes to maximize inter-group differentiation (see p. 189); add a couple more for outbreeding, plus a year or two for climatic delays, and the sum works out at about 9 years for the grouse and hares and 4 for the lemmings and voles.

It is necessary to add that pest outbreaks of rodents can develop in most regions of the world, and usually have quite a different cause—most often the temporary provision of edible crops by man. Even some cycles of voles are facilitated if not triggered in this way, so that there is a possibility of confusion. Gene-pool cycling, on the other hand, is quite autonomous. It is apparently confined to the north temperate and sub-arctic zones to which the tetraonid and microtine families and the one hare species are themselves naturally restricted. All the true cyclers are small herbivores.

6   Predators that depend on these same small herbivores as staple prey tend to reflect the latter's density cycles. Not only do lynx, foxes, hawks and owls increase their own litter or clutch sizes homeostatically when food is abundant, but some of them are noted prospectors and opportunists, especially in the high north, and habitually assemble where there are erupting demes. Even so the prey peaks normally swamp the predators' best efforts to exploit them. Causally therefore the herbivores' cycles have nothing to do with classical predator-prey oscillations, but just surge ahead of their own momentum, in spite of predatory mortality. The HBC records show the trade in lynx and coloured-fox skins closely tracking the rise and fall of snowshoe hare densities; but there are independent historical records from missions in northern Labrador, beyond the ranges of lynx and snowshoe hares, and where the latter is replaced by the non-cyclic arctic hare; and these show both the white and coloured

foxes following a four-year cycle, undoubtedly dependent on lemmings and other microtines.

It appears virtually certain that the predator cycles are always homeo-static and secondary, and do not perform any genetic structuring function.

CHAPTER 15

EVOLUTION IN STRUCTURED POPULATIONS

15.1 Vertebrates in general have adaptively structured populations

Two later chapters are devoted to examples of population structuring chosen from the invertebrate phyla, and particularly from those still in a presocial state of organization because they have important pointers to offer. In this chapter and the next attention continues to be focused on social animals, and mainly on the vertebrate classes. It seems safe to assume that wherever one chose to look among these groups one could expect to find evidence for adaptive structuring, comparable with what we have seen in the red grouse and microtine rodents.

In the majority of vertebrate species the members confine their movements, most of their lives, to a home-range that is small in relation to their powers of locomotion. Most individuals leave their birthplace and move away to establish a home elsewhere only once, and some may never change their residence at all. In a minority of species, mostly in groups with high mobility such as birds and fish, the members make yearly two-way migrations between alternative domiciles, and in that event they are normally faithful to the locality and deme in which they breed, and sometimes to their alternate locality as well. Living a roving life when not engaged in breeding can also be a viable strategy, found for example among many pelagic marine animals; and at the extreme development of vagrancy there are a few species, such as the snowy owl mentioned previously, some of whose members may never take up a perennial abode, and all of whom appear free to wander anywhere within their habitable biome. In contrast, for the vast sedentary majority of species, once in a lifetime is the norm for dispersal.

Dispersal is an important factor in maintaining spatial structuring, and as such it brings benefit to the population as a whole. In the event, it may often incur risks to the individual. Individuals must generally be more secure when they live in familiar surroundings, provided these continue to offer the necessities of life, and this presumably tends to minimize the number of residential changes they make. In many species the juveniles are allowed to stay in their natal home-range until the next breeding season approaches, and in some species they stay till the next but one or two; but the prospect of having a new cohort born into the habitat commonly requires that a cohort of adolescents leave beforehand in order to free sufficient carrying capacity. If, therefore, there is only one dispersal phase in the life cycle it is most likely to coincide with adolescence and the assumption of independence.

The sex difference in mean dispersal distance shown by the red grouse and many other birds and mammals is more than simply a device for preventing incestuous matings. Indeed, as the next example will show, the latter are apparently not always especially avoided. More importantly the difference is one of the common evidences of the existence of genetically-coded population structuring, in this instance serving primarily for regulating gene flow. There is a behavioural advantage in having at least one sex that remains close to where it was brought up, in the local experience it can gain and utilize, and the possibility of acquiring important traditions from its elders in the use of particular food resources, breeding and roosting sites, and the like.

On the genetical side, however, it appears that the classical assumptions about panmixis in mating systems are seldom if ever borne out by the facts, at least in the vertebrates; and in birds and mammals including primitive races of man, mating is generally regulated and contrived—often as in the red grouse being exogamous and patri- or matrilocal as the case may be. For this reason great interest begins to attach to the exact provenances of future mates, and their individual dispersals between birth and breeding sites. These, in the context of population density and dispersion, determine the effective size of the true interbreeding units in the population.

Another of the few species that have already been well documented in this respect is the great tit (*Parus major*). Here as in the red grouse both sexes are found to have skewed distributions of their dispersal distances, the males again being the more strongly so; but the difference in mean distance between the sexes appears to be less than in the grouse (medians 560 m and 880 m respectively; Greenwood *et al.* 1979: 127).

Van Noordwijk and Scharloo (1980) have analysed the degree of inbreeding in a relatively isolated, small population of great tits, occupying disrupted habitats on Vlieland, one of the barrier islands that skirt the Dutch Waddensee. This box-nesting population had been kept under surveillance for 20 years, and as far as possible all young birds raised had been ringed each year, with the result that the lineages of some 200 breeders became traceable at least to their grandparental generation. The island's effective breeding population had fluctuated between 38 and 94 pairs with a mean of roughly 50 (i.e. N ≈ 100). About a quarter of the pedigreed matings traced were between birds with a relationship coefficient of 1/16 or greater, and they included some sib matings. One interwoven family history reported in detail extended over 5–6 generations (back to the founders, A and B), and in the last generation the birds M and N each had two or three separate paths of descent from both founders. Given a closed randomly-mating population of 100 per annum the expected coefficient of relatedness between mates would be $1/2N = 0.005$; but a minimum estimate of the average coefficient actually found on Vlieland was 0.015, or about three times the expectation. The authors had reason for thinking this was not due to an effective subdivision of the population into smaller discrete units; but presumably it might have been explained (in part) by having a steno- or heteromictic dispersal system instead, programmed to produce a range of individual distances which reduced the effective choice of mates well below the potential maximum of 50; this is what the Oxford dispersal data just quoted appear to suggest.

It could also be partly accounted for, as the Dutch authors suggest, if the 'inbreeders' tended to leave more surviving progeny than the 'outbreeders'; and although evidence of inbreeding depression was found to result from some close-kin matings, and led to the failure of an increasing proportion of the eggs to hatch (a well-known consequence of inbreeding), it turned out in fact that the chicks surviving from these clutches tended to show a better than average, and thus compensating, survivability to adulthood. The data available are not sufficient to establish that inbreeding was significantly correlated with increased fitness, though they do indicate at least that inbreeding was not avoided. On general principles, however, it may be suspected that individuals standing high in the social hierarchy for genetic reasons would have a priority in the mutual choice of mates; and that such matings would have a correspondingly high probability of being fertile and producing good quality offspring, an appreciable proportion of which should succeed in repeating the process in their turn. As in human aristocracies, these are conditions likely to increase the probability of consanguineous matings. A revealing though perhaps irrelevant incident reported from the Vlieland birds, is that one matriarchal female was found to have contributed 37 adult grandchildren to the population, who constituted 10–20% of the island's breeding stock for about five years!

Rather different conditions of structuring in a breeding population have been described by Fisher (1976) for the Laysan albatross (*Diomedea immutabilis*) nesting on Midway atoll in the western Hawaiian Islands. These birds show a high degree of philopatry in returning as adult recruits, not only to their parent colony but to a strict location within it. Fifty-six males and 49 females whose exact birthsites on Fisher's study plot were known, returned in due course and established nest-sites. The sexes were found to have mean dispersal distances of 19 and 26 m respectively, between their birth and adult sites. (These averages do not take into account two birds banded on Midway although not on the study plot, which were subsequently found breeding in a colony on the next island, 100 km away to the west.) The nests are fairly closely packed in the denser parts of the colony judging from photographs; and if, to obtain a rough approximation, one assumed them to be uniformly spaced in a honeycomb pattern with 3 m between centres, that would produce an individual nest-territory of about 8 m². Assuming also that the distribution of dispersal distances is leptokurtic, it would imply that a large majority of matings took place within a philopatric circle containing 100–500 nest-sites. This can be set in perspective against a total breeding population on Midway of 180 000 pairs.

In this instance the structuring could readily serve for restricting gene flow between in-groups or demes, but it does not throw any light on the obscure problem that vagrant pelagic feeders of all kinds present, namely how their numbers can be matched with trophic carrying capacities in a limitless sea. It may eventually turn out that demes of breeders are associated much more closely than we now suspect with particular oceanic water-masses, and that a sufficiently effective structuring for selection to act upon is actually maintained on (and in) the high seas.

15.2 Wright's theory of the shifting balance process

The adaptive advantages of spatial structuring which accrue to populations as wholes can be separated into two functional kinds, namely the ecological and the evolutionary; both apply equally to cyclic and non-cyclic species. The ecological function is to ensure, by subdividing the population into small localized units each the hereditary tenant of its own local habitat, that the rewards of resource husbandry are transmitted to future generations perpetuating the same gene pools on the same ground. For non-cyclic animals this task is possibly less complex than it is for the cyclers because their population homeostasis is only concerned with tracking changes in habitat carrying capacity, and a single uninterrupted program may be all that is required for regulating inbreeding and emigration, and preserving the integrity of their demes. The evolutionary function cannot be summarized as briefly as this, and it forms the main subject of this section.

My own ideas about both functions have of course been reached by following ecological and sociobiological paths; but there is an older and more powerfully and persistently argued theory that has led to very much the same conclusions, with regard to the evolutionary function alone, in the work of Sewall Wright (e.g. 1931, 1940, 1945, 1949, 1968–78, 1980).

Wright's argument starts from the usual premise that by far the largest part of evolutionary change is mediated through changes in gene frequency. In any animal population such changes are due to four factors, in varying degrees, namely (1) mutation, which transforms existing genes; (2) selection, which exerts a directed bias on the transmission of alleles from one generation to the next; (3) the migration of individuals, which carries their genomes into or away from the population; and (4) chance or random events, which can modify the sample of parents who actually survive to reproduce in any given generation.

Attention must be given to the many largely heritable characters in multicellular animals (and plants) that exhibit continuous variation, which are controlled by multiple interacting genes. This is conspicuously true of metric characters—length, weight, proportions, colour saturation, numbers of replicated parts, precise timing of seasonal activities, and so on—in which any given sample of individuals will show a range of values, often distributed normally about the mean. If so, it implies that selection is having a stabilizing effect, in as much as the fitter majority are the ones that lie closer to the norm. Genes of small effect (polygenes) are involved in most or all these characters; and they are difficult to study experimentally because their individual contributions are undetectable in the phenotype. Many of them, and of the major genes as well, also have epistatic effects that modify each others' action. In the words of Mather and Jinks (1971: 32) 'polygenic systems provide the variation of fine adjustment, clothing the indispensible genetic skeleton and moulding the whole into the fine shape demanded by natural selection. They are the systems of smooth adaptive change and of speciation.' They must also be the main systems of demic differentiation.

Wright takes as an example a notional species of organism that has 1000 gene loci each with 10 alleles. The number of different genotype combinations that could theoretically be made from these would be vast beyond imagination; but in fact selective pressures would be so conservative, in a large panmictic population, as to eliminate almost all the aberrant genotypes that reproduction threw up. Of the four factors mentioned above that can lead to gene-frequency changes and thus to evolutionary progress, mutation is the producer of novel genes; but its rate is exceedingly slow, and selection in fact suppresses most of its products, excepting those rare ones that are favourable from the first. A large population made panmictic by free-ranging dispersal would consequently exhibit resistance to change, and could rarely if ever 'experiment' for more than one generation at a time by fostering off-centre genotypes.

If on the other hand the population is structured into many small units, containing of the order of 100 interbreeding members in any generation, with their inter-unit migration strictly controlled so that for some or many generations it is small or intermittent, the second and fourth factors, selection and random events, can be given enormously greater scope. In such small units the individuals that actually survive to breed will be rather few in number, and subject therefore to appreciable sampling errors, giving rise to corresponding variances between units even though the units may originally have been derived from the same base population. Within a few generations, as indicated in the previous chapter, gene-frequency variances between the units will be maximized, particularly if selective pressures are not strong enough promptly to extinguish those units that drift well away from the population norm.

The optimal conditions for such differences to develop have been modelled. Mutation can be neglected as an effective force, in what should prove to be a context of more or less incessant short-term change. Adaptive differences may however arise because of local variations in selective pressure (s), and these will have more influence in groups of relatively larger size (N). Conversely, non-adaptive random changes will become more marked as N is reduced. Both effects will be favoured by keeping the exchange of migrants (m) low. The greatest amount of differentiation is thus to be expected at some intermediate N and, it can be shown, of a size such that $4\,Nm$ and $4\,Ns$ are both in the neighbourhood of 1 (Wright 1940: 175).

To convey an objective impression of the 'field' of gene combinations that potentially exist, and starting with simple examples, Wright (1932) constructed a geometric diagram showing, in perspective, lines that connected up all the 32 homozygous combinations obtainable from five gene loci each bearing a pair of alleles. He explained that to represent these relationships symmetrically would in fact require five dimensions, and if all their respective adaptive values were also represented, a sixth would be needed as well. Each increase in the number of loci would add another dimension, so that if the pictorial concept is to be stepped up to the dimensions of the more realistic complement of 1000 multiallele loci, one must obviously be content with much less realism and more imagination in trying to comprehend the geometry of the field. He depicts it as presenting an uneven 'adaptive surface' of hills and valleys, with contours to represent differences in the adaptive values of the numerous viable genotypes that are certain to be inherent in a complex gene pool, if they could be given the chance to assemble and emerge (Figure 15.1). Each existing unit's gene pool occupies some appropriate point on the surface, according to its constitution; and any that is situated on a slope will be driven upwards by natural selection until it reaches level ground above. The units in most populations would eventually be clustered on a hilltop. How then would it be possible, by continual random or selective changes of gene frequencies, to descend again and cross one of the intervening saddles to reach another of the numerous upward slopes, leading perhaps to a still higher fitness peak?

Populations structured into thousands of local units each subject to drift, and sometimes to relaxed selection, would, he says, provide 'a trial and error mechanism on a grand scale'. Environments, both living and non-living, are always changing, and this may result in lowering or flattening out the adaptive hill the unit occupies. Rarely in such circumstances random events in a particular unit might cause the balance between interacting genes at two more more loci to shift, carrying the unit across a saddle and opening the way for a particular locus to play a bigger and more profitable role than before, and thus moving the unit's adaptive value to a higher level on a different upward slope. A 'mass selective' phase would follow. The successful unit would climb the new adaptive peak, expanding as it rose, and by emigration and crossbreeding might be able to pull the whole population, and perhaps the whole species, to the new position. The crossing of a single two-locus saddle 'is only an

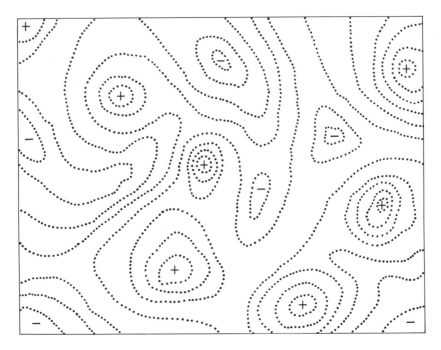

Fig. 15.1 *Wright's representation, in two dimensions instead of many thousands, of a field of potential gene combinations, with contours to show the relative adaptiveness of different situations in the field. From Wright (1932).*

elementary step in the differentiation with respect to a multilocus inter-action system' (Wright 1978, vol. 4: 524).

Populations structured on Wright's model could store a much greater genetic variability than panmictic populations of the same size; and not only might better adapted stocks emerge in the manner just described, but when environmental changes of wider than local importance occurred the probability would be correspondingly greater that pre-adapted units were already in existence, from which mass selection and expansion could take place. In his expressive phrase (Wright 1980: 841) 'creativity is raised to the second power' in populations structured in this way. In an earlier overall comment, especially relevant in the present context for its concluding words, he wrote (1940: 175): 'Under this con-dition (of N, m and s) neither does random differentiation proceed to fixation nor adaptive differentiation to equilibrium, but each local popu-lation is kept in a state of continual change. A local population that happens to arrive at a genotype that is peculiarly favourable in relation to the general conditions of life of the species, i.e. a higher peak combination than that about which the species had hitherto been centred, will tend to increase in numbers and supply more than its share of migrants to other regions, thus grading them up to the same type by a process that may be described as intergroup selection.'

He has observed (Wright 1980: 838) that his theory has frequently been misrepresented. The paper in question is intended to put the record straight, and particularly to emphasize the distinction between genic and organismic selection; the former is what most Mendelian theorizing is about; and the latter evokes adaptations that become universal property, benefitting all individuals alike and adding to the average fitness of the whole population. 'An organism' he says (p. 826) 'is a most extraordinary sort of entity from the physico-chemical standpoint. Starting from a minute bit of matter, it goes through an intricate series of transformations as it develops into its adult form, irrespectively, in the main, of ordinary variations in the environment. At all stages, all parts are integrated into a harmoniously functioning whole. There is nothing remotely comparable, in the nonliving world, of greater size than a molecule.' Individual genes are affected by the genic background in which they are situated, and the shaping of the phenotype results from a multifactorial network of inter- actions and multiple effects (pleiotropy) which integrate the genotype into a balanced whole. The extraordinary coadaptive nature of the pheno- type is not destroyed by meiosis and recombination; and the combi- nations themselves 'are broken up so rapidly in terms of the evolutionary process (except for almost completely linked genes), that it is only the average allelic difference that counts' (p. 825).

He concludes the paper (p. 841) by saying that several recent authors who have discussed group selection for group advantage at length, have rejected it as of little or no evolutionary significance, and seem to have concluded that natural selection is practically wholly genic. 'None of them discussed group selection for organismic advantage to individuals, the dynamic factor in the shifting balance process, . . . although . . . it is not fragile at all, in contrast with the fairly obvious fragility of group selection for group advantage, which they considered worthy of exten- sive discussion before rejection.' My task is obviously to show that the group selection I advocate is not essentially different from Wright's in raising average individual fitness, and is not fragile either. It concerns the organismic property of securing the most productive resources for the maximum number—an externalized group attribute as far as individual genotypes are concerned, like Wright's population structure. I have very little doubt that, if the shifting balance process is real, it was the mechanism through which cooperation, resource management and, later, sociality evolved.

15.3 Laboratory experiments on group selection

Wright's theory has gained long-delayed support in recent years, in part from the ecological realization that natural populations of animals *are* dynamically structured in space, on patterns that clearly resemble the kind he first predicted more than 50 years ago; and in part from a series of experimental tests initiated by Dr Michael Wade, in the same University of Chicago laboratories where Wright's (and Allee's) work had earlier developed.

In these experiments the flour beetle *Tribolium castaneum* has been the test animal. This species (and alternatively *T. confusum*) can be cultured indefinitely in controlled standard environments, going through their life cycle from egg to larva, pupa and breeding-adult in about 30–45 days; the adults may live considerably longer. Wade established small groups, to represent the structural units of a natural population, in separate shell vials containing the nutrient medium of wholewheat flour mixed with yeast. These were housed in an incubator. A beetle stock of high genetic variability had previously been created by a mass mating of four inbred laboratory strains, which were allowed to interbreed for several generations (Wade 1977).

Four experimental treatments were run in parallel, imposing different regimes of selective extinction and perpetuation on separate sets of population units. Each treatment was replicated 48 times, making 192 populations in all; and every population was started off in exactly the same way, by a 'propagule' consisting of 16 adults drawn at random from the stock. The trait under selection in all treatments was the rate of population increase, a multifactorial character that is affected by parental fecundity, egg, larval and pupal mortality (due in part to cannibalism), body weight and female development time—all being quantities known to contain genetic components. On the 37th day of the first generation the populations in each treatment were taken from the incubator and the young adults they contained were censused. This is the stage at which the beetles, if they were free to do so, would tend to disperse.

In Treatment A the one unit in the 48 replicates that had produced the highest number of adults was then selected, and divided at random into groups of 16 to serve as new propagules, and become parents for the next generation. Since the large number of adults required for the purpose was more than a single unit could provide, the second most populous one was taken as well and similarly divided, and so on until 48 sixteens had been obtained. All the remaining groups of adults belonging to that generation in Treatment A were eliminated from the experiment. The same procedure was followed for each subsequent generation.

In Treatment B the *least* populous units were similarly propagated and the rest eliminated. Treatment C was a control in which individual selection was left to take its course; each one of the 48 lines was therefore propagated independently by picking at random 16 adults to make one propagule for the new generation, and discarding all the rest. Treatment D was similar to A and B except that the units selected for making the propagules for new generations were chosen by reference to a table of random numbers, and the other units rejected: the treatment imposed here was thus a random group-extinction process. All four treatments were carried in their respective ways through nine generations.

Figure 15.2 shows that the different treatments led to rapid divergences in their productivities. The fact that the control populations continued to decline in productivity, and the other treatments did the same in

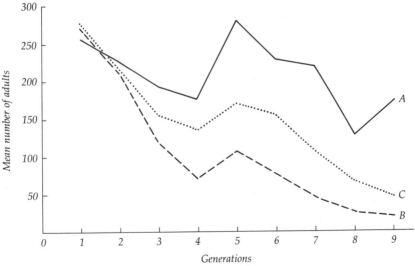

Fig. 15.2. *The mean numbers of adults for the High (A) and Low (B) group-selected populations and the Control (C) of Experiment I for all generations. Each point is the mean of 48 observations. From Wade (1977).*

varying degrees, and that all went through simultaneous irregularities in generations 4 to 6, was at first puzzling. Later work was to show that the decline was attributable in part to inbreeding depression and in part to the physiological effects of crowding during development (Wade & McCauley 1980: 803). The divergences produced by the artificial group selection, on the other hand, were strong and unequivocal: by generation 9 the A populations were averaging over 40 times as many young adults per generation as those of B. The variance among propagules caused by 'drift', on which these effects of group selection depended, was presumably large in comparison with the variance within propagules accessible to individual selection.

It had been widely believed by evolutionists at large that even a very weak gene flow from deme to deme would be enough to counteract the variance produced by drift; and to probe the truth of this assumption a second set of trials were run as 'Experiment II'. Surplus adults from the fourth generation of the Experiment I treatments were used to propagate parallel sets of descendants under the same four selection treatments as before, except that each new generation was initiated by mixed propagules, deriving their membership from two parental units instead of one. In Experiment I the random drawing of a number as low as 16 adults from a given unit to become a parent propagule had, as expected, led to genetic drift ('sampling errors') between propagules and a consequent divergence between lines within the same treatment. In Experiment II under Treatment A, instead of using up the whole of the most populous unit first, and then going on to the next largest, and so on, both were used together as a simultaneous pair; and to simulate recolonizations in which some of the founders were migrants from a second locality, the propagules were made up by drawing either 8, 12 or 16 individuals at random from the first unit and complementing them with 8, 4 or 0 from the second. Additional pairs of high population units were drawn on as required when these were exhausted, to fulfil the quota of propagules. This resembles a natural situation where members from two nearby structural units, under similar selection pressures, arrive to colonize a vacant habitat. The results showed that variation between units due to drift *was* in fact diminished to some extent; but even a 50–50 mix could still sustain, though perhaps not increase, the relative differences in production phenotype that Treatments A–D had produced.

In a review paper that appeared soon after these results were published, Wade (1978) compared and discussed the various theoretical models of group selection that had been advanced up to took as a definition of group selection 'that process of g which is caused by the differential extinction or proliferatio organisms'. The group itself might be 'any unit of popula ranging from families (kin selection) to whole populations (interdemic or interpopulation selection)'. The published models could be divided into two classes, the traditional and the intrademic. Those of the first type centred on Wright's shifting balance theory, and postulated local breeding units that were interconnected only by small numbers of dispersing individuals. Gene frequencies were altered in the population as a whole through the differential extinction of, and recolonization by, the almost isolated units. The second type was exemplified by D. S. Wilson's model (see p. 316), in which there are larger panmictic local demes subdivided into smaller, more or less transitory 'trait groups', containing a few individuals coming together at a particular stage of their life cycle. Trait-group survival and shared fitness depend significantly on the members' interactions with one another. When the trait group breaks up the members separate and mix with the rest of the deme, and the groups with higher survival rates contribute proportionally more to the population of the deme than the average. Kin selection (Hamilton 1964) is a special case in this class of model.

The authors of both types of model had made some or all of five assumptions, which, Wade points out, are inherently unfavourable to the operation of group selection. The *Tribolium* experiments had already suggested to him that these assumptions might not be essential and unavoidable, and that their validity could be experimentally tested. The first of the five was to accept unconditionally that the frequency of a single allele in a population can significantly alter the productivity or survival of a group, in the way it is generally held to do for the fitness of an individual. This assumption has figured prominently in debates on the conflicting advantages, for groups and individuals, of selfish and altruistic alleles, or of social and asocial alleles. But certainly there are important processes in group selection that cannot be reduced to the simplified context of single loci, because they relate to epistatic effects, and to interactions between different genotypes, including their collaborative responses in maintaining the homeostasis and genetic structuring of the population. Group advantage can arise from having genotypically different individuals within the same group, programmed to play complementary roles in their cooperation for resource management, for the production of young, and for exploiting their genetic resources. In the longer term the survival of larger populations may similarly depend on fostering differentiation between the gene pools of their local sub-groups.

The *Tribolium* experiments had shown that small isolated breeding units, drawn from a common ancestry and living under identical conditions, would readily diversify among themselves by drift when subject to random extinctions, and could be made to differ strongly within three or four generations when divergent group-selections for high and low productivity were applied. The selection process involved whole units without regard to individual genotypes as such, and acted on a trait that integrates the effects of a large number of loci and the interactions of many individuals. As will appear shortly, the group phenotypes it produced proved to be stable and have a high degree of heritability.

The second unfavourable assumption was that a structured population possesses a common 'migrant pool' from which all vacant habitats are colonized, and in which the gene frequencies are the same as the average for the population as a whole; and the third was that all units contribute migrants to the pool in proportion to their own numbers. Neither of these assumptions are supported by the ecological facts as we know them. The annual dispersal of individuals, at least among representative vertebrates, appears to be controlled and largely localized; and if any doubt remains that some local stocks have higher production rates than others for genetic reasons, and not merely because of local differences in resource availability, then again the *Tribolium* experiments demonstrate that large genetic differences in productivity can rapidly be evolved by selection between small units living in identical environments.

The fourth assumption concerned the origin of genetic diversity between population sub-units. The 'traditional' models attributed this primarily to genetic drift, and the 'intrademic' models to the random sampling of founder colonists from the migrant pool. These are different processes and occur at separate times, the first at mating and conception and the second at dispersal; but both appear to the model makers to be subject to the same limiting conditions, namely that the initial group size must be small, intergroup migration must be low, and individual selection must be weak. The models are severely constrained by these restrictions, which, though they are inferences based on theory rather than on experience, 'have become part of the common wisdom of evolutionary biology' (Wade 1982: 958–9). Wade (1978: 108) believes they have largely been responsible for the prevailing opinion that group selection is not likely to be a significant force in evolutionary change; and he has been particularly thorough in subjecting them to empirical tests.

Before describing these tests and their outcome, however, closer attention needs to be given to the method he chose for propagating the *Tribolium* generations, and the dependence of his results upon it. Each was founded in the manner just stipulated, by picking a small pre-determined number of adults at random from his laboratory 'pool', and repeating the process in order to set up a statistically meaningful number of replicate units, for subjection to a particular treatment. But once the first filial generation had matured the lines were usually propagated by taking propagule quotas of young adults exclusively from their own production line; and subsequent generations were established in the same way. Thus a sample of a gene pool already moulded by selection and drift in a particular unit was transmitted to the next generation; and on each occasion the most advanced units for the trait under selection were the ones chosen to provide all the propagules required for the treatment; and all other units were extinguished. If such a process occurred in the wild it would yield daughter colonist stocks resembling one another, but newly differentiated by sampling errors in the make-up of the separate propagules. The diversification and heritability of the group phenotypes that developed would be respectively much faster and greater than in comparable units drawn from, and thrown back into, a large panmictic pool at each generation instead, even if the selective pressures were the same.

Panmixis would homogenize any fine-grain variance that might appear at the end of each generation, whereas philopatric propagation of the kind familiar to us perpetuates it. Wade (1978: 110) emphasizes the point by likening the migrant-pool effect to blending inheritance, as conceived in Darwin's time, in contrast to particulate inheritance. Blending implies that, in the expression of any quantitative trait, the offspring must be intermediate between the parents; and thus any existing variance would on the average be halved at each generation until it eventually disappeared. Particulate inheritance, like philopatric descent, would preserve it.

Wade (1977: 143 and 1978: 109) drew attention to another effect of his propagation system. If the units were subject to a high level of random extinction (especially under Treatment D), the equilibrium variance between the units would not only be greater than could be attained by a panmictic population of the same size, but a large proportion of the

variance originally present in the stock could be shown to become converted into between-population, rather than within-population (i.e.
individual) variance, thereby making conditions more favourable for
group selection. So ideal might they become that only an occasional
group-selection event could result in a large genetic change.

Further experiments by Wade and McCauley (1980) first tested the
effect of varying the propagule size on the development of variance in
productivity. The methodology was the same as that used in Experiments
I and II, except that instead of founding new generations with propagules
of 16 adults, four different sizes, with 6, 12, 24 and 48 adults, were tried
instead. There were two selection treatments run in parallel. In Treatment
P, ten replicate units for each propagule size were perpetuated by making
up one propagule from their own production of young adults, and discarding the surpluses. There were no extinctions of whole units and no
group selection; but individual selection took its natural course within
each unit and generation. As a result the mean productivities of the units
rose to a peak in the third generation and as quickly fell back to about
60%, staying at about that level until the experiment ended with generation 14. This is the normal response to be expected for *T. castaneum*
cultured in this manner and, as already noted, it is due to a combination of
individual selection, inbreeding depression, and the physiological
consequences of crowding, the last two lagging initially but eventually
overtaking the first.

The units in this persistent-line treatment developed as much as
three- to four-fold differences in productivity, appearing most quickly as
expected in the lines perpetuated by 6-parent propagules, and successively more slowly with the larger propagule sizes. The increases of
variance with time in all four sets were curvilinear, but by generation 14
they had all converged at about the same level, suggesting (to me) that
wild stocks perpetuated in small in-groups of quite normal size (cf. 48)
need not present a serious obstacle to group differentiation and selection.
The authors analysed these P-treatment data to find out how far the
productivity of one generation was correlated with that of the next in the
same line; and the coefficients proved to be large, thus indicating a high
degree of 'population heritability' of their productivity phenotype, on
which group selection would have been able to act.

In the second treatment, called E because extinction was inflicted at
random on 15 out of each total of 20 replicate stocks at each generation, 20
new units at each propagule size were established with propagules taken

from the surviving five in the appropriate set. This treatment showed that although the total variance that each set possessed at the start of the experiment was depleted by the recurrent extinctions, between-unit variances still developed on a reduced scale. It did not appear therefore that group selection would have been prevented by random extinctions from taking effect, as had at one time been suggested. Random extinctions could actually lead to genetic changes in a structured population analogous to those produced by drift at the local gene-pool level.

The outcome of the P and E experiments was to demonstrate the resilience of between-group variance to conditions appreciably less stringent, as regards founding and breeding group size, than were previously thought necessary on theoretical grounds.

McCauley & Wade (1980) then used the between-line differences that existed in the P treatment, by the time the experiment ended, in an attempt to throw more light on the component reactions and responses that were causing the four-fold difference in productivity which had emerged between the highest ('H') and lowest ('L') lines. These lines were each allowed to multiply their stocks, by transferring them to much larger volumes of medium, in order to obtain adequate supplies of beetles. They were then used to make cross matings, both with single pairs and with mass matings between samples of 4, 8, 12 or 24 adults from each. The matings were designed to separate male effects on productivity from female effects, by making one or other of the four crosses H♂ ×H♀ (A), H♂ ×L♀ (B), L♂ ×H♀ (C), and L♂ ×L♀ (D). A second variable was introduced by establishing two regimes of population density. In one of these (CD series) the container volumes for the different-sized units matched the number of adults in the propagule concerned, so as to provide a constant density (and surface area) throughout, with two adults per gram of medium. In the other regime the units were all set up in vials of the same size regardless of the number of beetles, thus constituting a constant volume (CV) series. The CD series investigated the respective contributions of the sexes to breeding success in the absence of crowding, and the CV series their respective tolerances to crowding. Productivity, in terms of adult offspring produced per parent, depends on many variables, as previously stated, including female development time and fecundity, cannibalism on the younger stages, aggression and the secretion of a toxic gas from the adults' odoriferous glands.

The results showed there was a 'maternal population effect', made apparent in the closer resemblance of productivity responses between the

A and C crosses, both of which included H females, than between the A and B, which had H males. The average number of eggs laid by L females was only a little less than by the H females; but the survival rate from egg to adult was very much less for the L-laid eggs, which suffered heavy losses as larvae and pupae, largely attributable to cannibalism but with a smaller difference component in their intrinsic viabilities as well. The CV series, as expected, produced strong inverse density-dependent effects on the average production per parent; and these were opposed by the tendency for bigger propagules to give rise to larger collective totals of offspring, with the result that A and C crosses produced their highest totals from propagules of the intermediate size of 16. There was nothing surprising in this; but the D crosses, always much the least productive, yielded almost the same sized crops of offspring from all four sizes of propagule. The CD series also revealed anomalies in the productivity of the D propagules.

In sum, the data confirmed the enormous complexity of 'productivity' as a group trait, combining a network of behavioural, developmental and physiological variables, some of which are density dependent and others differently biassed between the sexes. There is much circumstantial evidence of homeostatic controls; and it is a foregone conclusion that these controls are tight in a species that utilizes not only cannibalism but also the emission of poison gas in the process. Such homeostatic balances clearly involve the coordination of very many loci, and integrate separate contributions to the process from paternal and maternal parents, and both juvenile and adult progeny—that is to say, they imply a common 'awareness' in the in-group as a whole. Wade has shown how responsive these particular balances are to group selection; and in practice there is no way in which to select for them at the individual level.

Craig (1982), building on Wade's foundations, chose another populational phenotype to select for, which was the tendency for adults to emigrate. Incidentally he worked with the other laboratory species of flour-beetle, *Tribolium confusum*. Young adults were presented with an opportunity to climb out of their population-unit vials, up a thread and into a receiving chamber; and, on the basis of variability in their responses, he applied individual and group selection concurrently. The individual selection was done by eliminating either 0, 5 or 25% of the emigratory adults in each successive generation, and putting the rest back into the unit they came from, thus imposing three intensities of bias against them. The group selection was varied in four different treatments.

In the first none was practised at all; the generations were seeded from own-product propagules, after 0, 5 or 25% of the emigrators had been removed. The 0% individual plus No-group-selection series was in effect a null control. In the second and third group-treatments, propagules for succeeding generations were taken from the units that showed the highest and the lowest rates of emigration respectively. In the fourth treatment a random group-extinction procedure was applied, the same as in Wade's Treatment D of Experiment I. In all, there were thus $3\times4 = 12$ permutations, and each was replicated 20 times.

Individual selection turned out to be ineffective. On no occasion in 14 generations did the average emigration rates differ significantly between the 0, 5 and 25% counterselected lines. That meant that the quantitative heritability of the trait was low, and, as one would expect of a variable with a partly homeostatic function, all individuals possess rather similar adaptabilities to the conditions they encounter. Thus they produce a largely non-heritable variance in phenotypic responses, reflecting their own individual experiences and feedbacks. The feedbacks are thought likely to occur at the larval stage.

The group-selection treatments on the other hand were very effective. In a final table showing the average numbers of emigrators produced under each of the 12 treatments in generation 14, the three High group-selected stocks were the top scorers, and the three Lows were together at the bottom. The three Nones were together in the middle, with two Randoms sandwiched between the Nones and the Highs and one Random between the Nones and the Lows. Only the 0, 5 and 25% Individual sub-treatments within each Group bracket were disordered. Statistically the results were highly significant.

To the layman, this is perhaps the clearest demonstration of all of the distinctive effects of individual and group selection. As a trait, emigration turned out to be positively related to population density, which does not alter our conclusions about the results of selection in the least, though it identifies a type of homeostatic adaptation with which we have become familiar in previous chapters—namely the density-dependent, variable, 'organismic' response, common to all individuals. It must of course be genetically programmed; and the individual responses are summated, and reflected in the productivity and survivorship of the group as a whole.

Reverting to Wade's 1978 review paper, there remains to be mentioned the fifth of the theoretical assumptions he identified as being made by the earlier modellers, which were unfavourable to group selection. It was that individual selection and group selection would normally be opposing one another. In practice, it appears, they are likely to be working towards the same end, to safeguard the survival of the stock. Individual selection is indispensable to the creaming off of a within-generation, high quality cohort of breeders. Group selection prefers the stocks that maintain favourable standards and statistics, while allowing individuals whatever flexibility this permits in responding to personal circumstances, in the manner we have just been considering. Conflicts between group and individual interest will automatically be minimized because they diminish group prosperity, though it may not always be possible to eradicate them entirely; but in the gene pool and genotype, group-adaptive traits are not chronically threatened by the interests of individual selection.

Wade (1984) has actually experimented with this possibility. At the close of his first experiments there were sets of 48 replicate units from both the high productivity Treatment A and the low productivity Treatment B, differing from each other by a factor of about 30 times in their average group productivities. He took ten of the most productive lines from A and ten of the least productive from B, and perpetuated all 20 by own-product propagules of 16 adults every 30 days, until the 33rd occasion (990 days). At that time a sporozoan infection spread rapidly through all the stocks; but they were successfully restored to health by sterilizing and rearing 100 or more eggs from each line, after which there was no evidence that significant changes in unit productivities had occurred during the epidemic.

The outcome of suspending group selection for three years, and an unknown but large number of overlapping generations, was not that the A and B stocks had converged to the same productivity level. Left to the forces of individual selection their initial genetic differences had been largely or entirely retained. In fact drift had led to further divergences among the 10 populations of each stock; but the A and B means throughout the period remained separated by more than three standard errors.

Wade's use of the propagule as the link from one generation to the next, and particularly his combining of adults from two lines in the same treatment to simulate 'migration' in Experiment II, may arouse doubts about the applicability of his results to the natural infiltration of one local stock by another (cf. J.M. Emlen 1984: 266). Nevertheless it appears to be a closer approximation to the natural reality we have had to deal with in this book than the panmictic matings of theoretical genetics. We have had reason to note in previous contexts that the concept of panmixis can be widely at variance with ecological facts, and now begin to see it as almost if not quite incompatible with group selection. In normal philopatric animals far the most numerous type of immigrant must come from close by, from another unit that is under virtually the same selective pressures and differs only through drift from the one the migrant is seeking to enter, much as in Wade's experiments.

As regards propagule size, it also seems reasonable to presume that an individual's effective breeding group is often no larger than 50. It should not be impossible to obtain an estimate of this in an amenable species of bird such as the red grouse, and it would be desirable to make the attempt because of its importance to evolutionary theory. Accurate measurement of the dispersal distances of perhaps 100 individuals from birth site to breeding site, and at the same time monitoring the general dispersion and population density in the study area, might provide data for a first esitmate.

The severity of group selection and extinction practised in the *Tribolium* experiments was almost certainly much more severe than that which normally obtains in nature, where considerable variances in gene frequency are likely to be viable, especially in favourable environments. This should in fact favour the probability that such variances would develop, and sustain moderate differences between neighbouring group phenotypes. The average productivities of the groups concerned could consequently be expected to differ among themselves, keeping the group-selective process active; and at intervals inviabilities and extinctions might occur. There seems little doubt that group adaptations, such as the regulated patterns of dispersal characteristic of many vertebrate species (see p. 238), have evolved by group selection of a type that does not differ fundamentally from the one that Wade has made to function so well in the laboratory.

15.4 A model of population structuring

The details of structuring in space and time can differ considerably from species to species, but the same principles appear to be common to many animals, and that makes it useful to present an idealized model as an aid to understanding its functions. As we saw, Wright (1931) was the first to appreciate that the local philopatric colonies or stocks of plants and animals, so long familiar to naturalists, could have an important evolutionary function; and he used his 'island model' (Wright 1943) to illustrate what the significance might be of small, nearly but not quite isolated sub-populations, which later came to be known as demes (Gilmour & Gregor 1939). My own views on structuring were brought into sharper focus largely through the work of D. S. Wilson (1980); but the model described here is based as usual on the vertebrates, and applies to cyclic and non-cyclic species alike.

The prime characteristic is the attachment of individuals to localities. Structuring is normally geographical, and distance is the most universal of isolating factors between the local subdivisions of populations. Geographical isolation is, for the same reason, the usual factor that allows one species in the course of time to become divided into two; and, taking a world view, we can find many examples of species groups that have thus split but still remain largely allopatric or non-overlapping in their respective geographical ranges. Taxonomists use the terms 'species group' and 'superspecies' to designate such constellations, depending on whether there are greater or lesser degrees of difference and geographical overlap between the component species. Taxonomists also use the terms species and subspecies, in effect, as lower grades in the same hierarchy. I mention this simply to show that different tiers of biogeographical patterning are recognized for practical reasons, far above the level to which one has to descend to examine the fine-grain structure of population ecology that concerns us here.

The higher tiers are all, to varying extents, artificial, in that they make use of separate steps to classify what is really a continuous process of differentiation, evolving in time and space. Where the taxa in question are allopatric, even the species can become a subjective category. Categorizing the degrees of relatedness and separation is what systematics and taxonomy are largely about; and as long as one accepts that the process of classifying organisms is one of simplification, in which amounts of divergence are expressed by employing a limited hierarchy of subjective grades, no misunderstanding need result. Conceptually the simplification process is illuminating, and without it classification would be impossible.

The same considerations apply at the demographic level, as to how many tiers of subdivision are desirable. Sewall Wright used only two, the *population* and the *deme*; but there are good reasons for introducing a third into bio-demography, namely the *in-group*, as a subdivision of the deme; and to complete the perspective the *individual* ought perhaps to be given a place as the lowest structural entity. My in-group is practically identical with what Wilson (1975) called the 'trait group', and I have decided to adopt another name only because it is so often required to convey a social and behavioural meaning, rather than one that is explicitly genetical. It is sometimes convenient also to interpose the term *meta-population* (Levins 1970) between the deme and the subspecies, when referring to a geographical province of intermediate size. This allows one to keep the versatile word 'population' free from being tied to a unit of specified size.

In practice, populations of organisms are commonly broken up into irregular fragments by discontinuities of habitat—fragments that can vary in size by several orders of magnitude. Their structuring, pro-grammed from within, has of course to adjust to the opportunities that exist. Some fragments are large enough to equate with more than one deme, while many have room for multiple in-groups.

The in-group can be defined as a set of individuals that live in every-day social contact and are more or less closely acquainted with each other, at least in the vertebrates. They interact as rivals, allies, consorts, parents or dependants in domestic life. In other words the in-group can be indentified with the social arena in which individuals cooperate and compete, and resolve such other important matters as feeding hier-archies, property rights, population densities, breeding quotas, and reproductive rates. 'Social group' might seem a more apt designation than in-group, but it seems better to choose a distinctive term *ad hoc,* and not accept one likely to be used in non-technical senses.

Some in-groups correspond exactly with a permanent flock or aggre-gation and, if so, the groups can be seen to be separate entities. A lek (p. 8) perhaps comes nearest to being an ideal example, if one includes the females that visit it for copulation, because it is normally small enough to contain a single social hierarchy among the males. But where the residents of large areas of habitat are spread out in a mosaic of all-purpose territories, or in personal home-ranges which they share in part with others, in-groups cease to be physically separable and their boundaries therefore can only be virtual. Nevertheless they do exist from the point of view of single individuals, whose regular contacts are limited to a circle of neighbours. In these circumstances the individuals' ability to com-municate and react, even with the various other members of their personal group, may be non-uniform, and tend to diminish to nil as the distance increases. This must apply for instance to territorial song-birds, or individual members of large flocks and shoals. One might reasonably guess that in the majority of vertebrate species, individuals would normally find themselves interacting with 10–100 other individuals, in ways that affect the group's survival value. Where much larger groups interact, it is a fair assumption that they are not all mutually acquainted.

In-group structure is often ephemeral, lasting only part of a generation or life-cycle, or effective only during some transient social exercise repeated each day. For example, the Atlantic salmon (*Salmo salar*) has different in-groupings for fry and for parr (fingerlings) in the natal streams, and presumably for adolescents and adults on their marine feeding grounds, and again for adults when they have re-entered fresh water and arrived on their spawning beds. Similarly we can expect summer groupings and winter groupings in migrant birds, and perhaps other transient groupings as well.

Breeding in-groups in some vertebrates re-form in the same places annually and thus retain their integrity for a period of years; and sometimes for much longer, as a result of the site fidelity which typically brings their members back, once they have been accepted into the group, to rejoin it as often as possible for the rest of their lives. Site fidelity helps to stabilize the genetic structure but does not preclude the exchange of genotypes between group and group. In the course of time the existing members die off and are replaced by recruits; and while some recruits usually settle and find mates in their own immediate natal area (that is to say philopatrically and endogamously), others leave home beforehand and mate in another breeding group (exogamously). We shall see that individual prospective breeders, born in the same locality and year-cohort, can be variously programmed towards these alternatives.

Dispersal is crucial to genetic structuring, and our model must take into account that it is normal for the individual's life-cycle to contain a principal act of dispersal, most often in the period between emancipation from parental care, and breeding for the first time. This results in a partial redistribution of pre-breeders. It may serve other purposes in the maintenance and regulation of populations at local level, but where the details of the dispersal process are known they suggest that it is primarily geared to genetic structuring, particularly to the control of gene flow and maintenance of favourable levels of inbreeding.

The deme, in contrast to the in-group, is usually defined as being a panmictic unit, largely but not wholly isolated from other demes. If, as I suspect, panmixis is rare or impossible in the real world on the scale that this implies (see p. 200), there remains no special attribute to characterize the deme, and it becomes merely a convenient cluster or collection of in-groups. Most often when there appears to be some semblance of population units of suitable size and mutual isolation to represent demes, such as one finds among the colonial coast-breeding sea mammals and birds, or in the sea-run fish stocks that return to spawn in parent rivers, the true explanation is almost certainly that the species concerned are simply accommodating to economic and geographical necessity, and maintaining bigger stocks where the feeding grounds are larger or richer, and smaller ones where the grounds are not so favourable.

More genuine examples of a two- or multi-tiered demographic structure exist, or formerly existed, in stone-age human races known from various parts of the world. The in-group level is represented in such populations by small, territorially separate bands, each united by ancestry or lineage of some kind, but all sharing a common culture and language and recognizing the other bands as members of a common tribe. There may sometimes also be an intermediate tier which can be described as a clan, between the band and the tribe. Tribal custom obliged at least some members of one sex to leave and marry exogamously, by joining a neighbouring band.

The best-documented of these primitive races are the Australian Aborigines (cf. Birdsell 1978). Before their population structures disintegrated under the impact of Europeans, it is known that in the arid interior of the continent their in-groups averaged about 25 individuals of all ages, living in face-to-face contact, and possessing rights to hunt and gather food on a large communal territory. About 20 bands, regarding themselves to be of 'one blood' and sharing the same dialect and totemic bonds and assemblies, constituted a tribe, which rather consistently numbered in the region of 500 persons. Marriage was exogamous and patrilineal; that is to say, the men of the band formed a hereditary kinship group who obtained wives from other bands, normally belonging to the same tribe. Tribes shared strict boundaries with neighbouring tribes; and if the tribe is equated with the deme, there were in fact interdemic marriages between members of peripheral bands on either side of the boundaries, but at frequencies estimated at between 40 and 10% of all marriages in such bands. Intertribal isolation was sufficient to allow tribal dialects to develop, and this together no doubt with some spiritual misgivings were apparently the (adaptive) obstacles to intermarriage, between otherwise friendly bands.

The scarcity of permanent water and the meagre productivity of food resulted in low population densities, closely correlated with mean annual rainfall. Homeostasis based on food production was effected primarily through infanticide (see p. 36), which discriminated against girls and left the adult population with a sex-ratio of roughly 3:2. Consequently men could not marry until they acquired high enough (which generally meant senior enough) status. Top-ranking males might nevertheless have two or several wives including young ones. Pregnancies were spaced so that mothers never had more than one child to carry about most of the day while they were out gathering food. If nevertheless the tribe thrived and grew, it is concluded that it must have split dialectally into two, since tribes of say 1000 are said to have shown signs of doing so, and larger ones seem not to have been known. If on the other hand it dwindled to 200 or less, its people and their territory were presumably amalgamated into a neighbouring tribe.

The parallels in structure and function with other species of the higher vertebrates are remarkably close. Perhaps the most significant new information to emerge, in the present context, is the real existence of two or three discrete structural tiers, and the numerical sizes of the units, which for the in-groups at least are similar to what we have been led to assume in other species. Another parallel but more modern human example from New Guinea can be found in Rappaport (1967).

Sewall Wright (e.g. 1969, vol. 2: 290) showed that demes (earlier called neighbourhoods in Wright 1946) have to be assumed as a necessary abstraction for the purposes of modelling, to represent more or less isolated and comparatively homogeneous local stocks, arranged like a patchwork quilt and constituting a metapopulation. Gene frequencies in the latter are conceived as slowly changing in evolutionary time through the different rates of productivity, and hence the export of migrants, between the component demes. The easiest structure pattern to treat mathematically was his hypothetical 'island model'; and starting with that he subsequently worked up to the most generalized type of structure, which would exist in a uniformly and continuously dispersed population, living in an extensive tract of habitat where isolation was solely related to distance.

From an ecological point of view this uniform 'continuum' must be considered the basic type of dispersion, which is liable to disruption by extraneous factors such as changes and breaks in the habitat, or by internal ones such as group territories or isolated colonies. If one can understand the structural dynamics in a continuum there should be little difficulty in adapting them to confined or insular distributions. Gene flow and the inbreeding coefficient in a continuum depend, as elsewhere, on how far the individual members of the population move away from their birthplace (or place of conception) to where they eventually mate and breed; and the degree of isolation that in-groups experience will be determined, at any given population density, by the restrictions their programs impose on free movement.

The subdivision of a metapopulation is of course the primary function of spatial structuring because this is what allows group phenotypes to differentiate and group selection to work. The units must be small enough to facilitate drift in the manner discussed in the previous sections, and isolated enough to retain gene pools more or less intact for a run of generations, while the descendent stocks continue to live on the same ground, reaping the rewards and failures to which their gene pools lead them.

I have put the emphasis on the in-group as the effective interbreeding unit, whereas Wright, who recognized no subdivisions smaller than the deme, necessarily put his emphasis on the deme. The choice is almost entirely a question of relativities. In-groups occupy smaller areas, are more expendable, and tend to lose their genetic and topographic identities sooner than demes. The former might remain essentially intact for perhaps 5 or 50 generations, the latter for 100 or 1000. The evolutionary progress and survival of the species will ultimately depend on the better-adapted gene pools supplanting the less good through a single, or sometimes stepped, mechanism of gene flow that starts in the in-group, spreads to the deme, and thence to the metapopulation and perhaps to the rest of the species.

Wright's demes were assumed to be panmictic, that is to say, to allow random mating; but as a rule in wild populations mating is more or less elaborately structured through programmed dispersal. For simplicity one can assume that dispersal is a single event, completed before an individual breeds, connecting the place where its parent gives it birth to the point at which, as a parent, it does the same. In some vertebrates there are other migrations as well, undertaken by pre-breeding adolescents, and by adults in non-breeding periods of the year, from which the migrants normally return to the permanent breeding area of their native deme; and the navigational miracles that many of them perform in accomplishing these journeys, bear testimony to the extraordinary survival value that must attach to the precise location of individuals within their structural systems. In addition to seasonal two-way migrations, some adults of some species may disperse again to breed at a different site on a second or subsequent occasion, and may also re-mate. One's impression, however, is that site and mate fidelity for life is the commonest practice in longer-lived birds and mammals, and this suggests that once a particular year-cohort has been programmed into position it is advantageous to alter the structural pattern as little as possible, consistent with efficient reproduction.

The dispersal of red grouse was shown to be far from panmictic, even at the in-group level. The movements of individuals are not only con-trolled, but their distance programs vary from one bird to the next. Moreover the means are different between the sexes, averaging two to three times as far in females as in males. The next chapter shows that similar differences are common in birds and mammals, implying that many other species also have their populations structured on the same principles as the red grouse.

The dispersal distances of young grouse sampled in Glen Esk (see Table 7.1, p. 94) show there was a wide spread between the shortest and longest movements of individuals in both the cocks and hens. Statistical treatment of the original figures of Jenkins *et al.* (1967: 105) reveal for both sexes that the dispersal-distance distribution curves are strongly skewed to the left, and more so for the males. The distance data are tightly clustered round the mode, making the 'bell' of the curve narrower and taller than it would be in a Normal distribution, again especially in the males. This is the phenomenon of positive (or lepto-) kurtosis (Snedecor & Cochran 1967: 86), and it is highly significant for these samples. As frequently happens in the circumstances, the tail of the hens' curve stretches further out to the right than would be expected in a Normal distribution, as though the intermediate values had been robbed to pro-vide a higher peak at the mode and a longer extension at the distant end as well. It may be recalled that the furthermost hen that was traced had moved 42 km. The furthest cock had gone less than 5 km; and some young cocks are of course known to have nested in, or close to, the nest-territory in which they were hatched 11 months before (p. 97).

It was not possible, unfortunately, to follow a large sample of birds between their precise birth sites and nesting sites, as had been done with these few telemetered cocks. The rest of the marked grouse were ringed either before they could fly, or when trapped for the purpose of tabbing at feeding sites at three months old; in neither case were the exact birth sites known. The great majority of the 'recoveries' were shot by hunters; and because of the practice in Scotland of driving grouse over a line of shooters concealed in butts, it is possible that some birds had been made to fly up to a kilometre from where they were first disturbed by the beaters. This results in concealing the actual effects of programmed dispersal on the size of interbreeding groups, and the distribution of degrees of relatedness between parental pairs.

The fact that such non-random distributions of dispersal distances exist, coupled with the strong difference in their modes and patterns in the two sexes, leaves no doubt about the dispersal system being adaptive, and moulded by group selection. The main functions are presumably to regulate inbreeding and gene flow, including for the latter purpose the proportions of individuals dispatched to greater distances and the approximate distances to which they go. The effects produced are statistical, benefitting population units rather than individuals. They are presumably controlled by polygene systems that can throw a similar array of individual-distance programs in every generation, as the genes are repeatedly segregated and reassorted through sexual reproduction.

The sexual difference may possibly bring additional advantages, through keeping the males, which have to decide what size of territory is required during the autumn contest, close to home, and giving them the maximum possible time in which to become conditioned to the present local plane of nutrition; and offsetting this by endowing the females, who have no corresponding opportunity for establishing a hereditary claim to real estate, with a bias towards moving further away and lessening the degree of inbreeding. The sedentariness of the males may also enable them to communicate their local knowledge, for example about feeding or refuge sites, to mates who have come from elsewhere, and in due course to sons who are going to succeed them. In mammals that show similar dual biasses in dispersal, and in many species of both classes that breed at traditional sites, the handing down of local information may be considerably more important than it is in the red grouse (above all in human aborigines). As to those female grouse, and possibly rare males, that have travelled into the unknown and find themselves among strangers in a populous habitat, it seems probable that only high-quality individuals are likely to become established, and that social selection by the resident males will be rigorously applied to them.

Group selection, by definition, must lead to the expansion of successful gene pools, and it appears that this would be most effectively accomplished by pushing the demic frontiers outwards *en masse*, little by little,
rather than scattering single genotypes far and wide, where the 'foreign'
social traits they possess, of which the special benefit will be frequency-
dependent, are likely to be isolated and of little effect. One might therefore expect that short distance dispersals would be the process by which it
was normally done. The arrival of single immigrants from elsewhere in
the deme or from another deme, on the other hand, might temporarily
restore some hybrid vigour to inbred groups.

It emerges from this discussion that population structuring is much
concerned with statistically planned rather than random mating. Even in
aquatic species that lack internal fertilization, like the bony fish, matings
can be elaborately contrived to confer statistical advantages, as will
appear later in the Atlantic salmon (see Chapter 16.3). It is characteristic
of fish, amphibia and reptiles that they grow asymptotically throughout
life. In the herring (*Clupea harengus*) for example, tens of thousands of
individuals gather to spawn at traditional sites each year. Females deposit
their sticky eggs on stones on the sea-floor, while males swim overhead
broadcasting spermatozoa. But it is the old large fish that contribute the
largest individual shares of genes to the next generation, because group
selection has made the eldest, most thoroughly tested survivors automatically the most fecund individuals. In birds and many mammals
(particularly the females) growth ceases at or soon after maturity,
although further survival is often differentially rewarded through
increasing rank, in the course of social selection, and can still result
in increased breeding success. Large salmon combine both these
advantages, and possibly herrings do the same.

It is relevant to one's concepts on structuring that very large aggregations, of from hundreds up in rare instances to the order of 10^5, 10^6, or
even more individuals, can be found in every class of vertebrates, for
example at the great communal roosts of some birds (e.g. swifts, starlings, icterids and ploceids) and bats; at breeding colonies of various
seals, fruit-bats, birds, sea-turtles and amphibia; or in schools of pelagic
dolphins and fish. Many of these aggregations give occasion for communal social displays, for instance massed flights by birds and choruses
of frogs, although it is clear (by human standards) that too many individuals are taking part for all of them to be mutually acquainted.
Breeding and roosting assemblies often require each member or mated

pair to possess and defend an exclusive site, which must automatically identify a personal in-group of neighbours for each site-holder, with which it interacts. The same must happen, at least transiently, in a manoeuvering shoal or flock. It was suggested in *Animal Dispersion* (Chapter 14) that the main function of the massed demonstrations is epideictic, indicating to the performers the size of the population currently inhabiting the area for which the display arena is the customary rallying point. The displays are typically conventional and social, and they show us that social interactions and bonds can extend sometimes far beyond the limits of the in-group, and of personal acquaintance, as of course they do in human experience.

In breeding colonies the individuals or pairs commonly return in succeeding years to claim their previous site by right. In some species we know that a proportion of their progeny also return to their native colonies, and some actually seek and obtain nest sites in the near vicinity of the one in which they were born. This applies for example to the Atlantic gannet (*Sula bassana*) (Nelson 1978: 75), and the common guillemot (*Uria aalge*) (Southern *et al.* 1965: 659), and it points to the possibility that largely isolated in-groups may subdivide the colonies.

At a particular colony of fulmars (*Fulmarus glacialis*) breeding on a 75 ha rock-girt island in Orkney, the data of Dunnet *et al.* (1979) have shown that roughly one recruit in ten was native to the island, the rest coming from outside. This estimate was made in the third decade of a 28-year study. The colony is not isolated: fulmars nest more or less continuously all along the cliff-bound coasts of the close-set Orkney island group, forming a diffuse population of some 100000 breeders (counted in 1969–70, Cramp *et al.* 1974: 64).

Comparable figures are available for a colony of nearly 30000 herring gulls (*Larus argentatus*) breeding on the Isle of May, off the coast of Fife, Scotland. Chabrzyk and Coulson (1976) arbitrarily divided this mainly gull-covered island into eight sections of about equal size; and they were able to estimate that on average 34% (or rather less) of the progeny that survived to breed, returned when adult to join the island's population. The remaining > 60% established homes elsewhere. The returning natives had a sex-ratio of 1.34 males to 1 female; and they did not scatter themselves at random over the island. 77% of the males settled in their own natal sections and another 11% in one or other of the next sections on either side; for the females the corresponding proportions were 54% and 20%. Putting these various figures together one can reckon that, on average, about 30% of all the males recruited into the island's breeding population took up positions within 200 m or so of their birthplaces. For females similarly recruited, the corresponding figure was 15%. The authors state that an average adult lives through 15 breeding seasons, so that, assuming a balanced sex-ratio and a static population (far from true at present, with the herring gull in Britain increasing at 12–13% a year), a colony of 30000 birds would need to admit 2000 recruits annually, on average; and of these, $(30+15)/2 = 22.5\%$, or 450 birds, would be closely philopatric. The remaining 77.5% would be less so, or not at all. Approximately the same proportions would apply to members of every year-cohort represented in the colony. Pair formation presumably occurs on the island, so that some of the birds might possibly mate with, or share in-groups with, near relatives.

In spite of having these unusually comprehensive data, there are still too many unknowns to allow us to calculate whether an effective differentiation of sub-units by genetic drift within the colony can, or cannot, be expected to occur. Complete isolation between small units soon leads to fixation and loss of resilience to environmental change; and to maintain both inter-group diversity and viability among demes or in-groups requires that their isolation be relieved by a compensating interchange of migrants. To produce such a result something resembling what is observed in these herring gulls would presumably have to be contrived; and this argument, in the absence of any conceivable alternative explanation of why such a positive non-random dispersal could have evolved, lends no little strength to our hypothesis. (For a helpful reference on the general problem, see Falconer 1981: 63–75).

Certainly the complex nature of the dispersal patterns makes it impossible to believe that their varying demands confer either advantage or disadvantage on the individuals concerned. Indeed Greenwood *et al.* (1979b) have shown that great tits (*Parus major*) nesting in close proximity to their own kin in Wytham Woods, Oxford, do not gain any significant advantage in breeding success over similar pairs of neighbours that are unrelated. The same authors (1979a) found that approximately 25% of the male great tits and 10% of the females established themselves on territories within a radius of two territory widths from the nest-box at which they were born—a distance arbitrarily chosen to define close philopatry; and also, very interestingly, that there was a significant genetic component of heritability in dispersal distances, when parents and offspring were compared.

Summary of Chapter 15

1 Much evidence from individual marking experiments indicates that most vertebrates have self-sustaining philopatric stocks, and that their dispersal movements are generally short in relation to their powers of travel. A common evidence of gene-controlled population structuring is the difference found between males and females in mean dispersal distances, mentioned earlier. There is an ecological advantage in philopatry, in the opportunity it gives for acquiring local knowledge, and transmitting information by tradition, about the use of particular food sources, breeding and roosting sites, etc. The reason why this function is often entrusted largely to one sex is probably genetical, in that it frees the other sex to utilize dispersal for optimizing gene flow and the inbreeding rate, on the same principles as human exogamy (i.e. taking a spouse from a different in-group).

A species as well or better documented than the red grouse in this respect is the great tit (*Parus major*). Both species have similarly skewed distributions of their individual dispersal distances, though in the great tit the mean sex difference is less. A small population of great tits living on the Dutch North-Sea island of Vlieland was monitored for 20 years. It averaged about 50 breeding pairs, and virtually all the offspring were ringed before fledging. Eventually some 200 individual lineages could be traced over 3 generations or more; and about a quarter of all matings between pedigreed individuals involved coefficients of kinship of 1/16 or greater. This shows that the effective mating-group size must, on average, be considerably less than the average total population of 100.

2 Wright's theory of the shifting balance process in evolution also involves populations structured in space. It does not take into account the ecological advantages of structuring which have been my central interest, but relates solely to the parallel genetical advantages, which arise from speeding the rate at which populations can adapt to changed or novel environments.

Initially genetic adaptation within a population depends very largely on altering the frequencies of alleles. Four factors can bring this about, namely (1) mutation, (2) selection, (3) the movement of individuals out and in, and (4) random events which, particularly in small populations, impose 'sampling errors' on the cohort of individuals that actually survive to become parents for the next generation. Another important consideration is that very many genetic traits exhibit a continuous variation, including all the metric or quantitative characters that show a statistical range within a population, distributed about a mode or mean. Such distributions imply a stabilizing selection process.

For argument a standardized species can be postulated, having 1000 loci with 10 alleles at each. The number of alternative combinations this implies is enormous; but in a large panmictic population, selection would be so conservative as to allow only a small fraction of them ever to survive if they appeared. Thinking next of adaptations to external change extending over a finite span of generations, one can rule out mutation as being too slow to be of general value. But in small almost isolated subunits (i.e. gene pools), consisting of, say, 100 individuals or less, with the migration factor low and selection light, random events through time could lead to rapid change, and the changes that took place would be different in different units. The optimal conditions for such variances to develop can be modelled mathematically in terms of unit size, and selection and migration coefficients.

Such random drift in a suitably structured population would subject the allele frequencies in the component gene pools to frequent if not continual changes. All manner of new balances among interacting genes might appear and be exposed to selection, with differing degrees of success. Wright draws us a picture, much simplified, of the potential field of genotypes that could emerge, as comprising a 3-dimensional surface— a sort of genetic landscape, on which the genotypes occupy imaginary coordinate positions, and there are hills and troughs that indicate their potential adaptive values. Existing gene pools can be assigned to their appropriate places. Because selection acts to preserve and if possible improve adaptive values, most of them would be found clustered on or near the tops of hills. Any environmental changes that occur are likely to alter the selection pressures, and that could cause an adaptive peak to collapse, bringing the units down to the plain (average level) or into a hollow. Those that survived would continue to drift about across the surface, and some might happen to reach another upward slope. If so, selection would propel them towards the top of it, possibly at a still higher level than before.

We are particularly concerned here with genotypes as integrated wholes, bonded together by numerous interacting effects and polygene balances which selection has tended to perfect and stabilize. The combination of random drift and weak selection, however, might occasionally allow a novel configuration to emerge in which a particular locus could play a much more prominent role than before, thus shifting the previous balance and starting the climb to a new adaptive peak. Wright has laid great emphasis on the organism as a harmoniously functioning entity, and on the 'organismic' adaptations that gene pools confer on all their viable members, which are the dynamic factor in the shifting balance process.

3 Wade has tested Wright's theory experimentally with laboratory cultures of the flour beetle *Tribolium*. This has a generation time of 30–45 days. By merging four strains, he prepared a genetically diverse stock, and made up 'structured' populations from it by establishing sets of parallel lineages, in the form of small units living in separate shell vials. At the end of each generation the new adults that had matured in each vial were counted, and a 'propagule' consisting of a specified number of them (e.g. 16) picked at random, was used to produce the next generation.

The trait under investigation was unit productivity; and by varying the treatments he contrived to subject it to selection, migration, and drift, and to quantify the results. Productivity is a very suitable group-trait, being compounded of individual fecundity, cannibilism, age-related mortality, development time and several other gene-based effects. Because of the compactness of the units all the treatments could be generously replicated, and the cultures kept physically constant in an incubator.

Four parallel treatments were applied in Experiment I. There was an A set where units were selected for high productivity. At the end of generation 1 the quota of propagules to keep the lines going in generation 2 were all made up from those units that had produced the largest crops of new adults; all other units were discontinued. In the B set the least productive units were selected instead; and in both the same procedures were continued for 9 generations. In the C set, as a control, there was neither group selection nor group extinction; each replicate was propagated by 16 of its own adults; and thus individual selection alone operated. In a D set a random-number table was used each time to select particular units for propagation and the rest were extinguished. The results in brief were to produce an A stock 40 times as productive as the B by the 9th generation. The C and D stocks remained intermediate between the A's and B's. Thus group selection turned out to be much more effective in producing changes than either individual selection or random group extinctions.

To simulate migration, Experiment II included a treatment in which some of the propagules were given a mixed origin, taking 8, 12 or 16 individuals from one unit and adding 8, 4 or 0 from another. The stocks used were surplus individuals from the 4th generation of Experiment I in which a large divergence had already developed between the A's and B's; and the question was whether mixed propagules coming from two different replicates of A would undo the success already achieved by group selection in advancing productivity. In effect it was like recoloniz- ing a habitat with migrants from two neighbouring in-groups that differed from each other only as a result of drift. The results showed that this 'migration', repeated for 7 generations, only slightly counter- acted the effects of the High and Low productivity selection, in com- parison with the unmixed controls.

Most previously published theoretical models of group selection had concluded it to be ineffective as an evolutionary force because of the restricted conditions in which it could work—in very small and highly isolated populations, where individual selection was weak. Reviewing them, Wade showed that these negatory conclusions depended largely on all or some of five assumptions the authors had made, that were inherently unfavourable to group selection; and his experiments were accordingly directed at testing their validity. Arguments had tended to be based on classical single-locus, two-allele lines, whereas his experiments showed the reality of 'group phenotypes' with a high degree of herit- ability. This supported Wright's views on multi-locus interaction and the wholeness of the genotype, and also the large part played by mutualist interaction between members of the group on which our own previous chapters have insisted. Single alleles are not the prime consideration in these contexts. Moreover the models had mostly ignored group structur- ing, by their assumption that panmixis is a real phenomenon; whereas in practice structuring sets up small persistent groups that function to prevent it, allowing between-group variances to develop instead. The authors' last erroneous assumption was to expect that group and in- dividual selection would work in opposite directions, whereas with few exceptions they are supportive.

Subsequent experiments showed that propagule size is not very critical after all, at least up to a limit of 50 individuals, which is within the probable range of natural in-group or mating group sizes. Also that the genetic diversity initially present in the form of individual variance between members of the founding propagule, as drawn at random from the laboratory stock, tends to be rapidly converted by the experimental structuring procedure into between-unit variance. This is presumably because fixation proceeds rapidly in the small breeding groups, but different loci happen to get fixed in different units; and it may also be this that simultaneously enfeebles individual selection and allows group selection to take precedence. The most convincing demonstration of the latter phenomenon has come from Craig's experiments on the variable tendency of adults to emigrate before they breed (see p. 216). Small unit gene-pools, transmitted to later generations, confer a high degree of heritability on group phenotype, which remains stable in the absence of further group selection.

In summary, Wade has thrown grave doubts on the validity of the assumptions on which the main objections to group selection as a normal evolutionary process have been based. Neither 'founder' units nor the frequency of intergroup migration needed, in practice, to be unrealistically small. He has found that structuring causes and perpetuates variances between component groups in the course of a series of generations, and tends to weaken individual selection whilst group selection grows stronger; and the structuring he devised for the experiments was acceptably close to that which has been shown to exist in the wild.

4   A model of population structuring is necessary as a conceptual aid. The phenomenon itself is, in effect, an extension of the grading process of taxonomy, reaching down to bottom of the scale. Both subjects relate to the natural process of genetic differentiation, to which isolation and time are major contributory factors. The factors are continuous variables, and so, virtually, are the gene pools themselves; and therefore to allow us to describe and compare the infinitesimal degrees of isolation-versus-mixing and difference-versus-similarity that result, we need a set of grades such as the deme and in-group, artificial and subjective though they may often be, which extend the genus, species and subspecies of taxonomy downwards.

Wright used just two grades, the population and the deme. In our context, in which social and cooperative behaviour are under review as well as genetic structuring, there is a clear need for another grade below the deme corresponding to the individual's acquaintance group and social arena. There is a ready-made and very similar category in D. S. Wilson's 'trait-group'; but because that expression has a strictly genetical meaning, the 'in-group' has been substituted here instead. In addition, 'population' has been set free for general use by adopting 'metapopulation' as the next grade above the deme.

An individual animal normally becomes linked with a home locality that it shares with other members of its own in-group. In a continuous habitat occupied at a constant density, in-groups can only be virtual, centring on each individual and extending to a perimeter where the frequency of personal interactions becomes negligibly low. But the sharing of a communal territory, for example, can give the in-group a much more objective definition. Habitats are commonly fragmented, and structural units must come to terms with divisions so enforced. In-group membership (and often size) can alter with different life stages and activity changes; and, because of dispersal, mating in-groups are not necessarily congruent with breeding in-groups. Dispersal, whether it is differently programmed in the two sexes or not, always plays a central part in structuring. In the red grouse some of the males matured and bred exactly where they had been born (dispersal distance zero), but they could still mate exogamously because the females dispersed more freely. Stone-age peoples have in fact provided our best examples of well-defined three- or four-tiered demographic structures.

Wright identified the effective interbreeding unit with the deme, his smallest subdivision, whereas here that role falls to mating in-groups thrown together by dispersal. Interbreeding in-groups are considered to be smaller, more expendable and evanescent than demes, lasting perhaps 5 or 50 generations compared with 100 or 1000 for demes. Survival and progressive change depend ultimately on the better gene pools supplanting the less good, by invasive gene-flow and/or recolonizations.

The individual dispersal distances found in the red grouse not only differ between the sexes, but their statistical distributions are non-Normal (skewed). This fact gives two separate lines of evidence that dispersal is genetically programmed as a group adaptation, for the statistical regulation of gene-flow.

Large aggregations of up to 10^5 or more individuals occur in many species. Some of the more permanent ones, e.g. at roosts and breeding sites, are internally structured; and transient in-groups may also exist within travelling flocks or schools. A large-scale marking study in a big herring-gull colony, for example, revealed that close philopatry within the colony itself was commoner in males than females, as one might now expect, although many adolescents of both sexes must have survived as emigrants, established in other colonies. In the great tit, although a hereditary component has been demonstrated in individual dispersal distance, no significant individual advantage accrues from nesting close to kindred birds. There seems no alternative to the conclusion, from either of these examples, that their dispersal patterns are genetically coded and find a statistical rather than an individual justification in demographic structuring.

CHAPTER 16
POPULATION STRUCTURING AS
A GROUP ADAPTATION

16.1 Sexual differences in philopatry in birds and mammals

About 20 years ago I began collecting from the literature examples of sexual differences in dispersal behaviour such as those mentioned in the previous chapter; but these efforts were later overtaken by P.J. Greenwood's more comprehensive review of the present state of knowledge on the subject (1980). The preliminary distinctions he makes are first worth noting, between 'natal dispersal' occurring before the individual breeds for the first time, and 'breeding dispersal' involving subsequent, secondary movements of individuals that have already bred at least once.

He was able to tabulate 45 species of birds and 69 species of mammals for which data on dispersal, separately attributable to each sex, were available; and from these tables a striking generalization emerged. If there *are* significant sexual differences, it is nearly always the female that is more philopatric than the male in mammals, whereas in birds the opposite is true. In Greenwood's terminology, dispersal is male-biassed in mammals and female-biassed in birds. In both classes the phenomenon is known to occur over a wide spectrum of types, including representatives of many families from penguins to passerines in the birds, and from marsupials and bats to ungulates in the mammals. In both classes there are a few exceptions where the bias is reversed.

In discussing possible causes for these opposite developments he first dismisses any possibility of a connection with the sex-chromosomes, which show a similar reversal. He could find no fundamental difference in the dominance relations of one sex towards the other in the two classes, which might have provided the cause; nor in the benefits, such as avoidance of inbreeding, that might be expected to accrue to one sex rather than the other from greater or less dispersal movement. Eventually he was led to ask whether the basic difference could be a question of role differences in the mated pairs—birds being mostly monogamous, with males that tend to be defenders of resources, whereas male mammals are mostly polygamous and tend to be defenders of mates instead. Putting this idea in another perspective, he saw the male bird as seeking possession of resources whereby to induce a mate to come and join him, while the male mammal, conversly, leaves home in search of mates.

This is not a very satisfactory conclusion, and in fact its premises will not bear close scrutiny. Mammals do not actually have any monopoly on polygamy: though less frequent it is nevertheless widespread in birds. By no means all male birds defend resources, though one sex at least needs to have the use of a nest-site (not always their own!) as a basic resource. Similarly there are some almost or entirely monogamous species of mammals like the red fox (in which males disperse further than females, cf. Corbet & Southern 1977: 317; Banfield 1974: 299), and the pine marten; and there are many mammals whose males possess and defend breeding sites, and even food territories. In short there is no consistent difference in resource tenure or the division of labour between mated pairs that would explain why birds should generally be patrilocal and mammals matrilocal.

The probability that the difference somehow enables one sex or the other to maximize its individual fitnesses is diminished by the fact that in most species there is not a clear-cut segregation in dispersal behaviour between the two. Instead, individual dispersal distances of males and females often broadly overlap, and it is only the average (and perhaps the shapes of the distribution curves) that are significantly different. At most, only a fraction of the more philopatric sex can expect to inherit an actual family holding; and in some species a majority of the recruit breeders settle at some appreciable distance. Thus in the herring-gull example (p. 229), 62% of the males and 72% of the females that survived to breed had left their native island and joined colonies elsewhere.

Dispersal distance can be regarded as a continuous variable (though broken by habitat discontinuity); and what we have learnt to expect is that it will be more strongly skewed towards zero for the males in birds and the females in mammals. It seems more likely to lead to an acceptable explanation if we view dispersal as a primary component in population structuring whereby one sex is assigned the predominant role in sustaining the continuity of local groups. At a later date, when mating and reproduction occur, this would automatically result in the second parent being drawn from the sex which has the more open pattern of dispersal, thus making it probable that mated pairs are seldom so closely akin as to depress their progeny's fitness. This advantage was earlier noted by Maynard Smith (1978: 139). A considerable spread of degrees of relatedness among mated pairs could result from such a dual lottery system, and a statistical benefit be derived from the inbreeding-outbreeding balance it produced in the deme as a whole.

If this view is correct the most likely explanation of why the birds and mammals have come to achieve similar ends by alternative means is that the difference is a phylogenetic one, and that it is simply an accident of parallel evolution that the parts the sexes play happen to have been reversed. The sex-chromosome anomaly already mentioned is another example of much the same kind.

Similar facets of social adaptation have frequently evolved in parallel in different major animal groups, and especially in the two great evolutionary series that have become socially most advanced, the vertebrates and insects. Social hierarchies, territoriality (usually dominated by the male sex), dawn and dusk epideictic displays, male songs and adornments, leks, and the use of traditional sites for social assemblies, have all evolved several times over in the Insecta. Similar parallels are still more numerous between one vertebrate class and another; and like sociality, population structuring in some form most probably existed in the early fish-like vertebrates and then spread by common descent to the later-appearing classes; and in each class it has obviously made advances along with the rest of evolutionary progress. The reptilian stocks that gave rise to birds and mammals diverged with the first great radiation of the reptiles; and subsequently each class 'took off' as they acquired (also independently) warm-bloodedness, and enjoyed a flush of diversification as they colonized the habitable world. There is nothing surprising in their developing parallel adaptations: the four-chambered heart is yet another example.

In the insects there are numerous families belonging to at least eight orders, in which the adults are flightless in one sex though functionally winged in the other. Well-known British examples are the common cockroach (*Blatta orientalis*) and the glow-worm (*Lampyris noctiluca*). It is usually the females that are flightless: they are in both the species just mentioned and in all the Lepidoptera in which this type of sexual dimorphism is found. In some of the latter the dispersal of females depends largely on what they can achieve as larvae. Where one sex is flightless in the Hymenoptera (e.g. in the army ants, Dorylinae), it is also usually the female; but the fig-insects (Agaontidae) have flightless males that never leave the inflorescence in which they came to life, whereas the females are winged. Among the Diptera, the interesting family of fungus-gnats (Mycetophilidae) contain species with wingless females, such as

Pnyxia scabei (some of their males too are wingless and others winged); and also species in which the male's wings are reduced and the female's are normal, such as *Sciara semialata* (Imms 1957: 618). Whether any of these adaptations are primarily connected with genetic population structuring is unknown; but it is worth mentioning that the larvae of other species of *Sciara* (the so-called 'serpent worms') form quasi-social aggregations containing enormous masses of individuals. Still another species of mycetophilid has luminous larvae and female adults, though apparently non-luminous male adults: it is *Arachnocampa* (= *Bolitophora*) *luminosa*, and its larvae are the so-called glow-worms of New Zealand. I have seen them living gregariously on a steep humid slope, on the base clay beneath a bushy vegetation, and each larva was stationed in its own recess in the bank behind a miniature barricade, consisting of 20 or more vertical, beaded, mucous threads. Their food is presumably fungal or bacterial.

The Argentine ant (*Iridomyrex humilis*) deserves special mention here, because ants are notable among insects for often having group-territorial structuring. The Argentine ant is a small, aggressive, destructive species that has been carried accidentally by man to many warm-temperate parts of the world. It is interesting to natural-selectionists because its colonies are always multi-queened (and often very large), and the workers are thus not necessarily sisters. New colonies are founded, not by a nuptial pair, but by budding off a self-sufficient detachment from an existing colony into an adjoining territory. Though males and virgin queens are both winged, only the males are known to fly (at least in the Cape Town neighbourhood of South Africa if not elsewhere). They do not engage in nuptial swarms, but fly out individually on warm nights after mid-summer and seek a female in her own nest, which may be the same as the male's nest or a different one. There mating occurs (Skaife 1961: 12). Viewed in the present context, the dispersal programs of the sexes are not only different, but bear a distinct resemblance, as far as the inbreeding-outbreeding balance is concerned, to those of the corresponding sexes of some microtine rodents (see Chapter 14.3, p. 180–1).

Returning to the birds and mammals, and to those few anomalous species in which the sexes' dispersal distances are biassed the opposite way to the norm for their class, Greenwood's data show that the only ones known in birds are in the Anatidae. In these geese and ducks, pair formation often begins well in advance of the breeding season; and in migratory species with separate winter and summer habitats it may therefore occur in localities far distant from the breeding site, and thus between birds not reared as neighbours. When philopatry is stronger in one sex, a mate from the other sex can consequently be seduced away to breed a long distance from its birthplace. (This has been called abmigration (Thomson 1923).)

Such a relationship, with the normal avian sex-bias of philopatry reversed, is well authenticated in the lesser snow goose (*Anser c. caerulescens*) which breeds in arctic and subarctic Canada. It is a bird with two colour-morphs, a 'snow goose' morph in which only the wing-tips are black, and a 'blue goose' morph in which only the head and neck are white and the rest dark coloured. The blue morph has multiplied and spread swiftly during the present century, showing that an unusually rapid gene-flow is in progress. It appears to result from the recruits' typical habit of forming pair bonds in their winter quarters in Texas and Louisiana, or during their northward spring migration. Goslings, soon after hatching, become imprinted on their parents, and when later they reach maturity the males normally select a mate resembling their mother or father. Ganders with white parents choose white females, and those with blue parents blue ones. The two morphs are largely controlled by a single gene, with the blue allele almost completely dominant to the white. Heterozygotes are therefore essentially blue phenotypes. A gander who, as a gosling, had one parent of each morph, is thought to be equally attracted by a white or blue mate. The male plays the dominant part in pair formation, and if he gets accepted by a white goose, she will normally lead him back to her natal colony, which may be one in which the stock is still largely white. This is the simplest hypothesis to explain the explosive spread of the blue morph, from its apparent place of origin in southern Baffin Island, right round to the western shores of Hudson Bay over 1000 km to the southwest (Cooch 1963; Cooke *et al.* 1968, 1975).

Another anatid with the females more philopatric than the drakes is the common shelduck (*Tadorna tadorna*). At a breeding colony near Aberdeen, under study for some 20 years, females outnumber males by three to one among the native-born individuals that return to breed when adult (Patterson 1982).

Whether there is any philopatric bias or not, it should be noted that an unusually high degree of panmixis may result in migratory birds that form pair bonds when living in large or randomly derived assemblages in the winter. This is perhaps reflected in the low rate of subspeciation that characterizes some species of ducks such as the mallard and shoveler

(*Anas platyrhynchos* and *A. clypeata*). The former breeds almost throughout the north temperate zone, but has split off only one local race, which is wholly isolated in Greenland. The shoveler has a similar north-temperate distribution, and no recognized subspecies. The common eider (*Somateria mollissima*), a sea-duck, is also circumpolar, but it has six recognized subspecies. At a very large breeding colony near Aberdeen with 1000–2000 breeding pairs, about a third of the adults are sedentary and do not leave the area in winter, while the remainder move about 100 km south to the Firth of Tay, where they join eiders coming from other colonies. When the migrants return in spring almost all are already mated (Milne 1965). Milne and Robertson (1965) showed that the frequency of an electrophoretically identified constituent of the egg albumen, which is either present or absent in any given egg, was substantially higher in the eggs of residents than in those of migrants. This demonstrates that a surprising degree of local reproductive isolation can exist in this species. The two moieties of the population were more or less spatially separated within the colony, revealing once again the effect of finely controlled philopatry, presumably in the female sex.

In the mammals the clearest exceptions to the normal rule of male-biassed dispersal are in a neotropical bat, *Saccopteryx bilineata*, in the chimpanzee and mountain gorilla among the great apes, in the pika (*Onchotona princeps*, a small lagomorph), and in the African wild-dog (*Lycaon pictus*); the literature references can be found in Greenwood (1980).

To these one may add many of the primitive races of man. The central tribes of Australian Aborigines, as they were naturally structured a century or more ago, have already been mentioned (p. 223). Their marriages were exogamous, patrilineal and patrilocal, meaning that kinship and territorial rights descended through the male line, and wives had to be imported from another clan and place, though usually belonging to the same tribe. Various other relatively primitive tribes, on the other hand, such as the Iroquois nation in eastern North America, were matrilineal, and thus more nearly resembled the typical mammalian plan. The complexities of exogamous customs found in different races were very great and proved extremely puzzling to past generations of anthropologists, who could not imagine how they had come into being or what their function might be; for a brief review see Marrett (1914: 152–181). The progressive emancipation of the human intellect no doubt made it advantageous to involve the program for genetic structuring in the mystery and superstition of totemism, in order to get it enforced in the face of a growing temptation to follow one's own desire in selecting a mate. In non-rational animals disobedience is much more easily blocked; but in any case, should disobedient mutants arise and begin to spread, the productivity of small communities would be likely to suffer, and that, in a competitive world, would keep the mutant frequency suppressed.

In the denser human societies of later days all that has normally been required as a safeguard against inbreeding is the prohibition of incestuous marriages; but the relics of our own former patrilineal and patrilocal feudalism also survive in the acceptance by brides of their husbands' surname and, particularly under European law, in the preference given to male heirs in the succession to noble titles and landed estates.

In man, of course, the divergence and proliferation of dialects and languages has long had a special significance as a barrier to gene flow. Our colloquial speech still incessantly changes, in response to fashion and novelty, and in times long past the rates of dialectal change with time and distance were no doubt subject to group selection. Accents are still mutually and accurately standardized among teenagers at school; but such fine nuances, though far from being lost on us, have ceased to maintain any biological integrity in local groups in the western world. I included a rather longer discussion of the evolution of property tenure in primitive peoples in *Animal Dispersion* (p. 187–191), and some views on the evolutionary change-over from compelling brain-programs to conscience and ethics, as reinforcements for social behaviour, in Wynne-Edwards (1972).

The fact that some of the primitive human races were matrilineal and others patrilineal points to the possibility that exogamy was occasionally allowed to lapse, and then at a later stage revived, but with the opposite bias. Alternatively the change could perhaps have been made deliberately at some point, and even been imitated elsewhere. In non-human mammals where a reversal of the norm exists, and in geese and ducks, it appears more certain to have arisen through the atrophy of the ancestral structuring program which controlled dispersal, and the subsequent evolution of a substitute arrangement, biassed at random the other way round.

16.2 Two ancillary adaptations that increase genetic quality and reduce effective group size

The balance between philopatry and dispersal that determines the viscosity of the population may be affected by one or both of two secondary factors, which have yet to be discussed in this context.

One is longevity, coupled with the more or less permanent residence of the individual once its phase of natal dispersal is passed. It is a widespread phenomenon, particularly common in the vertebrate classes. To illustrate its effect, consider two different populations, one belonging to an annual species of animal in which the members disperse and reproduce once during their single year of life. The other belongs to a perennial or iteroparous species whose members breed, on average, for ten seasons. Each species has only one dispersal stage, which carries the individual to a permanent home. It is obvious that, if the densities and mean dispersal distances of the populations are the same, the first will have ten times the turnover and ten times as much dispersal a year as the second. Increasing the lifespan and number of recurrent breeding seasons thus tends to increase population viscosity, and reducing them has the opposite effect.

Quite independently, as mentioned earlier, longevity tests the ability of parents to survive, and the longer the test continues, the greater the expectation that their offspring will share similar qualities and become efficient members of the local gene pool. Incidentally, therefore, it is an example of a trait that could be favoured both by individual and group selection.

The second ancillary factor is polygamy. If, through polygamy, a breeding stock reduces to one-tenth the effective ratio of males to females, the inbreeding rate must increase tenfold, compared with that of an equal stock of females similarly distributed and monogamously mated. In other words this simply means that the breeding in-group is effectively diminished by the amount to which other males are excluded from participating.

From the eugenic point of view, polygamy produces much the same group effect as longevity. For the most part, polygamous individuals owe their dominant social status to the excellence of their genotypes; and, to an extent which varies much from species to species, their status tends to be enhanced by survivorship and to increase with age. Polygamy, in whatever form it takes, carries the process of social selection a step further by creaming off a still smaller team of champion males (or rarely females) to provide the fertilization service for their in-group. It is interesting to find Charles Darwin's grandfather Erasmus conjecturing in his celebrated book *Zoonomia* (1794–6), on the subject of the horns of stags and their use in combat for the possession of females, that 'the final cause of this contest among the males seems to be, that the strongest and most active animal should propagate the species, which should thence become improved'.

For birds, male polygamy (polygyny) is not easily compatible with a system of dispersed territories except on a limited scale (e.g. bigamy), because (1) the male cannot defend enough ground to provide resources for numerous females, and (2) he cannot contribute an effective share in caring for dozens of dependant young. Consequently polygyny in birds usually involves a switch from a food-territorial to a lek system for the males, as the method of controlling the quota of breeders (see p. 65). This in turn is impracticable unless the females are capable of bringing up a family of adequate size on their own. The lek concentrates the local breeding males at one focal point for much of the day, and often for much of the year. For that reason, having access to foods that are easy to find and quick to gather is, for both sexes, a virtual prerequisite to polygyny, and makes it a feasible option mainly for fruit- or nectar-feeders or herbivores. For most insectivores it is ruled out (see p. 60–1).

It is perhaps necessary to comment very briefly on the inclusion here of lekking under the heading of polygyny, in view of Lack's (1968) three-way discrimination between polygyny, polyandry and prom-iscuity, and his classification of lek birds as promiscuous. The effect of lekking is normally to reduce the effective ratio between breeding males and breeding females to less than one to one; and whether the female's contact with the male is very brief or more enduring matters little or not at all for purposes of population structuring. A female lek is hardly a practical possibility, because even if the males undertook the incubation, the egg-layers would have to spend more of their time feeding, and would need to leave the lek whenever they laid an egg, quite apart from the social stress that lekking would impose.

According to Lack polygyny, as he defined it, has evolved about eight times in birds, and lekking nine times, making a total of 17 taxa, some of which, such as the grouse, hummingbirds, manakins and weaver birds, contain a number of polygynous species, as defined here. For a more recent reference see Payne (1984).

An even rarer departure from monogamy in birds (and other animals as well) is polyandry, which demands a very radical re-allocation of tasks between the sexes during reproduction. Here the females do take on the male's normal job and assume the brunt of the social competition. The result is that they typically become larger and more brightly coloured in the breeding season than the males. The males take over the nesting function, including incubation and care of the chicks, after mating with the territory-owning polyandrous female. The adaptive advantage of this remarkable reversal of the normal sexual roles has long remained obscure.

The best documented example is the American jaçana (*Jacana spinosa*), studied by Jenni and Collier (1972) in colour-marked wild birds. In this species the females average 1.75 times the weight of males, although their plumages are alike. Breeding females were found to hold territories large enough to accommodate one to four nesting males (average 2.2); and the males shared out the female's ground between them, each defending an exclusive part of it. The hen successively laid a clutch of four eggs for each of her mates. Jacanas' eggs are exceptionally small in comparison with those of rails and limicolines of similar size and habits (see Lack 1968: 216). Only the males develop brood-patches (as also in the phalaropes, mentioned below). Jenni and Collier found a large reserve of non-breeding females and males present in the habitat, many of which were caught and marked; and when the weights of breeders and non-breeders were compared, sex for sex, the breeders turned out to be significantly heavier, the females by 19% on average and the males by 11%.

A second instance of polyandry in the Jacanidae has much longer been known, in the pheasant-tailed jaçana (*Hydrophasianus*), and it is thought there may be still other examples in the family.

Elsewhere, avian polyandry is well authenticated in the bustard-quails and some of their allies (Turnicidae), the painted snipe (*Rostratula benghalensis*), the dotterel (*Eudromias morinellus*), and two species of phalaropes (Phalaropidae); and this is probably not an exhaustive list. All these species have the females larger and more handsomely adorned than the males. In none of them are the rates of polygamy high, compared with those of the more strongly polygynous birds. Where the ratio does not usually exceed two males per female (diandry), it is by no means easy to detect and establish the fact for certain, unless the birds have previously been colour-marked. Only in the last 20 years has the suspicion been confirmed that the dotterel (see Nethersole-Thompson 1973), and two of the three species of phalaropes (Hilden & Vuolanto 1972; Raner 1972;

Kistchinski 1975; Schamel & Tracy 1977, on colour-marked birds), though at times monogamous, are at least diandrous at other times.

All these polyandrous species are birds with nidifugous chicks, which can feed themselves and thus ease the burden that falls on the males. Even so, the scale on which polygamy is possible must generally be limited by the size of territory the female can hold, and still more by her capacity for laying successive clutches of eggs. As a means of decreasing in-group size and increasing the variance between groups it cannot therefore be pushed very far.

The rarity and sporadic occurrence of polyandry in the avian class as a whole suggests that it may be something of an evolutionary blind alley, which tends to lead to extinction. It adds heavily to the load on an already burdened sex and results in a very uneven division of labour. The reversal of the sexual roles, as I mentioned in *Animal Dispersion* (p. 239), has arisen in families or higher taxa in which little or no sexual dimorphism otherwise exists, or could have existed at the time polyandry arose. Morphologically, the same would have been equally true of the far commoner eventuality in which the males have evolved as the epideictic sex. In other words it is not in fact a 'reversal' of an outward dimorphism that existed before in either case, but an innovation; and the type we are now discussing has obviously moved against the run of the tide.

In both eventualities there must have been a previous state of something like parity or interchangeability in the behavioural roles of the sexes at breeding time, such as we can see in almost all types of colonially breeding birds, where either partner of the mated pair has to be able to defend the nest-site single-handed, and tend the eggs or young, while the other is away feeding. The move towards sexual dimorphism, in either direction, creates a more specialized division of labour; and the commoner and more equable trend has been to make the male very largely or wholly responsible for social competition on behalf of the pair, for securing the territory and regulating population density, while the female produces the costly eggs and takes the major share or exclusive task of incubation and parental care.

Even in the sexually-alike colonial species the male is often a little better armed or adorned, reflecting the physical fact that the maturation of testes is a minor physiological burden compared with the production of eggs, and he is consequently the one with more energy to divert into the social contest. This makes it easy to see why sexual dimorphism usually develops in the same direction, by building on what is already in incipient difference. Why then does it ever go the other way? Why do we find any species, even if they are only a few, with epideictic females?

It could only have happened if it somehow enhanced the survival value of the groups and species in which it has evolved; and it seems likely that this advantage is primarily derived from adopting it as a basis for polyandry. In Chapter 20 I stress the importance of breeding from the best phenotypes, because that is a main source of the genetic superiority that characterizes the fittest in-groups and demes. Introducing a polygamy rate of 1:2 (bigamy) halves the number of the polygamous sex that are required, to achieve a given quota of matings in any particular season, and this effect is obviously the same no matter which sex assumes the bigamous role. It must halve the quota of successful candidates in the social selection of breeders, and double their fertility if the same output of progeny is to be obtained. That will substantially raise the standard required for acceptance and ensure that winners comprise the highest flyers in their hierarchies.

Quite a modest polygamy ratio may produce a selective advantage for the group, sufficient to put the adaptation within the bounds of possibility for females, at least in species where a monogamous female role is relatively light. The fact is that the particular syndrome of adaptations identified with avian polyandry, namely females that take the principal territorial role and are bigger and/or more colourful than the males, and males that perform the main parental role, seem not to be found except in association with polyandry—with one remaining exception, Wilson's phalarope. It tempts one to predict that this species will ultimately prove to be polyandrous as well.

Polyandry is virtually denied to the placental mammals by the unalterable condition that their embryos have to be gestated and suckled by the mother. She cannot put her latest consort 'out of commission' as a sire by deputing these tasks to him. Nor can she come into oestrus herself while pregnant; and when she eventually does so, the males then available are still likely to include the same consort as she had accepted the time before.

In fact there do not appear to be any actually proven examples of polyandry (excluding balanced promiscuity) in the animal kingdom except in birds. Among the crustacea and insects there are species with females that are more ornate or better endowed in other ways for social competition than the males; but whether any of them are polyandrous is unknown. An added factor swinging the advantage towards having polygamous males is that females are often adapted to have their whole year's or life's stock of ova inseminated on a single occasion, and cannot thereafter repeat the process; whereas males can usually retain a store of semen. Only in very exceptional circumstances are the males 'inseminated' with eggs, e.g. in pipefish and seahorses (Syngnathidae) and, in effect, in those other fish where the male mouth-broods the eggs and young.

Polygyny on the other hand presents little difficulty to mammals and is correspondingly widespread and common. The female mammal is usually self-sufficient in raising the young until they are weaned, and even after that the father's help in bringing food to helpless progeny is commonly called upon only in some of the carnivores. Herbivory generally speeds up the finding and ingestion of food, and it is much more frequent in mammals than birds, because having to carry a bulky digestive system is not such a handicap to pedestrian animals as it is to those that fly. What is more, the females in polygynous mammals tend to participate in their own right, either in the home-range system by which the density of spaced-out populations is usually regulated, or in the female hierarchy if their in-groups are gregarious. Perhaps because of their generally inferior mobility, formal leks where a group of males gather to compete and have to disperse in order to feed, have rarely evolved in the mammals. Examples can be found, however, in the Uganda kob (*Kobus kob*, an antelope) and in some fruit-bats, e.g. *Hypsignathus monstrosus*, both of which are tropical African species (see Leuthold 1966, Bradbury 1981).

By far the most numerous class of vertebrates, at least in terms of species, are the bony fishes, and among them many of the social and reproductive adaptations of birds are paralleled. The majority of those that live in water sufficiently shallow to give access to a visible substrate have developed sexual dimorphism in colour and/or structure, at least when breeding. Many have territorial males, and a minority of these use sound signals or even electric discharges, analogous to the songs of birds, to advertise their claims (see *Animal Dispersion*: 72, 85).

In a most original and illuminating book Fryer and Iles (1972, Chapters 6 and 7) have given a vivid account of adaptations such as these, in the cichlid fishes of the African great lakes. The Cichlidae have undergone an astounding evolutionary explosion, which has produced 500 species in these lakes and the associated rivers. Lake Malawi alone, with over 200 cichlids plus 40-odd non-cichlids, has a richer fish fauna than any other lake in the world. Lake Victoria comes second. There is a high degree of endemism; that is, independent species swarms have originated in the separate lake drainages, with the result that the 126 cichlid species in Lake Tanganyika, for example, are all endemic. Clearly most of this prolific speciation has occurred within closed lake basins; and that appears to have been made possible, in large part, by a high degree of philopatry, coupled with repeated alternations of rocky and sandy substrates around the shores, which have served as ecological barriers to species already adapted to live on one but not on the other.

In many species on the sandy substrates the males construct a low circular platform of sand, often slightly concave on top; and over it the owner displays to attract mates. On hard substrates they may clean a patch of rock instead. In what are presumably the more primitive types, living on hard substrates, the eggs are attached to the clean bottom and guarded by a monogamous pair. More sophisticated types on both substrates pick up the eggs, and brood them in their mouths and pharynges; and they continue to do the same with the fry until they are old enough to feed. Protection is urgent because the communities are dense and there are many predators on the look-out for such morsels. In some species the males' courts are aggregated, and in particular instances these are known to be at traditional sites, and thus form typical leks. The courts may be crowded and contiguous: in *Tilapia macrochir*, for instance, groups of at least 200 are reported. In other species such as *T. karamo* or *Haplochromis heterodon*, the sparser grouping more resembles the lay-out of bower-birds' bowers. In *Lamprologus congolensis* the males hold very large territories in which they accommodate a harem of mouth-brooding females, each in her own sub-territory.

Some of these African cichlids appear to be promiscuous maters. In *Tilapia macrochir* the sexual encounter between a territorial male and a visiting female, including courtship, laying and fertilizing a batch of a dozen eggs, and sucking them into the female's mouth before she departs, may be consummated in less than a minute. Both partners are then free to mate again and again, as long as their motivation and supplies of gametes hold out. They may repeat the operation together or form new partnerships. Possibly in this instance, as also in some of those known in birds where promiscuity is proved or suspected—e.g. in the brushland tinamou (*Nothoprocta cinerascens*) (Lancaster 1964)—a transient liaison between the sexes allows several females to mate with one or more of the higher ranking males. If so the net result, as far as genetic structuring is concerned, may again be polygyny, that is, raising the breeding sex-ratio to more than one female per male.

Attention should perhaps be drawn to the differences in emphasis I have made in discussing longevity and polygamy here, as compared with the parallel discussions in *Animal Dispersion* twenty years ago. There my main concern was to bring out the contributions these traits could make to population homeostasis. Thus longevity was seen as strengthening the transmission of traditions, by which homeostasis is sometimes partly achieved; and polygamy, as concentrating in the hands of a few males, and thus facilitating, the control of the annual breeding quota of females. Without wanting to belittle these, my chief emphasis now is differently placed, on eugenics. If the total number of breeders in the in-group remains the same, changing the sex-ratio among them has no lasting effect on the size of the gene pool (though it does alter the inbreeding coefficient). Practically all populations of domestic mammalian livestock, from racehorses downwards, are deliberately bred on the same polyga-mous basis, and for the same reason, that top winners are extremely scarce in proportion to the merely good or very good material with which they can be crossed. It would lead to an undesirable level of inbreeding, as far as the gene pool is concerned, if sires were regularly to inseminate their descendants; and in the wild, harem size is usually less than 20 and often under five. There are also numerous species in which a few sub-ordinate males also participate, both to reduce the inbreeding rate and, on occasion, as will appear in the Atlantic salmon (next section), apparently to raise the level of philopatry. In leks it is usual for the matings to be shared, more or less unequally, between two or several males. Young or secondary males also 'steal' copulations in various gregariously breeding seals (Pinnipedia); and in the unusual mating system of the ruff (*Philo-machus pugnax:* Aves) there is apparently a hereditary caste of secondary males (van Rhijn 1983). I discussed the existence of 'high' and 'low' males in arthropods in *Animal Dispersion* (p. 259–70); time may show whether or not their function is the same, and concerned with genetic structuring also.

What has become much more obvious than in 1962 is that the phenomena of longevity and polygamy make sense when viewed as optional group adaptations, capable of contributing to population structure and quality. Variations will arise between in-group and in-group and between deme and deme, in mean longevity and breeding sex-ratio, just as in mean dispersal distance and its standard deviation;

and thus group selection will eventually reveal, and favour, the balances that prove most productive and viable. No such coherent explanation of either phenomenon has ever emerged from classical Darwinism.

16.3 Genetic structuring in the Atlantic salmon

The Atlantic salmon (*Salmo salar*) appears to exemplify many of the adaptations for population structuring already discussed, and to introduce a remarkable new one, namely the precocious spawning of minnow-sized males which has presented an insoluble puzzle for 150 years. Like all other Salmonidae, it breeds only in fresh water. An important phase of its life history is typically spent in the sea, and involves a two-way migration that takes most individuals away to distant oceanic feeding-grounds, where they remain during adolescence. They return as adults to spawn, commonly in the same river, and even in the same tributary, as that in which they were born. Once in fresh water their population structure becomes increasingly easy to visualize, because of the geometry that branching rivers impose on their habitats, and the linear chains that their contiguous in-groups necessarily form. They control their home densities territorially—not only the spawning adults, but also the fry and juvenile stages that develop from the fertilized eggs; and the territories form normal two-dimensional mosaics where the streams are wide enough. The adults are sexually dimorphic when mature. Male polygamy is thought to occur as a rule, and the mean dispersal distances of the sexes are different, as will shortly appear.

In the British Isles salmon are still indigenous to all sufficiently unpolluted rivers, although their numbers have recently diminished, probably through overfishing in the sea or through the overfishing of their marine prey. They return to breed after one to three or more years at sea, spent by many of them in Baffin Bay, off the west coast of Greenland, on feeding grounds which they share with their conspecifics from the rivers of eastern Canada. They suffer a high mortality during their sea life so that only a small fraction survive to become adults. Most of these find and re-enter their native river system; but an appreciable minority, of the order of one in ten, finish up in another river. A higher proportion than this fail to appear in their own natal streamlet, yet breed elsewhere in the same watershed. Nevertheless it has been highly exceptional for a British-marked salmon to turn up in Norway, and so far none are known to have been exchanged with Canada. The North American and European stocks thus appear to be genetically isolated (cf. Thorpe & Mitchell 1981). The metapopulation native to the Baltic rivers is also largely if not wholly isolated; its members confine their sea life to the Baltic Sea.

It seems reasonable to suspect that the scatter of adults into fresh waters other than their own is not primarily due to a malfunction of their homing ability, as has often been supposed, but is a normal part of dispersal. It presumably provides recruits for colonizing new or vacant habitats with fish of more or less local provenance, and maintains a desirable rate of gene flow from one river stock to another, within the same deme or metapopulation.

Salmon in Scotland spawn in late autumn, in flowing water where the bottom consists of coarse gravel and waterworn stones, sufficiently large to allow a current to percolate through the chinks between and beneath them. When the female is ready to spawn she holds her position head to current, just clear of the bottom, and with flips from her strong tail, swirls or skips the stones out from under her; the current carries them to rest just downstream. Persistently she 'cuts a bed' some 20–30 cm deep, big enough to receive a large batch of eggs. At the end of each mating sequence (described below) she moves a length or two forward and, by the same motions as before, fans more stones backwards to bury the eggs and fill the bed. The eggs are then secure from predators until they hatch, and even after that the larvae remain concealed until their yolk-sacs are absorbed before they swim up and enter the open stream overhead, usually in May.

For 1–3 years or longer the young fish stay in their natal stream. In the first summer the fry grow into parr, which have a row of 10 or more fingerprint-like dark patches along their sides—the parr-marks. From the outset they are territorial, enlarging their holdings as they grow and displacing outcasts, on which socially-induced mortality falls (see Gee *et al.* 1978). There appear to be genetic differences in individual growth rate, such that a single batch of eggs produces parr which reach the critical size of about 10–14 cm and are ready to migrate to sea in two, three or even more successive years (Thorpe 1977). Each spring the leavers drop downstream, undergoing a striking change in appearance as they descend, and emerge into the sea as silver smolts.

The spawning itself merits close attention. In the short late-autumn days, mature adults are dispersed wherever the stream bed is suitable for cutting, from the headwater streams just deep enough to float a salmon, down almost to the tide. The males have assumed a reddish cast and temporarily developed a conspicuous 'kipe' or upturned hook at the apex of the lower jaw, with which to butt or threaten rivals (Fig. 16.1). It varies in size and formidability from one male to another. A small assembly of both sexes are typically present on each of the 'redds' where spawning occurs; and among them a dominant male keeps other males off, while he courts with females one at a time, which come and select a site in his domain. The dispersion of males somewhat resembles an extended lek. The female is also territorial while spawning and will attack other females if they try to supersede her. With the male swimming beside her, she labours for hours to hollow the nest, and when it is finally deep enough the pair work up to a climax, at which moment the male quivers, and batches of eggs and sperm are simultaneously ejected into the bed. The orgasm over, they move forward, and the female soon begins cutting again, starting a second bed upstream of the first, as mentioned above. The same laborious sequence is repeated again and again, perhaps for more than a day, until her eggs are finished.

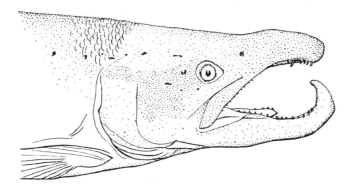

Fig. 16.1 *Head of spawning male Atlantic salmon* (Salmo salar) *(Scotland, 26 Nov. 1957). The 'kip' or hook on the lower jaw is an aggressive weapon or adornment, present only while the male is mature. The upper jaw is also temporarily overgrown. (Standard length of fish = 93 cm. Reduced to ¼) (From Wynne-Edwards, 1962).*

She then leaves the redd, but the males hold their ground. My main authority for this account , J.W. Jones (1959: 130), felt convinced that some males spawned with more than one female, an opinion that is generally shared by experts. He made prolonged observations in an artificial channel through which part of a river (the Welsh Dee) was diverted, and into which he could see from the side through underwater windows. The channel was constructed only 1.7 m wide so as to keep the fish near enough the windows for accurate observation and photography. Consequently there was not enough room to attract many wild fish at a time, and actual confirmation of polygamy was not obtained.

Jones did confirm a more singular fact. A century earlier it had been proved by John Shaw (1836) not only that some of the diminutive male parr became precociously ripe, but that the sperms they produced were capable of fertilizing salmon eggs *in vitro*. Still more, eggs so fertilized could be grown in artificial channels until, at two years after their own conception, some males among them became similarly fertile. Under some conditions the most precocious parr will even mature at age 0 +, meaning the end of the calendar year in which they hatched. Jones and Orton (1940) collected samples of parr from 20 British rivers and found that in November–December when the spawning season is at its height, 75 % of them were sexually mature or nearly so.

In the observation channel Jones watched them on the spawning redd. Male parrs have a social hierarchy too, and the alpha one in the immediate vicinity may succeed in slipping in beneath the courting adults, hiding close under the female's vent. Being a little fish he is out of sight beneath her large bulk, and not molested. When she ejects eggs the parr lets go his sperm.

Jones wished to confirm that parr sperm are effective in 'field' conditions; but he found if he shut the adult male out of the observation channel the female would neither cut a bed nor ovulate, in the presence of the parr alone. But he succeeded in castrating a big fish so painlessly that it was not inhibited from acting as the female's consort, and thus he finally proved that the parr are effective parents. Incidentally Shaw (*loc. cit.*) had discovered that in the small stream where his juvenile captives were kept, he could take a 1½-ounce (40 g) parr and milk out of it enough sperm to inseminate all the (artificially stripped) ova of 'a very large female salmon'.

There has been prolonged speculation about the function of this male paedogenesis, without eliciting a single promising suggestion; but it now looks probable, in the light of what we have already seen in this chapter, that it is another device for increasing the inbreeding rate. The parr, still living in their original nursery stream, are closely philopatric, whereas the spawning adults homing from the sea are on average less so, as shown by evidence that has now become available.

The Freshwater Fisheries Laboratory of the Department of Agriculture and Fisheries for Scotland have conducted a protracted experiment that throws light on this, at a site only 40 km from my home. In 1966 they built a pair of salmon traps in the Girnock Burn, one of the upper tributaries of the Aberdeenshire Dee, about 80 km from the sea. The burn is a hill stream 9 km long, and the traps are not far above its confluence with the river. They are massive structures, placed one above the other, designed to withstand floods and catch descending and ascending fish respectively. Two scientific papers on the results have been published at the time of writing (Buck & Youngson 1982; Buck & Hay 1984). Others are in preparation, and I have been given additional information through the courtesy of the director of the Laboratory and Mr. Alan Youngson, one of the researchers taking part.

During an 11-year period of continuous daily operation, about 25 000 descending well-grown parr and smolts were caught and tagged. Those that eventually returned as adults had mostly spent two years in the sea. The incoming run of adults varied in number from year to year between 38 and 269 fish, with an average of about 150; but of these, over the 11 years, only 15% were found to carry tags. This was a lower proportion than previous smolt-marking experiments elsewhere had led the researchers to expect, and it was suspected that some tags had been lost during the fishes' hazardous absence and enormous growth. So in the two final years, 1976 and '77, all the descending young fish over 9 cm long were also clipped on the pelvic fins. The adults with two years of sea life returning in 1978 and '79 numbered 130, and of these 21 had tags and clipped fins, 31 had clipped fins but no tags, and one had a tag only, having escaped the clipping (or regenerated the missing fin-rays). That gave a sample estimate of 53/130 or just over 40% as the proportion of ascending adults that were native to the Girnock Burn.

Additional Girnock-tagged fish were reported elsewhere in British waters, considerably more of them in fact than eventually reached the Girnock. The great majority were caught in the lower reaches of the Dee or in nearby coastal nets. Obviously by being caught they had been prevented from returning to the Girnock, if that had been their programmed destination; but the probabilities of death or diversion into an alien tributary are the same for all the returning fish, so that the 130 that did come through the trap still provide a random sample, as far as their own provenance is concerned. No fishing was done in other similar nursery streams in Aberdeenshire or elsewhere, but a few of the dispersed adults were caught far up-river in other watersheds in eastern Scotland. It is perhaps necessary to stress once more the tremendous mortality that salmon suffer, amounting to about 95%, between the time the smolts are marked and the time the adults return from the sea to spawn, one to three years later.

The sample suggests that rather less than half the adults ascending the Girnock are native. Since the Girnock redds probably make up less than 1% of the total spawning ground in the Dee watershed, the 40% or so that did return were far from being just a random sample of Dee fish. They had proved their amazing capability of finding the exact place, from a starting position perhaps 2500 km away in Baffin Bay. Their homing faculty within the river is thought to depend on remembered olfactory cues with which they became imprinted as young fish, in their nursery stream and on the way down to the sea; and it seems fairly safe to assume that if 40% of individuals can be programmed to perform this feat, 90% or more could be similarly endowed if that were to result in a commensurate gain in the survivability of the deme. In other words the observed figure of 40% is unlikely to signify incompetence on the part of the other 60%, and more probably exemplifies, once again, the usual range of programmed variance in individual dispersal distance which we have found in every previous instance.

The potential that exists in the Salmonidae for precise homing has been demonstrated in another species, the sockeye salmon (*Oncorhynchus nerka*) in British Columbia, a fish that shows conspicuous differences in life-history and population structure from *S. salar*. In one of the earliest experiments on homing, all the descending smolts from a particular lake were marked in two successive years. The survivors returned four or five years later; and of 2881 adults that came up and entered the same lake in those years, all but 17, or 99.4%, were marked fish (Foerster 1936, quoted from Ricker 1950).

The reasons which would make a partial abmigration of the returning fish advantageous are probably much the same for the Atlantic salmon as for the red grouse, and numerous other species. The first is to ensure that new and depopulated habitats are filled up. Failing that, the species' range would shrink by default. It is likely to demand that dispersers be rather generously broadcast, because with mortality on the scale the salmon's sea life entails it might chance quite often, if perpetuation of the stock depended solely on returning natives, that a small stream like the

Girnock received no incoming adults at all. Such a risk would be much reduced if a substantial proportion of homecoming adults each year were programmed to enter any comparable stream that was vacant or sparsely populated in preference to those that were better stocked, including their own parental burn. (Quite possibly the salmon's keen chemoreceptors can inform a prospecting fish, when it enters a side stream, roughly what concentration of adults already exists upstream, and whether parr are abundant or scarce.)

There are other responses which work in parallel with dispersal to diminish the risk of temporary local extinctions. These are (1) that the metamorphosis of each generation of parr into smolts is normally spread over two or more years; and (2) that there is a similar spread of years during which a generation of adults return to spawn. Both must result in damping the accidental fluctuations in natality and mortality that occur from year to year, and preventing the extinction of small local stocks simply by the failure of adults to arrive and breed in one particular year (see *Animal Dispersion*: 575–6).

The second function of a partly wider dispersal is genetical, to alleviate the close inbreeding that would occur in completely philopatric in-groups and allow a modicum of gene flow and outbreeding instead. For this purpose dispersal needs to be controlled in order to produce a viable balance between isolation and gene flow. In the red grouse the genetical function appeared to be the one that dictated what the mean dispersal rates and distances should be, most of the time, whereas the colonization function was fulfilled intermittently, by means of a spontaneous population cycle which produced a major exodus (see pp. 154, 190). In the salmon the dispersal and recolonization functions have both been appended to the return migration from the sea; but philopatry has come to depend also on enlisting some of the male parr (which have never left their natal stream) as members of the breeding in-groups. This turns out to be a relic of an ancestral habit in the Salmonidae of having stay-at-home males, as I shall show later. Its net effect is to reduce the mean dispersal distance of the male parents, though by how much we cannot know because the shares of insemination due to parr and adult males may never be disentangled. There will also be a corresponding increase in the skewness and positive kurtosis of the distribution of male dispersal distances, in a manner clearly parallel to those of the distribution we were able to quantify, even if roughly, in the red grouse (p. 226).

Ryman and Ståhl (1981) reported that highly significant differences exist among the salmon in a set of seven neighbouring, pristine rivers in northern Sweden in the frequencies they show of particular allozymes; that is to say, there is evidence of genetic divergence between adjacent river stocks.

The Girnock Burn data suggest that around 60% of the adults are non-natives. Some of them may of course have been born in other streams not far away; but there seems no doubt that a regular contribution of gametes from philopatric male parr could add to the inbreeding rate. The figures we have for the proportions of them that ripen precociously (> 0.5), and of the smolts that survive their sea life and return as adults (0.03 or less for males), suggest there might on average be 20–30 ripe parr for every male adult on the spawning redds. The 100-fold difference in size between the two (50 g : 5000g) makes it probable that the adults (meaning the grown-up fish) could yield the bigger combined total of spermatozoa; but, as always, the vast majority of sperms go to waste. The spawning parr may have some advantage of position where he lies right under the female's belly, so that his sperm might reach the eggs ahead of the adult male's. Anyway, that is a possible explanation of why the female is inhibited from ovulating except in the presence of an adult male, namely that if she were not, the parr might steal altogether too much of the act. The coefficient of inbreeding and the frequency of genes identical by descent among the parr will depend on the average shares that fall to the parr, the native adults, and the alien adults in the parentage of successive generations.

The adult males make valuable contributions to the survivability of local stocks in at least three other ways. (1) The more dominant of them presumably control the density of breeding females on the redds (which the parr would be too feeble to enforce). (2) They are possessors of long-tested genotypes themselves; and (3) they can accompany the females as pioneers in colonizing streams in which there are no native parr.

In his studies of the descending smolt run in a small Swedish river, Österdahl (1969) found by histological examination that only 35–40% of them were males and the rest females. Assuming a primary ratio of 1:1 (confirmed by hatchery data), differential mortality between the sexes must thus have brought about a shortfall of 33–46% among the males, and this he attributed to the costs of precocious spawning, inferring that the shortfall was due to failure by many of the spent parr to survive and turn into smolts. Campbell (1977: 226) records a corresponding preponderance of females among the smolts in the Tweed, and cites other authors who had observed the same imbalance elsewhere in British rivers. This disparity among the juveniles probably accounts in part for the similarly unbalanced ratio in returning adult fish. The figures for the Girnock between 1966 and 1981 totalled 1029 males and 1250 females, or 1:1.2, indicating an 18% shortfall of male adults. Hutton (1924: 48), on the other hand, sexed over 15 000 adults in the River Wye by dissection, and found a ratio of 1:2.5, implying a 60% male shortfall.

In Scotland almost all adult males die after one spawning, whereas a minority of the females re-enter the sea and 'mend', and are later able to make a second spawning migration, usually the following year. A few exceptional specimens are known to have returned as many as five times—a fact that can be read from their scales. In the Aberdeenshire Dee, however, even second-timers amount to only about 2% of female spawners, so that the sexual differences in mortality among adults can have little effect on the 18% difference in numbers mentioned above. In some west-coast Scottish rivers, on the other hand, one adult in three has spawned on a previous occasion (Menzies 1931). Such local differences in vital statistics, which appear also in the local frequencies of maturation among male parr, are thought to be at least in part hereditary, and again to reflect a genetic diversity between stocks. If and when further data about philopatry are obtained from other equally small stocks for comparison with those from the Girnock, the average balance between natives and aliens among the adult spawners seems likely to show a similar variance between rivers.

Not unexpectedly, although the males survive less well, the largest salmon that come in to spawn are males, returning for the first time after spending an extra (usually a third) year feeding at sea (Menzies 1931). In Hutton's Wye sample mentioned above five out of six of the very large fish were males. Such a superiority in male size is, of course, characteristic of species in which males regulate the breeding dispersion of females, and still more of those where they are polygamous as well.

A brief look at the similarities and differences in life history found in other members of the Salmonidae is enough to suggest how the paedo-genetic spawning of parr in *Salmo salar* came to evolve. The three genera the family contains must all be descended from a common freshwater ancestor; and typically they inhabit cool northern rivers and lakes of rather low productivity. They tolerate salt water, and the majority of species of salmon, trout and charr that make up the family have at least some populations whose members spend part of their lives feeding and growing in the richer waters of the sea. A still larger majority of species also have some wholly landlocked relations which, being isolated in their present watersheds, tend to diverge and become endemics. It is invariably true of the family as a whole that they spawn and undergo their early development in fresh water.

The brown trout (*Salmo trutta*) of Europe and the rainbow trout (*S. gairdneri*) of western North America are two examples of species with very successful populations of exclusively freshwater fish; yet both have, in some parts of their range where rivers run cool enough right to the sea, sea-run forms that can interbreed with the inland stocks, known as sea-trout and steelhead trout respectively. Exactly the same is true of the North American brook-trout, which is technically a charr (*Salvelinus fontinalis*). In the European sea-trout more females than males tend to migrate out to sea, no doubt because the extra growth they gain affords a corresponding increase in fecundity. Less advantage would accrue to sea-run males because even small males produce superabundant gametes. When sea-run females return to spawn they not infrequently pair with a brown trout that has never left his native water (Campbell 1977). The same predominance of females among the sea-run migrants and/or the participation of resident freshwater males in their spawning have been reported in the rainbow/steelhead trout, the kumzha trout of Black Sea rivers, the chinook, sockeye, coho, and masu salmon (*Oncorhynchus gorbusha, nerka, kisutch* and *masou*) of Pacific rivers, and in the brook trout and Dolly Varden trout (*Salvelinus fontinalis* and *malma*) of North America. In most of these instances, sea-run females outnumber sea-run males by between two and five to one (for references to original sources, see Campbell 1977). The participating freshwater males tend in every case to be smaller and younger than the sea-run females, and they are no doubt fulfilling just the same function as the spawning salmon parr, by contributing philopatric genotypes to their local gene pools. The opposite parental mating, of sea-run males with home-grown females, has not been reported.

The Atlantic salmon differs in population structure from these other species principally by not having any permanent resident freshwater adults that are left behind when the smolts go out to sea. The landlocked populations of Atlantic salmon *sensu lato* that do exist, both in North America and Europe, are physically cut off from the sea, and consequently from all intercourse with their more numerous sea-run relatives.

It would be interesting to· know whether parr spawning has already disappeared from their mating systems, as our hypothesis leads one to expect it should, given time. In sea-run populations of *Salmo salar*, on the contrary, the ancestral function of topping up the inbreeding rate with the aid of stay-at-home males, has evidently persisted.

What may be a parallel development of parr spawning has long been known to occur on a small scale in the largest of the Pacific salmon species, the chinook (*Oncorhynchus gorbuscha*). It has been reported in the upper, inland tributaries of the Sacramento, Klamath, and Columbia rivers which are in the southern part of the chinook's geographical range. It involves only a small fraction (e.g. 2–3%) of 0+ parr, which mend after they have spawned. In the same or other upper tributaries, non-migratory or 'residual' males occur, which mature at 1+ and die after spawning—presumably merely as an extension or alternative to parr spawning. Parr spawning does not occur in the stocks of chinooks in the smaller coastal rivers, and it has been shown by cross-breeding these with upland stocks that the difference is genetically determined (Ricker 1972: 79–80).

In the two remaining species of *Oncorhynchus* not so far mentioned, *O. tshawytscha* and *O. keta,* it appears that all the mature fish of both sexes have been to sea. Presumably for them the inbreeding/outbreeding balance is maintained by raising the level of adult philopatry instead, with a corresponding sacrifice of dispersal capacity. The figure already quoted (p. 258) for the percentage of accurately homing adults in the sockeye (*O. nerka*) suggests this is true of them; and it may apply therefore to other species in the genus *Oncorhynchus*.

Summary of Chapter 16

1   The disparity in mean dispersal distances between males and females is considered further. Greenwood (1980) showed that it is a common phenomenon in bird and mammal species and, curiously, that the roles of the sexes are generally reversed in the two classes; so that in mammals the females and not the males are the more philopatric sex. In both classes there are a few anomalous taxa that have the opposite arrangement from the others in their class, e.g. geese and ducks amongst birds and some higher primates amongst mammals.

Greenwood concluded that the reversal of polarity between the

classes had no functional connection with the sex-chromosome complements, which are similarly transposed, and with that I agree. In the end
he did not find any obviously acceptable explanation, either for the sex
differences themselves or for the bird-mammal reversal. I suggest here
that the sex differences are simply a division of labour, in the building and
preservation of population structure through the usual medium of dispersal. Very close philopatry can be an advantage ecologically because it
enables progeny to inherit the resources their forbears have husbanded,
and acquire traditions about the use of particular sites. But beyond a
certain intensity, philopatry becomes a diminishing asset genetically; and
by assigning it to one sex, the other sex can be freed to follow a slightly
more liberal dispersal program, which regulates the effective size of
mating in-groups. The groups can thus be shaped to procure the great
advantages that flow from genetic structuring and the promotion of
variance between local gene pools.

The bird-mammal reversal I take to be evidence that their respective
dispersal codes have evolved in parallel and independently, and that
which role became linked to which sex was largely if not entirely an
accident. These classes show many similar parallels, e.g. in homeo-
thermy, and the four-chambered heart; and some of these have also
assumed opposite polarities, e.g. the XY-XX chromosomes, and the survival of the the right or left aortic arch.

It is noted that some duck and goose species are migratory, and form
pair-bonds when far from 'home'. Thereby males have been seduced to
enter allopatric breeding grounds in violation of the normal structuring
practice. Also, sex differences in dispersal occur sporadically in insects.

2   Two attributes are next discussed which are common among vertebrates, and both diminish effective dispersal rate and in-group size,
namely longevity (iteroparity) and polygamy. Longevity allows the more
efficient survivors to amass higher personal fitnesses, and this eventually
benefits in-group productivity as well. Polygamy can likewise have a
eugenic effect on the group. Polygyny is generally more practicable than
polyandry: it is fairly common in birds and more so in mammals. An
uncommon but specially interesting form of it occurs in species in which
the males display at a lek and are visited there by females for fertilization.
Polygynous matings and leks are also known in fish. Polyandry though
less convenient achieves the same eugenic advantage, and may have led
to the evolution of the reversed secondary sexual roles found in a few
birds, e.g. phalaropes.

3   Genetic-structuring benefits may have led to a remarkable mating phenomenon found in the Atlantic salmon (*Salmo salar*). First demonstrated in 1836 but never explained, it is the participation of minnow-sized male parr (post-larvae) in inseminating the eggs of adult females over 100 times their weight. All salmonids breed in fresh water; and in *S. salar* there is a juvenile downstream migration to the sea and thence to distant feeding grounds, at age 1–3 years. Return to inland waters occurs 1–3 years later, but there is a huge mortality in between. About 90% of those that come back enter their native river and, in an example detailed, about 40% reached the same branch or stream. Others arrive to breed in adjacent rivers or tributaries. The dispersal function is thus appended to, and forms the final stages of, the homing migration. Because very few (usually < 5%) of the original emigrants return, local stock extinctions would be common if there were not a substantial scatter; i.e. the colonization function looms large.

It is suggested that philopatry, weakened by this necessity, is augmented to a level favourable to genetic structuring by the spawning parr, themselves closely philopatric, not having left their parental stream.

In the event, an individual parr of sufficient status among its peers moves in, hides under the female's belly, and ejaculates when she ovulates. There is always also a full-sized, dominant male beside her doing the same; but the parr spermatozoa are known to be fully efficient, and their net effect can hardly be other than to reduce the mean dispersal distance of males.

In several other salmonid species only a fraction of the males join the females in migrating to sea, and share in the richer feeding, greater growth and fecundity that ensue; many or most of them mature at a smaller size at home instead. The Atlantic salmon, which frequents rather unproductive northern rivers, has apparently preserved the enhanced male philopatry which such divergent habits probably provided for its remote ancestors, by advancing the males' precocity so astonishingly as to allow them to spawn philopatrically as fingerlings, and still have the chance of feeding and growing in the sea.

CHAPTER 17
POPULATION STRUCTURING IN
THE INVERTEBRATE PHYLA

17.1 Two gastropod molluscs

My hypothesis suggests that ever since the earliest vivivorous animals evolved they must have been capable of limiting their own increase, otherwise they would have eroded the primary production of plants so severely that they themselves became extinct (see Chapter 4, p. 53). In the vertebrates and several classes of arthropods the hallmarks of sociality are more or less abundant, and remarkable parallels in social adaptation and behaviour have independently evolved in the two phyla. Especially in some crustacean and insect orders it is common to find species in which the males are endued with bright colours, penetrating voices or impressive weapons, and compete among themselves for conventional property or rank in just the same manner as vertebrates. Investigations have shown, at least for some insects, the expected faculty of controlling population density and thus safeguarding food resources (see p. 23).

The fact that population regulation is the norm among animal species is witnessed by our everyday observation that plants, the primary producers, are not all obliged to be toxic (which would make herbivory impossible), nor are they chronically abused by natural consumers, whether vertebrate or invertebrate. Instead, with transient exceptions, the producer and consumer industries co-exist in a stable and sustainable relationship, of forbearance on the part of the consumers and sufferance on the part of the consumed. Below the arthropod level in the evolutionary scale the signs of possible sociality are rare or absent. Many of the species are hermaphrodite, but in those that have separate sexes, sexual dimorphism and the secondary sexual characters that distinguish competitive males are very scarce. One example is the Bermuda fire-worm, *Odontosyllis enopla*, a marine polychaete which, each spring, has synchronized communal spawnings. They are held in the surface water layer, after sunset and only in the third and fourth weeks of the lunar month—conditions which ensure that night will fall fast. Both sexes have luminous organs, but whereas the female emits a more or less continuous glow, the male gives sharper intermittent flashes. Fertilization of the eggs is external. There is little evidence that the spectacular display is anything more than a device for assembling the mature worms for their annual reproduction, nor that the different types of luminosity are other than signals to the sexes, enabling them to get close enough for successful fertilization; nevertheless, some competition between rival males has been observed (for references see *Animal Dispersion*: 256 & 349).

The Mollusca are a large and versatile phylum in which the individual organisms are either hermaphrodite or of separate sexes, according to species. Where they are dioecious, sexual dimorphism of the competitive and social kind seems to be unknown, except in the highly advanced cephalopods (squids, cuttlefish and octopuses) which possess good eyesight and all have separate sexes. Even here, though the animals are visually demonstrative and their signals are ritualized, social competition appears to be weak. The study of their social behaviour is hampered by the fragility of their skins which precludes individual marking (cf. Moynihan & Rodaniche 1982).

Thus it seems proper to describe the molluscs as presocial animals. Notwithstanding this it is known that some, and probably most or all, are capable of population homeostasis and habitat conservation. For example, the population dynamics of the pond snail *Lymnaea elodes* have been studied experimentally by Eisenberg (1966), who found that their numbers were rapidly adjustable to changing rates of plant and detritus food production, so that overexploitation was forestalled. In the pond near Ann Arbor, Michigan, where the experiments were carried out, the water level followed an annual cycle, being highest in spring, falling through the summer, and reaching a low in the autumn and winter (when the surface froze). Each year, reproduction took place about midsummer, and the progeny quickly hatched out. The water was already beginning to recede from the inshore habitat, so the young had rather a short time, depending on their distance from the shore, in which to feed and grow before it became necessary to burrow into the substrate and aestivate to escape being left high and dry. Once dug in they remained dormant till the following April. Then they emerged and renewed their rapid growth, and were ready to breed at 12 months old, in June and July. After breeding all but a small minority died. Their food consisted chiefly of living vegetation, rooted in the substrate in a wide uniform zone running parallel to the shore.

For experimental purposes part of the snails' habitat was enclosed by erecting snail-proof partitions to make many large pens of uniform size. In the 'basic' series of experiments there were three alternative treatments, each of them replicated four times. In one set the populations present in June were reduced by hand to a quarter of their natural density. (The original design, to reduce them to a fifth, was shown too late to have been based on an underestimate of the natural density.) In a second set the natural stocks were increased to five times their initial density, and in a third the initial densities were left unaltered to provide a control.

In addition to this series there were four other pens in which the natural snail populations were left unchanged, but extra food in the form of a 10-ounce package of frozen spinach was added every three or four days.

In all the pens the adult survival rate was similar throughout the breeding season. Notwithstanding the 20-fold difference in population density between the 'high' and 'low' pens, *all three basic treatments resulted in similar average totals of eggs and young being produced;* and this was attributable to the different average sizes of the egg mass that each adult deposited, which varied inversely with the density of adults present. It was a striking result. But no less so was the effect of providing extra food: the fed populations produced 25 times as many eggs as the controls at the same population density. Before long however this difference had been diminished by high mortality, to leave at the time of aestivation only a little over twice as many surviving young as in the control pens.

The author concluded that all the stocks, except in the fed pens, must have been under severe food limitations, and consequently unable to realize their full potential fecundity. Yet it was apparent in all the pens that there remained an abundance of vegetation and coarse debris of a kind that the snails could and did eat. He thought it was perhaps just high quality food that had been in short supply. There was little indication that any of the snail stocks had actually damaged the vegetation in their quest for food, and he had to conclude that population regulation in relation to food must be 'relative rather than absolute'. Regulation there certainly was, sufficient to fulfil his definition of it as 'the maintenance of the numbers in a natural population within a certain range, around some level that is determined by the whole of the environment, and is brought about as a feedback from population density'. Thus he seems to suggest that the regulation was both adaptive and self-imposed.

If so, Eisenberg's conclusions differ little in principle from those we derived from the study of the red grouse. The pond snails could evidently manipulate their density and, faced with mismatching between density and food resources, could rectify it within a single annual generation. The machinery for regulation must operate not through sociality and conventional behaviour but via some other feedback process, most likely dependent in part on substances that the snails secrete or excrete into the water and can monitor as a density indicator. The possibility that the growth and fertility of gastropod molluscs might be influenced by a pheromone has been suspected for some time (see C. A. Wright 1960).

Many types of freshwater snails can be cultured in the laboratory and subjected to experiment, and it was well known that overcrowding tended to stunt their growth and make them infertile, even when they received adequate food. This could simply mean they became short of some resource other than food, such as oxygen or calcium, when confined in a limited volume of water, or that the water became polluted with metabolites, including ammonia, which is their principal excretory product and can become toxic, especially when the pH is high.

There has been a powerful motive in recent decades for discovering the true explanation for these effects, because water snails are the intermediate hosts of parasitic worms that complete their life cycles in man and domestic animals. The most serious of these are four species of the trematode *Schistosoma* which, in tropical and subtropical countries round the world, sap the vitality and burden the lives of millions of people: it is said indeed that the malarias now contribute less to the sum of human misery than the schistosomes. If the snails really control their own populations by means of a pheromone secreted into the water it might be possible to manufacture and use it as a specific molluscicide without inflicting injury on other forms of aquatic life. That might open the way to controlling the spread of schistosomiasis by suppressing the intermediate hosts. The largest research effort in this quest hitherto has concentrated on the small planorbid snail *Biomphalaria glabrata*, a common South American host of the most widespread schistosome species, *S. mansoni* (see Thomas 1982 for a brief review).

After a decade of intensive study, Thomas and his colleagues state that they have found 'no good evidence' that crowded snails secrete an inhibitory pheromone. Instead, the effects of crowding seem directly attributable either to the depletion of resources, especially calcium, or the accumulation of metabolites, especially ammonia. Unexpectedly, however, the research has disclosed evidence of an opposite adaptation,

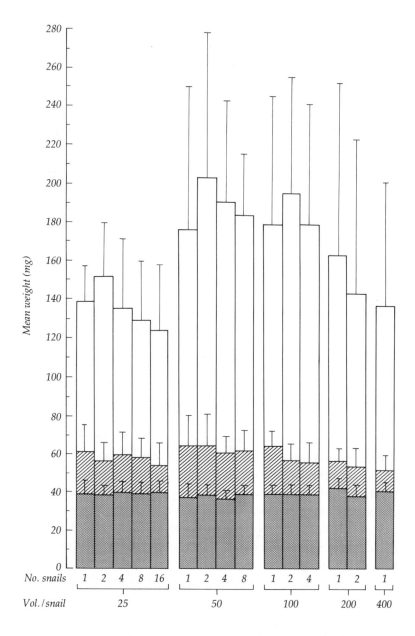

Fig. 17.1 *Comparison of mean growth rates of snails* (Biomphalaria glabrata) *in beakers with different numbers of individuals and different volumes of water per snail. In the first three sets of columns, snails in cultures of two achieve the highest individual growth rates, and solitary snails grow more slowly. Positive standard errors of the means are shown. For details see text. From Thomas & Benjamin (1974).*

namely organic chemical factors that stimulate growth and fertility in other snails; and these of course are just as relevant to the subject of this chapter.

When snails are placed in a vessel containing 'standard water' and fed on discs of lettuce in excess of their needs, their metabolism progressively alters the solutes originally present in the water by removing some and adding others. The water is said to become 'conditioned', and in conducting experiments the status quo has to be restored at intervals, usually by replacing it with standard water every three days (Thomas 1973). The rate at which conditioning proceeds depends, when temperature is kept constant, largely on the number and size of the snails present and the volume of water the vessel contains. In a broad-based experiment on the growth of young snails, 15 permutations on these two variables were set up in the necessary number of beakers. Population was varied by putting 1, 2, 4, 8 or 16 weighed snails to a beaker; and water volume, by allowing 20, 50, 100, 200 or 400 ml per snail. All the treatments except the one requiring 16 snails were replicated 10 times, and that one five times. The snails were allowed to grow for 12 days, when the experiment was ended and they were re-weighed.

The results are shown in Figure 17.1. Each single column in the graph is divided into three parts, the bottom part showing the mean starting weights of the snails in that particular one of the 15 treatments; the middle piece shows the amount of growth up to an intermediate weighing after three days; and the top part, how heavy they had become by the 12th day.

Different amounts of growth were achieved in all 15 combinations. The first set of five columns, in which 25 ml of water was allowed per snail, show the fastest growth as occurring in the two-snail series. At snail densities lower and higher than two, it was slower. The second set, at 50 ml per snail, contains four columns all indicating considerably faster growth than their counterparts in the first set, but again putting the two-snail groups ahead of the rest. In the third set, with 100 ml per snail, the twos also lead, but here and in the two remaining sets growth was progressively poorer than in the second set in spite of the increasing volumes of water per snail.

Performance by these young fast-developing snails was obviously very variable between individuals within the same treatment, and even between individuals in replicated treatments with isolated snails. This is brought out by the large standard errors of the means that characterize all the results alike. The aims of the research were not primarily concerned

with behaviour and ecology and no explanation of this particular phenomenon was looked for; but it is potentially important as possibly offering
some kind of functional substitute for the social hierarchy in a presocial
species, by which some individuals are given a head start and others are
held back for a time. Such an effect could result if necessary in a palliative
mortality of surplus population, with little or no need for competition.
The large variances seem to have been responsible for an experimental
side-effect, namely that the authors' statistical tests could only demonstrate significant differences between the growth-rates of snails in the
first set of five treatments; and even in that set, only between the 2, 4 and 8
snail groupings (Thomas & Benjamin 1974: 35). Fortunately there were
other nearly parallel experiments in which either one or two snails were
placed in 200 ml of water, and their growths compared. These were each
replicated, and also repeated four times for snails at different ages and
sizes. Collectively they gave a highly significant confirmation that the
average differences in performance between singles and doubles were
genuine (Figure 17.2).

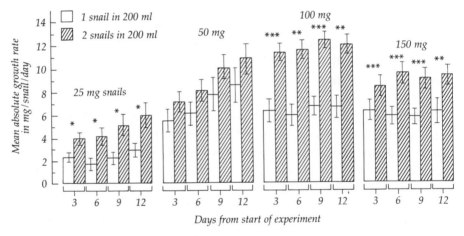

Fig. 17.2 *Mean growth rates of snails of different sizes, cultured alone or in pairs, with
other conditions held constant. In all four sets, two together both grew faster than the
solitary snails. P-values based on t-tests are indicated by asterisks:* *<0.25, **<0.01,
***<0.001. From Thomas (1973).*

Closely parallel results were obtained in the original permutation
experiments for the age of attaining maturity and for subsequent fertility.
Companionless snails took longer and had to reach higher weights before

maturing, compared with two living together and occupying the same amounts of water per snail. Their fertility rates (eggs per day) also tended to be lower when they did mature (Thomas & Benjamin 1974: 41–2). In other experiments where 'large' snails were used (300–700 mg), crowding effects had presumably begun to show, because solitary individuals then appeared equally productive to the pairs (Thomas 1973: 334–5).

So far the general outcome of the work had been to show that not only the volume of water but also the number of snails in the group can influence an individual's growth and fertility. In respect of each of these variables an optimum could generally be demonstrated, with two as the most favourable number and 50–100 ml of water per snail as the optimum volume. The graphs often hint that three snails might have done even better than two, but this seems not to have been tried. One should not lose sight of the vast difference that separates the laboratory conditions from those that exist in the wild, where *Biomphalaria glabrata* usually lives in far larger bodies of water and is by habit a gregarious animal; but there seems nothing unexpected about such creatures suffering harmful effects from crowding when the natural remedies of dispersal, and mortality from predators, are denied to them. Intuitively it may seem more anomalous that being solitary should be a disadvantage to growth, seeing that it implies the individual has the resources of the habitat entirely to itself. The latter discovery did in fact lead Thomas to explore by what means two individuals were able to promote one another's welfare—for instance by secreting a supportive pheromone into the water.

To this end he employed conditioned water as the snails' medium. It was being produced in quantity in the tanks where the snail stocks were kept, and was routinely drained out each week and replaced by clean standard water. The previous work had indicated that any stimulant effect would be demonstrated most easily on isolated snails, and these were therefore taken at the most impressionable age and placed, one in each of a series of assay chambers of 200 ml capacity, and interconnecting so as to keep them in aqueous continuity with a reservoir of conditioned water through a set of four doubly mesh-covered windows. These windows precluded bodily contact between snails on either side of the partition, but allowed a free enough exchange of water, which was kept stirred by the aerating system. For some experiments a flow of conditioned water was circulated through narrow tubing to the test animals; and for all experiments controls were provided by having duplicate sets of assay snails living in a medium of standard water, renewed at frequent intervals.

These experiments confirmed that there was indeed a water-borne factor or factors that could exert growth- and fertility-promoting effects. A concentrate of the organic content of the conditioned water was prepared by means of ultra-filters, and the first steps taken to characterize the active factors, which showed that their molecular weights must lie between 500 and 10^5, and that polypeptides, glycoproteins, lipoproteins, or small growth factors bound to larger molecules might be implicated (Thomas *et al.* 1975: 21).

In the same paper (p. 20) the authors emphasize how far from simple the action of the stimulant factors appeared to be. The effects were not in linear proportion to the concentration at which the factors were administered, and indeed at low concentrations they were found to *retard* growth significantly instead of promoting it (same ref., p. 6–7). In other words, the extract could have both positive and negative consequences.

We need to think here about whether, and if so how, such pheromones might serve the purposes of population homeostasis and structuring. For homeostasis the standard requirements are inputs of current information on population density and on the state of the food resources. These are fed into the central nervous system and integrated there to produce responses by the individual. The responses should result in adjusting the density as required in order to forestall any immediate risk to group survival through food shortage and, in the longer run, any impairment to the regeneration of future supplies.

To be effective as indicators of current population density, chemical messengers such as pheromones must presumably decay with characteristic half-lives, so that their information keeps up to date; but their information value could be increased if they contained several factors with different decay rates, so that they 'aged' with time, for a matter of hours or days perhaps, as well as attenuating with distance from source. In some of the mammals at least, pheromones probably incorporate a personal identity as well. In general, persistence allows their concentration to rise as long as production is temporarily exceeding the rate of decay—as may have happened in the *Biomphalaria* fertility experiments (p. 273) in which the oviposition rates were found to rise for three days running. Part at least of the *Biomphalaria* pheromone complex was shown to be stable enough to survive the characterization procedures of ultra-filtration and dialysing, boiling, and pronase digestion for 3 hours (Thomas *et al.* 1975: 3).

A perspective on the cybernetic complexity of the feedback is provided by considering three beakers, two of which each contain a single young snail and 50 ml of water, and the third, two young snails and 100 ml. We have learnt to expect that the two living together will be growing faster than the isolated ones; and that all it would be necessary to do to even things up would be to pour one of the isolated snails, water and all, into the other one's beaker. If all four snails were producing an ephemeral pheromone at constant rate throughout, its concentration would be the same in all three of the original beakers and both the final ones, after the merger. All the snails would of course continue to experience an indentical density of 'snails per litre'. Only if the individual can distinguish 'own pheromone' from 'alien pheromone' could the pheromone system sufficiently inform, and thereby control, the snails' growth and fertility rates in the way that has been observed. If this faculty does not exist, then the pheromone concentration will do no more than indicate snails-per-litre (or snail biomass per litre), which would be immaterial in this imagined experiment, and leave the snail to derive its information about the presence or absence of companions through some different perception channel. But given the three qualities I have postulated—transience, ageing and individuality, a water-borne pheromone could go a long way towards providing an index of in-group size or population density.

In practice one pheromone may not as a rule be sufficient. Like most snails, *B. glabrata* slides over the substrate on its ciliated foot, secreting a mucus trail as it goes. Townsend (1974) has shown that, as was already known of some other kinds of snails, *B. glabrata* possesses a keen sensitivity towards the slime trails it comes across, distinguishing conspecific from foreign ones, and often tracking conspecifics for a short way in the forward direction. Like secreted pheromones, the trails can probably provide useful current information, some of it, as Townsend found, having a persistence of less than half an hour, though other characteristics, such as the breadth and thickness of old trails, stay readable for many hours. Trails could yield another index therefore, of 'snails per unit area' rather than unit volume, and thus feed a separate additional dimension into the nervous system's estimate of density.

17.2 The Allee effect

These speculations do not throw any light on why, in Thomas's experiments, the isolated snails were so often found at a disadvantage. The phenomenon itself is far from new, however, and has appropriately been named the 'Allee effect' (Thomas 1973), after its well-known discoverer and exponent, W. C. Allee (who lived 1885–1955) of the University of

Chicago (Allee 1931; Allee *et al.* 1949: 393–419). Allee referred to the handicap as undercrowding, to associate it and contrast it with the better-known consequences of overcrowding.

The Allee effect appears to be widespread though far from universal, and to conform closely to a pattern. The classic example is provided by the flour-beetle *Tribolium confusum*, in which it was first identified nearly 60 years ago, and later exhaustively investigated, with great benefit to our understanding of population homeostasis in insects. R. N. Chapman (1928) was the originator of the laboratory bottle-culture technique for *Tribolium* which was used in Wade's experiments (see Chapter 15.3). Like *Biomphalaria* the flour beetles condition their medium, and to maintain a habitable environment it has to be renewed at regular intervals. The old medium can then be sieved through millers' bolting silk, which allows all the growth stages from eggs to adult beetles to be found, extracted, counted and transferred as required to the new medium. It was Chapman who made the then novel discovery that the populations were, as we say, homeostatic, in as much as however many pairs of beetles he used to seed a culture, the surviving progeny always reached the same equilibrium density per gram of flour after two or three generations. For the beetles the culture bottles with their renewable food supplies had a specific carrying capacity, and, as later researches were to show, they as consumers possessed a battery of density-dependent adaptations for preventing their numbers from exceeding it, including variable fecundity, cannibalism, and secreting their toxic pheromone. No mortality occurred from starvation; and no other organisms were present to cause predation or disease.

Allee (1931) used Chapman's data to show that the reproductive rate, during the first four weeks of rapid population growth in newly seeded cultures, was highest not in the cultures with the minimum number of founders (one pair of adults), as might have been expected since they had the lowest starting density and the greatest scope for multiplication, but in those that had two founding pairs instead. When the numbers of progeny (eggs and larvae) present were counted in cultures 11 days old, the two-female colonies held roughly five times as many as those with only one female; and in terms of eggs and larvae produced *per female* they also outdid the females in cultures with 4, 8, 16 and 32 pairs in the same volume of flour, which suffered increasingly from the effects of crowding (Figure 17.3).

Chapman's work was confirmed and extended by Park in Allee's

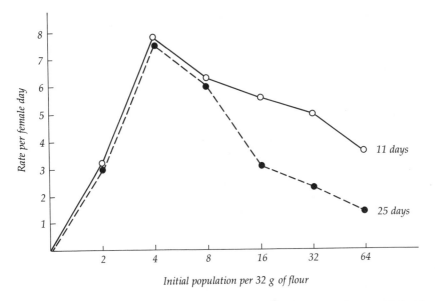

Fig. 17.3 *The Allee effect. Egg-laying rates of* Tribolium *females in cultures seeded with 1 to 32 pairs of adults in 32 g of flour. The highest individual fertility occured in cultures with two females: it was about 2½ times that of the 'undercrowded' lone females. Cultures at higher densities were increasingly affected by crowding. (Eventually all the cultures reached the same equilibrium density.) From Allee (1931), using data from Chapman (1928).*

laboratory in Chicago. He concluded at first (Park 1933) that there were two simple opposing factors at work. He found that repeated copulations stimulated the females' egg-laying, and assumed this effect would be weakest in the one-pair cultures because the rate of encounter between individuals would be lowest, with only two per standard volume of flour; and that it would grow in strength with increasing numbers. He also found that the adults ate eggs, and assumed that the higher the adult density, the greater the egg mortality from cannibalism would be. The optimal compromise must therefore be, as his experiments seemed to show, in cultures with two founding pairs.

Later work at Chicago and elsewhere was to show that, while the copulation factor was responsible for the undercrowding handicap at low densities, the conditioning of the medium caused female fecundity to fall, making it inversely dependent on density, and this was a much more effective homeostatic factor than cannibalism. Conditioning could also affect the rate of larval growth and survival.

The Allee effect and the inversely density-dependent fecundity effect were independently discovered by MacLagan (1932), who went on to perform similar experiments with other cereal-eating insects, one of them the grain-weevil *Sitophilus granarius*, a species that did not show the Allee effect. The female inserts each of her eggs into an intact grain of wheat, and there the larva afterwards develops; but MacLagan found she 'never uses the total number of grains at her disposal, for purposes of oviposition, no matter how hard pressed she may be for space'. In fact crowding could not be made to induce the females to use more than half the grains; and though he did not say so, the conclusion is plain that they are programmed to leave enough seed-corn for next year's crop. He appears to have been the first population ecologist to appreciate just how profound a truth these culture experiments reveal, namely 'that the organism itself imposes the ultimate limit to its own abundance when all other factors (biotic and physical) normally inhibiting population increase, have failed'.

When writing *Animal Dispersion* 25 years ago I could make no sense of the Allee effect, and merely said it seemed premature to conjecture about its adaptive significance (1962: 560–2). It was only when preparing the ground for this and the two preceding chapters on population structuring that I thought seriously about it again and saw what its possible role might be. The *Biomphalaria* snails, for instance, are hermaphrodite, and no doubt they normally copulate with another snail before laying their eggs; but in the absence of a mature partner they can probably fertilize their own. To do so is presumably a last resort because self-fertilization is the fastest possible road to gene-fixation: only half the heterozygosity the parent possesses is bequeathed to each offspring, and 50% more double-recessive genotypes appear in the progeny than would result from an outbred mating. The mortality falling on some of these double recessives is likely to be fairly heavy, and the loss of alleles that have little chance of survival to maturity except as heterozygotes, correspondingly speeded up. If the progeny were then to interbreed among themselves the same elimination would continue at a slower rate into the following filial generation. Self-fertilization is no doubt valuable in an emergency, especially when, in a sparse population, it is only a minority of adults that fail to find partners, and they are still able to contribute something towards recouping numbers. Nevertheless the inherent dangers are attested by the commonness of adaptations to prevent 'selfing' that have evolved in hermaphrodite animals and plants.

Similar consequences, though with a lower fixation rate per generation, would follow from the inbreeding of a stock founded by a single pair of *Tribolium* or any other two-sexed species. Considering the long-term survivability of the average deme it is probably an advantage that individuals which have found themselves cut off, and with an extremely limited choice of mates, should be programmed to go slow on reproduction for the time being, on the chance that immigrants may yet arrive and improve the prospects. Very isolated in-groups are perhaps most often likely to be affected by this kind of restriction, and their best course might be to maintain only the minimum stock needed for keeping possession of the habitat until immigrants appeared. A possible 'straw in the wind' is Park's (1933) finding that repeated copulations are a strong stimulant to the rate of egg-laying in *Tribolium* females. Crombie (1943), who knew this and deliberately seeded his *Tribolium* cultures with well-copulated females failed to encounter the Allee effect at all.

Many of the higher vertebrates, as exemplified by the red grouse, carry their pursuit of genetic excellence far beyond the avoidance of close inbreeding, through the process of social selection which normally permits only the socially successful genotypes to breed and excludes all others. Whether an Allee effect applies to any of them when the choice of mates becomes unacceptably small we do not know; but it sometimes looks as if vertebrate species threatened with extinction, like the California condor or the whooping crane, once their populations have fallen sufficiently low, lose the capacity for making a comeback. A few great auks and passenger pigeons lingered for years after their commercial exploitation had ceased to be profitable; and the Greenland whale has hung on as a tenuous remnant for much of the 20th century. There may of course be alternative explanations of their failure to increase, among them a continuing but surreptitious attrition at the hands of man.

Our certain knowledge of the Allee effect is in fact confined to a variety of laboratory animals. It could conceivably be a rather extreme manifestation of the individual animal's reaction, in the wild state, to a lack of quality when faced with choosing a mate, or to being in an in-group of suboptimal size. Given that population structuring exists and serves a genetic function, it is predictable that mating-group size and gene flow must be regulated at or near the optimum level for retaining potentially useful variance between individuals and between in-groups; and that the programs for achieving this will contain prohibitions as well as preferences about behaviour.

With these reflections in mind I went to check on Allee's views about the possible adaptive functions of the undercrowding handicap, in his last publication on the subject (Allee *et al.* 1949: 403–7). To my delight and surprise (considering it happened so long ago), I found he was cautiously feeling his way towards an explanation in exactly the same direction as I was. He was closely in touch with the work of Wright who had long been his colleague in Chicago (Wright, personal communication); and here is the essence of his thoughts. 'The relation between population size and density is mentioned here because various ecological aspects of evolution are both neglected and important'. He gives some references to Wright's publications, and continues: 'We are especially interested in the slower rate of evolution in oversmall populations as contrasted with those that are somewhat larger. . . . Omitting all details, the important point for general ecology is that one of the primary controlling factors in the rate of evolution, under many conditions, is the number of animals in the inter-breeding group. . . . As with many phases of individual survival, the rate of evolution is highest, other things being equal, in populations of optimal size, as contrasted with those that are overlarge or oversmall. Here the rate of evolution becomes a criterion for the existence of natural cooperation.'

Nothing more explicit than this is said: the connection he sensed between the Allee effect and Wright's hypothesis was just a hunch, not ready to be defined and clarified. It still needed thought and investigation that he did not live to undertake. But it was no shot in the dark: rather, a remarkable intuition.

Summary of Chapter 17

1 Our hypothesis predicts that all vivivores, whether social or not, ought to require self-regulating populations, otherwise they would sooner or later overexploit their food resources, and/or decline to extinction. As introductory examples of animals in a presocial state two species of gastropod molluscs are chosen, one of which is the pond snail *Lymnaea elodes*, studied experimentally by Eisenberg (1966). These snails live at high density in shallow inshore water which is liable to recede in summer; but before this happens they mature and breed, and afterwards die. The young develop rapidly as long as there is water, after which they burrow and stay dormant till the following spring. Their food consists of rooted vegetation and detritus.

Part of the study-habitat was partitioned into large pens of uniform size, and sufficient in number to permit 4 replicates of each treatment. There were also 4 treatments. In (1) the natural populations were reduced $\times \frac{1}{4}$; in (2) they were increased $\times 5$; in (3), as a control, no changes were made; and (4) was like (3) except that extra food was provided. In spite of the $\times 20$ density difference between (1) and (2) both they and treatment (3) all produced similar average totals of eggs and young. The fed populations produced 25 times as many eggs as the controls but mortality had reduced the progeny to only $\times 2$ at aestivation time.

Eisenberg concluded that population regulation had taken place, in relation to food supplies; but since there was no evidence of food exhaustion or serious damage to vegetation, it was 'relative rather than absolute'. With our hindsight we can confidently conclude, therefore, that the snails displayed a normal, competent homeostasis.

The second gastropod is *Biomphalaria glabrata,* an intermediate host for the debilitating human parasite *Schistosoma mansoni.* It is cultured and studied to aid in the search for preventive remedies, possibly including a pheromone for controlling the snails themselves. Most aquarium snails make slower growth and have lower fecundity when overcrowded; but Thomas (1973) and his colleagues showed that these effects in *Biomphalaria* are responses attributable to depletion of resources and/or the 'conditioning' of the water by excretory products. Unexpectedly he found a pheromone that actually stimulated growth among snails in an uncrowded group, as compared with snails in isolation. When they were grown over a standard period in cultures of 1, 2, 4, 8 or 16 in a similar multiple series of water volumes, with sufficient food at all times, it turned out that 50–100 ml of water per snail would produce the best results; but in fact 2 snails together would grow much faster than 1 by itself, and stepwise faster than those grouped in higher multiples.

Analysis shows that the pheromonal information necessary for this phenomenon must include (1) a predictable decay rate or half-life, and preferably a mixture of components with different half-lives, serving to indicate its age or distance from source; and (2) a distinction by the recipient between own and others' pheromones.

2   This phenomenon of handicap to growth, maturation, etc. in isolated individuals or isolated sexual pairs among small laboratory animals is well known. It was discovered by Allee (1931) and is called the Allee effect. He found it when analysing Chapman's (1928) results from the first *Tribolium* population experiments. Chapman had not only established

the standard culture technique for *Tribolium*, but had shown that however many adult pairs he used to seed his cultures the latter always reached the same equilibrium density per gram of flour within 2–3 generations. As described in Chapter 15.3, *Tribolium* is efficiently homeostatic with respect to the carrying capacity of its habitat; but, being an insect, its feedbacks are at least in part social, and it possesses a variety of density-dependent control mechanisms.

There is no accepted explanation of the Allee effect. I suggest it is a by-product of genetic structuring, i.e. of optimal mating-group size and rate of gene flow. Breeding in too small groups, e.g. by single hermaphrodite molluscs, or single-pair cultures of *Tribolium*, may therefore be inhibited, or deferred (by slowing growth, maturation and resource consumption), pending the possible later arrival of immigrants and averting the side-effects of rushing into it at present. In Allee's own last publication on the subject (1949) he expressed an intuition that his colleague Wright's structuring theory would eventually offer a solution to the puzzle.

CHAPTER 18.
POPULATION STRUCTURING AND
HOMEOSTASIS IN PLATYHELMINTH
PARASITES

18.1   The trophic status of parasites

This chapter takes a restricted and necessarily cursory look at one of the main phyla of metazoan parasites to see whether, and how, the structuring and self-regulatory hypotheses apply to them. I have chosen the Platyhelminthes because in a book of this kind one can only sample the vast animal kingdom, and various members of this phylum have been subject to highly informative experimental research in recent years. Not without reluctance, I decided not to include the Nematoda as well, on the grounds of the extra time and delay their study would have entailed. In their phylogeny and morphology both groups are more primitive than the annelids and molluscs; in their behaviour they are both unquestionably presocial.

Parasites are usually defined by a combination of three criteria. They must be dependent for a least part of their lives on a host organism, not only (1) for nutriment, but also (2) for their habitat; and (3) they must impose a net burden or cost on the host. The definition is not a rigorous one, partly because the popular conception of parasitism is more intuitive than logical. We are inclined to apply different standards to animals and plants as hosts, and we also appear to pay some heed to the relative degeneracy of the dependent partner. Fleas that inhabit the coats of birds and mammals and feed on the host's blood are seen as ectoparasites, whereas aphids that suck the juices of plants are merely specialist herbivores. Plants like fungi and broomrapes that lack chlorophyll and draw their nutriment from living host-plants are recognized as parasites, whereas insect larvae, nourishing themselves while concealed in fruits or taproots, are not. Elsewhere the definition breaks down for quite a different reason, because parasitism at some points is found merging into predation (see p. 39–40).

Parasitism is immensely successful as a way of life, chiefly because it affords to animals, in particular, the means of feeding on other animals without actually killing them, on the same principle that herbivores adopt in consuming the tissues of living plants. The trophic parallel is close, and undoubtedly the same danger of overtaxing the food resource exists for parasites as for any other vivivores. Viewed from a trophic standpoint, animal parasites appear to be a logical extension of the guild of carnivores, supposing for the moment that 'carnivore' can be made analogous to 'herbivore', and broadened to include not only predators but all consumers of live animal food.

Zoologists do not generally appreciate the immensity of this all-permeating and unjustly despised 'industry' of parasitism. Parasitologists on the other hand have come to assume, from what they already know about parasite distribution and taxonomy, that the vast majority of free-living metazoan animals, if they could be searched, would be found to harbour parasites, and most species of hosts would harbour several kinds. The difficulty of evolving a viable symbiosis between two different kinds of organism is such that all parasites are to some extent specialists in their exploitation of hosts, and some narrowly so; and their specialization must often have been sharpened further through competitive exclusion between one parasite species and another. The conclusion that emerges from these simple premises is that the animal parasite industry, were it fully explored, might be found to have reached a species total not greatly inferior in size to the total of all free-living animal species put together. In total biomass, the carnivores, with parasites included, would obviously be very much less than the herbivores. For one thing, parasites are of necessity smaller than their hosts; and for another, they belong to the secondary or higher trophic levels of consumers, and on that account might comprise only 10–20% of the mass of herbivores, if the usual conversion ratio for free-living carnivores holds true for parasites as well.

I shall not be considering here the microparasites, which include viruses, fungi, and the parasitic bacteria and protozoa. Viruses in particular tend to undergo rapid multiplication soon after infecting a host, and to be transmitted from there to new hosts before their temporary infection dies out. In contrast to them, the larger metazoan parasites including the helminths have longer generation cycles, during which the members of some species need to pass through two, three, or rarely more different kinds of host. A full cycle in these digenean or multigenean forms also includes an initial asexual phase of reproduction (see Clark 1974), undergone by larval stages within the body of the first 'intermediate' host (and sometimes repeated in succeeding intermediate hosts). This is followed by a potentially longer period of sexual reproduction after the final, 'definitive' host has been reached.

Sexual reproduction does not usually result in the multiplication of helminths within an individual host through the retention of their progeny. Instead, the parasites produce thousands or even millions of resistant eggs which pass out of the host into the environment, and start new life-cycles from 'square one'.

Prolific reproduction is of course a response to the enormous odds against survival for minute, often inert particles, scattered randomly by the host yet capable of development only if they reach another receptive host; and there are often quite long odds as well against subsequent transmissions of larvae from one host in the series to the next. It is a strategy for organisms with very little control over the density of their offspring. Helminths are nevertheless capable of varying their ovulation rates in response to changing circumstances, either in the host or the environment outside. Many have evolved annual or seasonal life-spans and generation times. But the reason why vertebrates are so popular as definitive hosts is probably that they offer the chance of longevity, plus, in birds and mammals, a homeothermic habitat ideal for egg-production.

It is in their own interests for parasites to minimize the harm they do to the host, and not to inflict damage that brings no compensating gain to their own fitness. The lighter the basic load on the host's vitality, the greater its carrying capacity will be as a habitat for egg-production. Well-adapted parasites can be expected therefore to produce the minimum of toxicity, limited to unavoidable waste products, in forms that the host can dispose of easily without being obliged to let anything accumulate. A highly beneficial feature of egg-production as a method of propagation and dispersal, compared to the alternative of proliferating within the existing host, is that only a fraction of the nutriment and energy the adult imbibes from the host is catabolized within the host's body; most of it is converted into body-growth or exported in the eggs. Thus the dominant item of cost that falls on the host in a mature symbiosis, and the factor limiting the host's carrying capacity, is not toxicity but nutritional demand; and if that were to exceed the sustainable maximum, the host would starve and its carrying capacity would diminish.

One must not lose sight of the fact that parasites are by definition a burden to the host, and the latter would be a fitter individual if the parasite were absent. Infection always depresses the host's fertility or life-expectancy in some measure, making it more likely to succumb to a predator, or a concomitant disease, or malnutrition in hard times, or diminished social status. Some helminth and crustacean parasites regularly take such a levy of nourishment that the host can afford to produce no gametes of its own, and remains sterile. That can be a viable strategy only if the parasites somehow limit the incidence of their infection so that a sufficiency of unaffected fertile hosts are always spared to perpetuate their stock.

It is rather common in helminth life cycles for the intermediate host to be eaten by the definitive host, which means that the latter has to be a predator. Where this happens, the parasites are normally taking advantage for their own transmission, of a predator-prey relationship that would exist anyway, whether the prey species were parasitized or not. It is to the parasites' advantage therefore if they cause their host to fall an easy victim at the appropriate time. In a stable predator-prey-parasite relationship the predator should assume the main responsibility, through its homeostatic program, for limiting its toll of prey to a sustainable level, taking account of the fact that disadvantaged prey are the easiest ones to catch. If so, the parasite may not be putting the survival of its intermediate host population at any increased risk. Adult helminths living in a *definitive* host on the other hand, if they cause the host's death, put a premature end to their own lives, and only if their demic population were running at an excessive level could that be an advantage to the deme.

Parasitic helminth populations have two clearly distinct structural tiers, first inside an individual host and second, collectively in the host population. In the majority of species it would be reasonable to equate the group of individuals that share a single host with the in-group (see p. 220), and the collective population with the deme and the higher structural tiers beyond, in parallel with the tiers of the host population. The only incompatibility would arise in parasite species in which one or two (or any very small number) of parasites are capable of preempting a host-individual and preventing additional recruits of their own species from gaining entry. The critical question is how many individuals it takes to make an in-group. In the limiting case where most or all individuals are isolated singly in individual hosts, the in-group tier effectively disappears from the fabric, and homeostasis must get along without it. In recent parasitological literature these two normal tiers have been designated the infrapopulation and the suprapopulation respectively (Esch, Gibbons & Bourque 1975).

Parasites stand to benefit by regulating their numbers at both levels. Ideally they need (1) a high potential fecundity, sufficient to overcome most if not all the local extremes of risk that must occur, because juvenile mortality is so great and uncontrollable, and because the host populations may fluctuate independently for reasons over which the parasites have no control. They need to couple a sufficient egg-production with (2) the regulation of their own established populations during each parasitic stage of their lives. Like other potential killers they must guard against overtaxing their host species to their own detriment. They need homeostatic programs just as much as any other vivivores if they are to keep their demands within sustainable limits, and to be able to respond to demographic and nutritional feedbacks from the hosts.

A host can afford to have a limited amount of nutriment consumed by parasites, and there must therefore be at any given time a particular rate of consumption that maximizes the productivity of the parasite in-group living in a single host, and/or the productivity of the parasite deme that is exploiting a population of hosts. The parasites are presumably programmed to reduce their food demand and cut down their egg-production as soon as the host shows signs of malnutrition or of suffering any reduction in its food intake. Without that precaution the symbiosis would be inherently unstable: the parasites would exacerbate the host's starvation and precipitate a rise in host mortality, and a still sharper rise in their own on account of the fact that hosts with the largest parasite burdens would normally be the first ones to die. Small initial abuses would thus be magnified in their consequences. Some such knock-on effect may at times be unavoidable; and if it were not damped by the parasites' homeostatic responses their populations would often be caught out, and would crash. On principle, if a host population is in decline the well-adapted parasite should react to minimize its own contribution to the damage. Conversely, when improvements occur in the host's nutrition the parasite should exploit them by increased fertility or growth in size or both.

For feedback information on the current in-group size within a single host, the parasites could well rely on emitting a pheromone which circulated in the host's internal environment and could be quantified by chemoreception. There is evidence as we shall see that reductions of numbers or biomass can be brought about within the in-groups of various types of helminths; and suggestions that pheromones may be involved have already been made, for instance in certain cestodes (Halvorsen & Williams 1968) and parasitic nematodes (Schad 1977). It is generally assumed that *nutritional* feedbacks also occur in helminths and are responsible for various well-known crowding effects; but unless it becomes possible at some future date to culture helminths *in vitro* instead of in a living host, actual proof of the existence and nature of homeostatic feedbacks may not be obtainable.

That there may also be a feedback from the parasites' suprapopulation or deme is perhaps more difficult to credit because it would have to reach the senses of internal parasites from outside the host; but it seems absolutely essential all the same. The necessity is most clearly apparent with parasites that castrate their hosts, or encompass their deaths by predation, making it vital that the incidence rate (proportion of hosts infected) should be buffered to a level the host population can stand. A demic feedback then becomes all-important. But it is also desirable even for species that actually conserve and protect their individual hosts from death by overexploitation. Reproductive output normally budgets for a surplus, otherwise populations would decline. If there were no feedback from the deme, therefore, the tendency would be for the incidence rates of parasites to rise to saturation, and for hosts to become loaded up to their maximum capacity. In a minority of helminths something like this may actually occur, but in the majority it does not, and their population structure provides strong evidence for a demic feedback, as will shortly appear.

The most likely density indicator to reach the adult helminth from outside appears to be the arrival of infective larvae, attempting to enter and establish themselves in the infected host. The demographic information such larvae are capable of carrying may sometimes be a bit out of date: the larvae may have grown from eggs laid weeks before, or even the previous year. But if they arrived often enough to provide reasonably reliable samples they could conceivably convey a 'transmission' feedback from the deme to its incarcerated adult members. As explained later, such larvae reaching an infected vertebrate host are often destroyed, because the host is already immunized and can suppress them with its antibodies. But the surge of antibodies directed against the parasites' own species which the host releases into its blood-stream, seems very likely to be detectable by the established adult worms inside the host as well. The size and frequency of such discharges would appear to provide the type of quantitative signal required. If the adult helminth had an innate standard and could tell whether the present transmission rate were above it or below, it could respond homeostatically by cutting down or speeding up its egg-production. Conceivably too, if the arrival of larvae were a rare event and the particular host had still some carrying capacity to spare, the adults might stop their own secretion of antigens which serves to keep the host's defence system constantly on the alert (see p. 294); and that would give larvae arriving later and finding the host unprepared, a chance to slip in and reinforce the in-group already established. But I anticipate the development of my theme.

It may be no coincidence that many years ago the carcinologist Geoffrey Smith (1909: 99) discovered that the crustacean semi-external parasite *Sacculina* attracts larvae of its own species, which fix themselves round the entrance to its muscular mantle cavity; but they never produce sperm, as the minute 'complemental' males of some other cirripedes do. Instead they rapidly degenerate *in situ*. This observation has never to my knowledge been explained; but it could be a close parallel to the helminth demographic feedback postulated here (Figure 18.1).

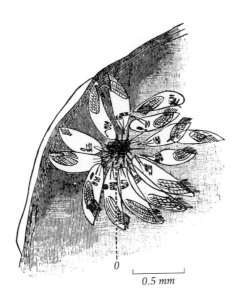

0

0.5 mm

Fig. 18.1 *Fourteen cypris larvae attached radially round the reproductive opening of a young* Sacculina—*a partly external crustacean parasite of crabs. The larvae rapidly degenerate, and may possibly be bringing a population-density feedback to the mature parasite. The parasite and its host are illustrated in Fig. 18.2 (p. 305). From Smith (1909).*

Parasite populations can survive only if they have the upper hand and play the controlling role in their symbiosis with at least one species of host for each of their parasitic life-stages. But like most organisms that serve as food for consumers, the hosts have evolved defences for their own protection, and the successful parasites have of course been able to overcome these. The defences depend on a basic faculty, apparently common to all animals, of being able to distinguish between the component materials or parts belonging to one's own body, and similar but foreign materials that have found their way into it from outside. Differentiation between self and non-self is prerequisite to taking defensive action against infection; and successful action means killing the foreign matter if it is alive, and afterwards disposing of the remains.

Identification of self requires that one's own cells carry on their outer surfaces a particular signature of macromolecules which uniquely encodes one's own species, or even each member of it individually. In the invertebrate phyla the body tissues incorporate scattered policing cells; or, if the animal possesses a circulatory system, similar but more mobile cells carried to all parts in the blood-stream. Their function is to recognize unauthorized alien matter, and having done so to take action for removing or encapsulating it. In the vertebrates the system is more elaborate. It recognizes the macromolecular codes carried by alien organisms as 'antigens'. When antigens of any kind are discovered by the circulating lymphocytes a train of action is set off, the progress of which is complex and still partly obscure. It results within three or four days in the manufacture of quantities of an antibody, put together as a particular permutation from an assortment of immunoglobulins so as to match and bind with the antigen under attack, rather as a key matches a lock. When delivered to the site of the infection the collective result is to arrest and agglutinate the invading organisms, whose remains can then be removed by phagocytes.

An important ingredient in the vertebrate system not known to be paralleled in any invertebrate species is that when such an episode is finished and the intruding organisms have been cleaned up, some of the left-over antibody-secreting cells are retained in storage to serve as templates in case they are needed again; and if the same variety of parasite enters the host on subsequent occasions the host can muster the appropriate antibodies more quickly and use them with more lethal effect. If everything has worked out perfectly the host is in fact made wholly or partly immune to reinfection, sometimes for years if not for life. The system is consequently known as the immune system. (For a general reference see Barriga 1981.)

Substances and objects other than parasitic organisms can also act as antigens. Allergies, for instance, are another manifestation of the activity of the immune system. But the main drive behind its long evolution has presumably been the increasing defence it has given against parasites. Yet in spite of it, parasites have continued to abound. Each new bid the hosts have made to get rid of them has been circumvented, at least by some species of invaders.

The reason for the success of this evasion is that selective pressures are always stronger on the parasites' side. If they are prevented from infecting their host they become extinct; whereas if the latest dodge the host invents to keep the parasites out is evaded yet again, life for all concerned merely continues as before. An exactly parallel evolutionary relationship holds for predators and their prey. If the prey manage to foil them by some new method of defence the individual predators may come under intense selection to get the better of it; but the future survival of prey stocks is scarcely ever threatened by well-regulated populations of predators. Just the same, vertebrate hosts do reap a great advantage from their immunity because it means in many cases they have only to suffer from a particular parasite once in their lives, or perhaps once in several years; and when they have recovered from it they are free of that particular burden as long as the immunity lasts.

18.2 The schistosomes

Schistosomiasis depresses the health and happiness of an estimated 200 million persons in tropical countries, and has been the target of intensive research for many years. The interaction between helminth and vertebrate is consequently better understood for the several *Schistosoma* species that infect man than for any other metazoan parasites. There is a vast literature, ultimately bearing on the prevention or cure of the disease; and the rapid advance of immunology has now raised the possibility of being able to immunize people at risk by vaccination. The discoveries made in this direction are of particular relevance to our subject here.

All the species contributing to the disease have similar life-histories and similar clinical effects. The one chiefly referred to here is *Schistosoma mansoni*, which is widespread in tropical Africa and South America. They are digenean trematodes which when adult normally live inside the host's hepatic portal vein, or in its mesenteric tributaries, this being the part of the venous system that gathers up dissolved products of digestion from the intestines and carries them to the liver. Man is not the schistosomes' only definitive host; *S. mansoni* occurs naturally in other primates, and laboratory populations of it are maintained in monkeys, mice, rats and hamsters.

The eggs pass out of the host chiefly by dissolving their way through from the blood-stream into the gut and then being evacuated in the faeces. If they fall or wash into water, a minute ciliated larva emerges which has enough fuel to keep it going for a few hours while it swims in search of a snail. If successful, it again uses histolytic enzymes for penetrating the host tissues, and makes its way into the snail's big digestive glands. There it becomes a sporocyst, which grows and undergoes a complete reorganization in preparation for fission. Eventually it breaks up to start a second generation of sporocysts, and they repeat the growth and fission but produce this time large numbers of a second swimming larva, the cercaria. The parasites have been living in the snail for about two months by the time the cercariae emerge, in enormously increased numbers. Individually, the larva has a forked tail for swimming; it is just visible to the naked eye, and can live for a couple of days. If it finds a human or other primate host while it still has sufficient energy left, it can penetrate the skin in a matter of minutes, using enzymes from its penetration glands and strong muscular contractions with its body and oral sucker. The tail is left outside when it enters the host.

The larva is now a schistosomulum. It reaches a lymph capillary in the skin, and is carried via the heart to the lungs where it spends several days, and can be found if necessary for experimental purposes. From the lungs it passes (somehow) to the portal vein.

Unlike the majority of platyhelminths, *Schistosoma* has separate sexes. They reach maturity and mate about five weeks after entering the definitive host. Pair coupling is permanent, and the female generally remains held in a wide groove on the ventral side of the male's body. Both are small, the male about 10 mm and the female 17 mm long; the male has a thicker body, and both are essentially cylindrical and vermiform (discounting the male's groove). In experimental hosts, worms can be recovered and counted at autopsy by a technique that washes them out of the portal vein. Schistosomes can live in man for many (perhaps even 20–30) years.

Much of the account that follows stems from research done at the National Institute for Medical Research in London. It was conveniently reviewed by the principal workers Smithers and Terry in 1976, where references to the original papers can be found.

Experiments by Cheever and Powers (1972) compared sets of laboratory hosts that were given light and heavy infections of 100 and 600 cercariae per host respectively. Up to about the 12th week similar numbers of eggs per worm-pair were found in the hosts' faeces in both sets; but between the 12th and 27th weeks two-thirds of the adult worms died in the hosts that were heavily infected, and the eggs laid per worm-pair decreased. In the lightly infected hosts there was no parasite mortality, and the eggs laid per pair actually increased. This confirmed a phenomenon that had been known for some time, namely that there is often a self-cure by the host in the early phase of schistosomiasis, after which a reduced number of worms survive and may preserve a steady state more or less indefinitely. It had been assumed initially, as the term 'self-cure' indicates, that this regulation was induced by the host and was attributable to the action of the host's immune system, though it appears difficult, considering the persistent selective pressure on all hosts to get rid of harmful parasites, to explain why, if it could kill some of the worms, the 'cure' stopped short of killing them all.

If an experimental host carrying such an established schistosome in-group is 'challenged' with a re-infection of cercariae, it is found that the latter fail to survive the schistosomule stage; and it can be shown that this *is* due to the host's immunity reaction. The host has in fact become immunized against reinfection, although it continues to harbour adult parasites. This condition was described as concomitant immunity (Smithers & Terry 1969). It was a puzzling situation; but the authors were quick to remark that it was more beneficial to the parasites than the hosts. From the parasites' point of view it leaves the host supporting an in-group sufficiently small to maintain an effective egg-production without serious risk to the host's life, and at the same time it prevents the entry of parasite recruits that might upset what seems to be a stable equilibrium.

The most puzzling feature was how the adult worms serenely survived in a host who possessed antibodies that would kill schistosome larvae; and the most obvious solution seemed to be that the adults had somehow acquired a disguise, and were displaying antigens similar or identical to those of the host.

The laboratory kept stocks of three definitive hosts of *S. mansoni*— rhesus monkeys, mice and hamsters; and an important clue was obtained by transplanting adult worms surgically from one species of host to another. The experimenters found that uninfected monkeys would readily accept worms from infected monkey donors, and the transplanted worms would be so little affected as to continue producing eggs in normal quantities. But worms from the other species of donors, implanted into monkeys, suffered setbacks which depressed their egg-production for the first five or six weeks, after which it returned to normal levels once more.

Another equally important result showed that the transplanted adult worms induced concomitant immunity in their new monkey hosts, proving that though they themselves might be disguised, they were still secreting schistosome antigens into the blood-stream and thus stimulating the host to go on making antibodies, and so exclude secondary infections.

If the adult worms really did disguise themselves with coatings of antigens matching those of a variety of host species, and if they could change their disguise to match a new kind of host within 5–6 weeks after being transplanted, the most obvious explanation was that they were acquiring new antigens direct from their host. The next, more ingenious step in the research was therefore to sensitize 'virgin' monkeys against mouse antigens, which could be done by giving them small injections of homogenized mouse tissue; and then to implant in them adult schistosomes taken from mouse donors. If the surmise that the worms in a mouse host were coated with mouse antigens was correct, the sensitized monkey ought to react against them with its anti-mouse antibodies. The experiment was done, and the worms were duly killed by the new host. Worms similarly transplanted (as controls) into normal non-sensitized monkey hosts, from mice and other donors, survived as before (Smithers, Terry & Hockley 1969).

These were most skilful and convincing experiments. It has since been shown that invading schistosomules acquire their first host antigens during the days they spend in the host's lungs, and simultaneously they lose the capacity to bind to anti-schistosomule antibodies. These discoveries are a product of new techniques that allow schistosomules, recovered from the lung, to be cultured in media for long enough to prove, for example, that they can pick up the antigen determinants of the A B O blood groups from human blood and, if tritium labelling is employed, that molecules from the medium have actually been transferred to the parasite tegument.

With insight derived from studying population homeostasis in other animals, one can without hesitation attribute these adaptations wholly to the initiative of the parasite, exploiting the immunity system of the vertebrate host to its own advantage. The stable symbiosis that can be established with an individual host shows how successful the parasite program can be when it works smoothly, and how powerless the host is to resist or intervene. There are obviously considerable gaps in our knowledge still, the most important perhaps being how the 'self-cure' process is effected without obvious resort to competition and the hierarchy. The worms have limited sensory faculties, but competition and the suppression of some individuals by others could conceivably be induced by pheromone signals. There is some evidence from other helminths to support this possibility, as noted earlier. All the pheromone would need to do would be to impair selectively their ability to maintain an adequate personal disguise, so causing the weaker parasites to succumb to the host's antibodies. If so, and if the pheromone could be isolated and characterized, it is also conceivable that it might act as a curative purge when injected into human sufferers from the disease. Under natural conditions its effect would be expected to cease as soon as the parasite in-group had been reduced to a size appropriate to the carrying capacity of the host; but for therapeutic purposes it might be persistently administered until, one likes to hope, every adult worm had been dislodged.

18.3 Tapeworms

A so-called crowding effect has frequently been observed among tapeworms, inhibiting normal growth, maturation and egg-production of some or all members of an in-group; and it is obvious that nutrition could be implicated as an underlying cause. As far as is known, tapeworms are narrowly restricted in their ability to metabolize carbohydrates to the sugars glucose and galactose; and this has been represented as a possible dietary factor limiting their growth and development (Read & Simmons 1963: 282).

When feeding conditions are good, on the other hand, their growth rate can be no less phenomenal than their fertility. Avery (1969), in experiments with tapeworms that are parasites of ducks, fed some cysticercoid larvae of *Dicranotaenia coronula* to previously uninfected domestic ducklings 4–5 weeks old. Only 9 days later he found that some of the larvae had grown to maturity and begun to lay eggs. The largest one weighed 345 mg (wet weight) and was 230 mm long. He calculated, assuming a constant exponential growth, that it had been doubling its weight every 9 h since Day 1. Another species, *Sobolevicanthus gracilis*, similarly administered to a parallel group of ducklings ranged up to a maximum of 126 mg during the same 9 days, and some of them had also matured. Both species are short lived, and 48 days after the experiment began no worms at all were left in either set of hosts.

A single inoculum of 20 larvae had been fed to each bird at the start. Not all became established and some of those that did soon died. The resulting in-groups discovered at autopsy, while they varied in size, were mostly small; but it was notable in both species that great differences had occurred in the rates of growth of individual worms so that the largest was often ten times the weight of the smallest in the same host. This did not appear to Avery to be an effect of crowding because the smaller worms were just as backward whether the group they were in contained many or few. The facts suggest to me that a type of feeding hierarchy had developed among the in-group members which had the effect of concentrating the resources on the stronger growers and speeding their development.

In the next experiment he used mallard ducks (*Anas platyrhynchos*) as hosts. At midsummer 1961, 24 ducklings 5–7 weeks old, and reared in tapeworm-free conditions, were put into a wildfowl enclosure containing a small pond. The pond was often visited by wild ducks and the small crustacea in the water carried light infections of tapeworm larvae belonging to 7 species. After the ducklings had been feeding in the pond and eating some of the crustacea for 17 days a small sample of them were sacrificed and examined for parasites, and the procedure was repeated at varying intervals for the next 3½ months. The commonest tapeworms found were *S. gracilis* and *Diorchis stefanskii*, and the course their infections followed was roughly the same for each species. On Day 17 a small number of adult worms were already present, but in the next few weeks they were all replaced by what appear to have been roughly equivalent biomasses of stunted immatures. At that stage it might have been thought that the worm in-groups were actually growing to fill the permissible carrying capacity, after which 'crowding' inhibited individual growth, although there was no apparent limitation placed on admitting recruits to the group.

At the halfway stage, on Day 60, no *S. gracilis* and very few *D. stefanskii* were found in the sample; but after that the infections of both picked up again, to levels higher than before. No more *S. gracilis* grew into adults, and egg-production by that species ceased completely. The few adult *D. stefanskii* found also dwindled to zero before the 3½ months were up. Taking the ducks individually during the second period, their worm burdens varied from 1 to nearly 100; but there was no inverse correlation between numbers and average worm size. All the worms were similarly stunted, whether there were few or many.

Such conflicting data suggested there must be other unidentified factors strongly contributing to the results observed. Hoping to throw light on them Avery began a more comprehensive experiment the following year in which he attempted to follow the flow of infection through the intermediate hosts as well as the ducks. A flock of adult mallards were brought into the enclosure in June 1962, but sampling was not started for the time being. An apparatus was constructed with which to make regular estimates of the population densities of the intermediate hosts, *Cyclops strenuus* (a copepod) and *Cypria ophthalmica* (an ostracod); and large samples of these were examined microscopically at the same time to quantify the incidence of cysticercoid infection among them.

Collection of data began in February 1963 and was repeated by sampling every 3 weeks thereafter. The first ducks sampled carried an average of 120 worms each, belonging to four cestode species, the commonest being *S. gracilis*. Between February and July the *Cyclops* population doubled (to reach over a million) and the *Cypria* population increased 30-fold (to 350 000). In July the infection rate of *Cyclops* was still much as in February, at less than 1 in 200, but in *Cypria* it had gone up from $< 0.5\%$ to $> 30\%$, and it reached 50% in August. The menace of this vast increase becomes clear when it is revealed the *Cypria* individuals are castrated by having a cysticercoid to carry.

On 3 July, in order to simulate the natural increase in the duck flock due to breeding, 22 uninfected ducklings were brought in. As naive hosts they picked up burdens of several hundred tapeworms immediately, many of which became large and mature. No doubt this contributed to the massive build-up of the infection in the crustacean hosts, though wild-duck visitors may also have helped. The worms in the old birds on the other hand, belonging to the three species routinely monitored (*S. gracilis*, *D. stefanskii* and *Dicranotaenia corunula*), had not been producing eggs for a long time; and when the *ducklings* were sampled the second time, in mid-August, their fertile adults had disappeared.

What were then left in the host flock were hundreds of sterile stunted worms. In the same weeks the *Cypria* population collapsed to 30 000, in mid-August; and no *Dicranotaenia*, which were solely confined to *Cypria* in their cysticercoid stage, were found in the ducks in September. Temporarily at least, the species had reached the point of extinction, or passed it.

It is possible to offer a speculative explanation of this dramatic story, and it seems worth doing so even if it is partly wrong, in order to present a population ecologist's view of the crowding problem, a view which contrasts sharply with some of the other conjectures commonly found in the literature. The fact that young birds, brought in to join old ones that already harboured nothing but sterile worms, were found to produce fertile adults suggests, as Avery pointed out, that it was not a defect in the ducks' nutrition that held back the stunted parasites in the adult ducks. Both ducklings and old ducks enjoyed the same diet, and moreover the ducklings averaged heavier worm burdens than the older ducks.

One needs to keep in mind that the pond was only 5 m across, and was much enriched by the ducks' droppings; it held a large and seasonally increasing crustacean fauna. And also that in this whole series of experiments, fertile worms were only produced in numbers on the occasions when naive ducks were infected for the first time. Few or none were found on any other occasion. The birds went dabbling in the pond day by day, consuming crustaceans along with much else besides, and thus picking up cysticercoids. If what seemed true of the schistosomes (p. 294) applies also to cestodes, these cysticercoids would have been repeatedly 'knocking at the door' trying to gain acceptance by the hosts. Many would have been killed by the birds' antibodies, and the frequent or continuous flushes of antibodies in the host's blood-stream could have been sensed and integrated by the established worms, and have given them an index of the current parasite transmission rates. The rates appear to have been high enough, long before the ducklings arrived, to have arrested the growth and reproduction of all but a few of the original worm infections the ducks acquired in February. Those could have produced half a million eggs apiece before their short lives were over, enough perhaps to merge the 'knocking' into a daily tattoo, and to switch all the fertility signals down to 'stop'. The first worms the naive ducklings picked up in July presumably got off to an exuberant start before their hosts had antibodies ready, with which to create a feedback; but before the next 21-day sampling date came round the fertility of new worms had also been summarily stopped.

Avery shared my opinion that the so-called crowding effect is a phenomenon that 'develops after ducks have been continuously exposed to a high level of infection'. Supposing the rest of my interpretation of it as a normal homeostatic response is broadly correct, it implies that the tapeworms react not only to safeguard the definitive-host population on which they depend, but to preserve their intermediate hosts as well. Judging from experience one would not expect any other conclusion.

A corollary arising from it is that several species of cestodes sharing the same host can apparently act in unison when it comes to controlling their growth and fecundity for the common good. They must be programmed to respond as members of an animal community, even though they are in direct or nearly direct competition with fellow species for the same food resources. Presumably they have no viable alternative but to cooperate. The different species would oust and exterminate one another if they had the means, and no doubt they do so when they can, as their tendency to segregate into different sections of the intestine suggests. But helminths appear ill-equipped for competition with rivals, particularly in the use of physical or chemical weapons. Perhaps any resort to 'force' proves inimical to the tolerance of their sensitive and vulnerable hosts.

18.4 Dispersion of parasite populations

If one could map the dispersion of individuals belonging to any species at any instant of time, few or none would be found distributed at random. For example, red grouse plotted on a map-area 1000 km^2 in extent in eastern Scotland would appear clumped because they are confined to heather moors, and the distribution of heather moors is disrupted. But if on the other hand one mapped the breeding dispersion of pairs on a 1 km^2 sample area of moor on a larger scale, their spacing would prove to be more uniform than a random scatter because of their territorial system (see Figure 8.3, p. 104).

Statistically speaking both these types of pattern depart from a Normal distribution. The first, with clumped spacing, is said to be over-dispersed, and the second, with too even a spread, to be underdispersed (cf. Cassie 1962). Both are frequently exhibited by the same species at the same time, as for example by schools of fish or flocks of birds which form separated clumps, but with each clump consisting of more or less equally spaced individuals. Both types of departure from randomness are brought about by the adaptations and behaviour of individual organisms. Large animals are often mobile enough to disperse themselves almost entirely by their own volition, in spite of interference by the elements, and have thus an almost free scope for optimizing their choice of sites and in-group companions. Plants typically rely on external agencies for their dispersal and have to provide for a large mortality of spores or seeds that fail to arrive at a habitable destination. Other organisms come somewhere in between these extremes in their ability to control individual destiny during dispersal. But the result is always the same: the organisms survive only in viable situations, and what is viable depends on physical and biotic factors, few or none of which are random variables in space or time.

One cannot therefore expect to get much insight into the adaptive functions of dispersion by mathematical modelling and statistical analysis of the density patterns that particular living species construct in space. Life processes frequently succeed only through their ability to overcome the vicissitudes of chance. But models based on random principles can be useful at least in showing where non-random processes exist, and thus pointing to biotic adaptations that might otherwise go unnoticed.

Both over- and underdispersions of metazoan parasites have attracted attention of late, and it appears that different species, when their demographies are analysed in terms of parasite individuals per host in a local population, tend to be polarized one way or the other, the overdispersions being much the commoner of the two. This is equivalent to saying that the majority of species are gregarious, and it gives us a useful clue to their population structure. Individual helminths spend the mature period of their lives immured in a host from which they cannot move; and the effect of overdispersion is to ensure automatically that the majority of them will be members of in-groups of larger than average size (an illustration of Lloyd's 'mean crowding effect', see p. 318). Most platyhelminth species are sexually hermaphrodite, and self-fertilization has frequently been proved; but there are signs that cross-insemination may also be fairly widespread. For instance, most of them are also protandrous, an adaptation well known in plants for promoting cross-pollination, and no different explanation has been advanced for the helminths. Experiments have been performed in which digeneans have been transplanted at separate sites, or inoculated singly on different occasions into a host, and have subsequently been found relocated as partners. In other experiments, radioactively marked flukes have been implanted into hosts that already carried light unmarked infections of the same species; and whereas 28 out of 37 worms that had no available partner inseminated themselves, only one out of 33 of those that found partners had done so; the rest were cross-fertilized (quoted from Crompton & Joyner 1980: 130).

Rather more tenuous evidence of outcrossing has been obtained for some hermaphrodite tapeworms. An instance is known where three individuals of *Acanthobothrium quadripartitum* were found in simultaneous cross-insemination (Crompton & Joyner 1980: 132). Many cestodes have protandrous proglottids (segments), and a few have protogynous ones (e.g. *Progynotaenia*). Some have proglottids that become detached from the end of the strobila before their sex cells actually ripen, and these have been found forming copulating pairs. A few species actually have separate sexes (see Baer & Joyeux 1961: 435). In the Digenea (trematodes), the differentiation of permanent males and females is confined to the schistosomes (Joyeux & Baer 1961: 609).

Gregariousness among free-living animals is one of the devices that can establish breeding groups with favourable inbreeding-outbreeding balances; and this gives another hint that overdispersed helminths do indeed take advantage of intermittent cross-breeding. We have seen that some kinds of established helminths are able to prevent the ingress of recruits to their in-groups, by keeping the host's immune responses on the alert; so that larvae of their own species get killed if they attempt entry. It is equally clear that if a massive infection enters naive host and becomes established and disguised before the host's immunity has had time to build up and stop them, the phenomenon of self-cure (which expels part of an established infection) and the inhibition of growth and fertility are likely to ensue, affording a strong presumption that these too are self-regulatory reactions on the part of the parasite in-group. A similar type of self-regulation appears to cause the 'arrested development' phenomenon well known to occur in some parasitic nematodes (Schad 1977).

Whether an in-group can also *promote* recruitment to its own number is not established for any helminth. However, it has been suggested in recent years that not only protozoan but also helminth parasites are able to depress the host's immune responsiveness, that is, to interfere with its ability to produce antibodies, together with the 'complement' that reinforces them (Barriga 1981: 327). If this necessitated a general lowering of all the host's immune defences it would seem a dangerous reaction for the parasite to invoke, because it would give the opportunity for competitors or pathogens to enter at the same time and threaten the parasites' or the host's well-being. If on the other hand the inhibition were specific it would allow just the sort of control that could benefit the parasites, by allowing them to supplement in-groups of suboptimal size.

Overdispersion could have other advantages if it meant, for example, that a particular parasite could also build up its numbers when it found itself in a host in which it was relatively free from competing species; or if a larger proportion of the host population could be left relatively free of parasites, and hence healthier and more vigorous (see below). Certainly there has to be a positive action on the part of the parasite to produce overdispersion—either a density-dependent attraction which draws recruits into an established group until it has built up to a suitable size, or an elimination of groups that fail to reach a threshold number, or both. Helminths living in aquatic hosts might be able to use pheromones excreted into the water as attractants; but since overdispersion occurs in helminths with terrestrial hosts as well, the demise of unacceptably small groups might be a more likely means of producing it.

Underdispersion is the contrary kind of anomaly, characterized by small infections in many hosts and usually limited to one or two adults in each. The in-group as a structural unit consequently vanishes. A good example is found in the primitive platyhelminth *Gyrocotyle,* investigated by Halvorsen and Williams (1968); its host is the equally archaic cartilaginous fish *Chimaera.* The latter are ovo-viviparous, with the result that the young fish grow to a considerable length while still provisioned with their embryonic yolk, before they begin to feed. The authors' collection of specimens, young and old, showed that 64 out of a total of 92 *Chimaera* were infected with *Gyrocotyle.* When the fish were graded by size into seven growth stages, it was found that more than half (25/45) of the youngest fish groups, 13–25 cm long, had so far escaped infection (these figures correct a misprint in the authors' Group II column of Table II, p. 134); and such very young fish account for all but three of the uninfected specimens they found. The remaining three were in the next two size groups, up to 35 cm long. Of the 20 that *were* infected in the two smallest size groups, 11 fish held one worm, 4 held two, 2 held four, and 3 had more than four. Only one other fish had more than four worms, and it again belonged to the next larger size.

A few young fish apparently took a long time to acquire more than one worm; but of the 19 fish that fell into the older groups, ranging between 35 and 84 cm in length, *all had become infected;* 1 held one worm, 16 had two, 1 had three, and 1 had four.

From these data the authors concluded that most hosts became infected quite soon after their yolk was absorbed and feeding had begun, at about 20 cm long. If a young fish chanced to ingest several post-larvae at once the number was usually whittled down quite soon to two. This was borne out by the average sizes of the worms themselves which tended to increase with the size of the host. But worm size was also affected by the number of worms present, and in an in-group of four, one or more of them were always stunted (usually 0.5 mm long), suggesting to me that here again there was a nutritional hierarchy by means of which the in-group could adjust its final number to two. The only possible means of establishing a hierarchy seems to be the production of some kind of external secretion circulating in the parasites' microhabitat, to which the individuals are differentially susceptible.

Gyrocotyle appears normally to be protandrous. The pair of adults that remained in the host were sometimes uneven in size, the smaller being still in the male phase while the larger was producing eggs. There was not enough material to demonstrate exactly how the sexual cycles operated, though cross-fertilization seemed the obvious motive. The largest *Gyrocotyle* found appears to have been only 12 mm long; and the authors' drawing, showing two apparently mature worms lying in the folds of the intestinal spiral valve, does not suggest that they imposed much of a burden on the host. This is consistent with the high apparent incidence rate in *Chimaera* adults.

Halvorsen and Williams mention in their discussion five tapeworm genera where similar eliminations of newly established post-larvae are known to reduce the final quota of adult worms per host to one or two. In three of these the sexes are dioecious and the normal infection leaves one male and one female together in a host.

Joyeux and Baer (1961: 420–421) mention that *Taenia solium* and *T. saginata* occur solitarily, as a rule, in the human intestine; and that if the established worm is expelled by medication the host some time later becomes susceptible again, so that if he remains exposed to infection another worm becomes established. This is a parallel situation to the concomitant immunity that occurs in primate hosts of *Schistosoma* (p. 293); and it can also be included under the wider term of premunition, which means that the presence of an established infection in a host precludes the addition of a secondary one. These *Taenia* species can reach a length of several metres and presumably it only needs a single full-grown worm, laying 10^5 eggs a day or more, to fill the carrying capacity of the host.

This underdispersion departs equally clearly from the Normal distribution, solely on account of the vital activities of the parasite. In the case of the large tapeworms it appears to be an adaptation appropriate to their size. The same dichotomy between over- and underdispersion is mirrored quite independently in parasites in another phylum, the Crustacea. In a very interesting discussion, Fryer (1966) draws attention to the dispersions of certain copepods the adult females of which are external parasites of fish. In a sample of 560 specimens of a large predacious fish in Lake Nyasa, *Bagrus meridionalis,* 333 carried at least one female parasite of *Lernaea bagri.* But among the fish there were six that bore 12 parasites, and one each bearing 24, 30 and 55 respectively. From the statistics available he calculated the probability that a fish could amass 12 parasites by chance alone, and it worked out at 2×10^{-6}; for one with 24 parasites it was 1×10^{-18}; and for the higher aggregations the chance became so astronomically remote as to be meaningless. It was also observed that when there were two or more copepods on the same fish they were often close together, even in species that had no marked preference for a particular part of the host's integument. Evidence from different sources about the dispersion of these and other parasites suggested to Fryer (a) that the larvae must sometimes gather into dense swarms which make it possible for several or many to settle simultaneously on a fish; and (b) that, because a host sometimes carries parasites belonging to different age-groups, the presence of established adults provides a magnet (presumably biochemical) for prospecting larvae. He concluded that in any case 'some factor other than chance usually decides where settlement will occur, and that a settled parasite is often capable of attracting others to the site'.

In contrast, certain crustacean parasites that are large in relation to their host's size are almost always found in ones or twos because there is no accommodation for more. Like other Crustacea these parasites are almost certainly social animals, and would have no difficulty in devising the means of pre-empting a host. One example is the rhizopod *Sacculina* whose externally visible body lies hidden under the reflexed flap, formed by the abdomen of its host which is a crab. It occupies the space where the female crab normally carries her eggs until they hatch; and incidentally if it parasitizes a male crab which is developing the usual narrow male abdomen, it causes the host to develop a broader one in time for the next moult (Figure 18.2). The parasite sends roots into the host's body to abstract nutriment that would otherwise go into the crab's maturing gonads, and the host is thus castrated. Another example is the parasitic amphipod *Bopyrus,* which occupies one of the paired gill-chambers of a shrimp. That allows a maximum of two of these rather large parasites to

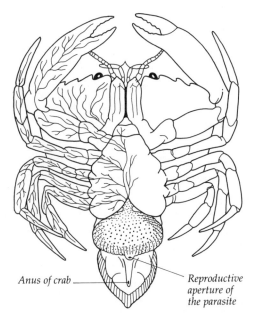

Anus of crab —————

Reproductive
aperture of
the parasite

Fig. 18.2 Sacculina carcini *parasitizing the shore-crab* Carcinus maenas. *The external part of the parasite is shown stippled, and its penetrating roots are diagrammatically represented on one side of the crab's body. From C. M. Yonge (1949)* The Sea Shore, *Collins, London. (Originally after Delage).*

be carried, one on each side; and when in place they cause the gill covers to form a conspicuous bulge. Because the infection is so limited in numbers the parasites are underdispersed, like the human tapeworms, and apparently for the same reason.

A final illustration of underdispersion comes from the cysticercoid larvae of the duck tapeworms, growing in the small crustaceans *Cyclops* and *Cypria* and mentioned earlier in the chapter. They are also large in proportion to host size, and one per host is their normal complement. Their adult stages in the ducks were of course overdispersed (complements of 1 to nearly 100).

From the *Animal Dispersion* hypothesis one might expect over-dispersed parasites to gain advantage from belonging to in-groups of appropriate size, particularly through cooperating in resource management and sharing in the maintenance of a varied gene pool. But in the circumstances peculiar to endoparasitic life these may not always be such valuable advantages. Free-living organisms are assumed to carry recessive genes and multiple alleles in their gene pools as an insurance against changes in selection pressures, which might be disastrous to a stock lacking in genetic adaptability. For helminth parasites on the other hand the environment to which they are keyed is no doubt exceptionally immune to change in the long term. They must be adapted to cope with normal diurnal or annual cycles in the host's dietary habits for example; but these can be counted upon to recur, if not actually *ad infinitum* then for another millennium or so at a time. Changes there may be in the community of competing parasites, but there are few habitats in the biosphere as predictable as the *milieu interieur* of a homeotherm mammal or bird. In the very long term the survival of parasite species will depend most of all on that of the host species, and they have got to rely on the latter's ability to adapt and change when major environmental challenges occur. It may consequently be far less dangerous for endoparasites in general than it is for free-living species to resort to self-fertilization and asexual reproduction as norms, and easier for them to survive in the homozygous straitjacket that results, especially if they ease it occasionally by crossbreeding.

18.5   Homeostasis and structuring for helminth populations

The evidence briefly presented in this chapter leaves no reasonable doubt that population homeostasis and structuring exist in the platyhelminths. The grounds for reaching this conclusion begin with the presumption that, being vivivores, parasites could not otherwise avoid the danger of overexploiting their food resources—a danger as real for them as it is for herbivores and carnivores. When we looked for confirmation we found it firstly in two widespread and partly overlapping phenomena, namely (1), the regulation of in-group biomass through differential *growth* among the individual members which can result in only a proportion of them reaching maturity; and through differential *mortality* which can dislodge supernumeraries from an established in-group ('self-cure'). The once-prevalent belief that these regulatory changes were imposed by the host is based on false assumptions and cannot stand up to critical examination. (2) The second tell-tale phenomenon is the non-random distribution of parasite species within their host populations, which commonly exhibit a degree of over- or under-dispersion that is characteristic of each particular species. These departures from chance cannot be attributed to the host either, because hosts' reactions to helminth parasites are to treat them all the same, and try to resist their entry.

To these may be added the general observation that helminths are not lethal to their hosts, except where the definitive host is a predator and picks up the parasite in the normal course of exploiting an intermediate host as prey. Homeostatic predators generally present no threat to the survival of prey stocks, either parasitized or parasite-free.

Lastly, we saw in the duck tapeworms a particular instance, under field-experiment conditions, in which supposedly 'informed' adults had had their egg-production completely inhibited (as I suggest, through feedback from the external environment outside the host). At the same time imported ducks just introduced in the same enclosure were colonized by naive tapeworms that grew and rapidly produced eggs. Only three weeks later, when a sample of the hosts were examined, feedback information appears to have arrived; many of the worms were already sterile and stunted. At the next sampling (six weeks) all were so. The potential trouble developing in the outside world was that 50% of one of the intermediate hosts, *Cypria*, had already been castrated by tapeworm cysticercoids, and their population was about to collapse. It would be difficult or impossible to explain these events on any basis other than programmed homeostasis. (The excessive incidence of cysticercoids, which should have been forestalled if homeostasis had worked properly, can possibly be attributed to the long unnatural confinement of numerous ducks in the experimental enclosure, and the small size of the pond it contained.)

Some parts of the feedback system on which the homeostatic responses of individual worms are based are not difficult to predict, since they must necessarily involve the state of the food resource (host digesta etc.) and the population density of the parasite, both in its own in-group and in the local population of hosts. It is safe to assume that parasites react sympathetically to their host when its feeding and nutritional plane become reduced. Conditions of hardship for the host tend to be multiplied in the danger they present to the parasites, because in a crisis the hosts with the heaviest parasite burdens are the ones most likely to die. It is no doubt easy for parasites to obtain nutritional feedbacks; but information relating to the population density of hosts, and the average incidence rates of conspecific helminths they carry, must sometimes be complex and confusing for them to interpret. I have suggested how one type of signal may get through from the outside world, and give the adults an indication of the current transmission rate of their last-stage larvae. Trying to picture how meaningful information could be conveyed in the circumstances just outlined above, for example, where one of the intermediate hosts is being castrated to excess, is much more difficult. If the apparent transmission rate of larvae is rising, how can the adults tell whether it means that the intermediate host is increasing in density while the incidence rate of cysticercoids stays constant, or that the incidence rate is going up, while host density remains constant? The first condition is benign, the second potentially disastrous; and egg-production needs opposite adjustments to adapt to them.

It seems certain that larvae searching for intermediate hosts, or reproducing asexually inside them, also require homeostatic programs; and hence a last-stage larva entering its definitive host might conceivably bring some useful demographic conditioning with it. Ten years later of course (e.g. to a schistosome in man) the juvenile experiences of an aged adult are unlikely to be relevant.

If it should appear at first sight beyond belief that such complex programs as these could be coded and integrated in the microscopic central ganglia of small helminths, the reader should perhaps recall that programs for constructing a human phenotype can be packed into one zygote nucleus.

Parasites depend largely on their hosts for dispersal and spread, within and beyond the home-range of the individual host. It is conceivable that, on an evolutionary time-scale, they have been able to adapt their own needs for genetic structuring (e.g. for balancing isolation and gene flow) to having them fulfilled at second hand, through the genetic population structuring maintained by their hosts. Even though their parasitic habitats are so changeless in the long term, a Wrightian demic structure must presumably still provide the key to their evolutionary advance.

Summary of Chapter 18

1 Parasitic animals are exceedingly numerous, feeding as vivivores on other animals much as herbivores do on plants. The platyhelminths are introduced here as representative presocial animals because many of them infect man and domestic animals and for that reason their ecology has been much researched. When adult their main activity is egg-production, in quantities sufficient to surmount the difficulty inherent in the infection of new hosts; but earlier life-stages commonly inhabit one or more intermediate hosts in which asexual proliferations take place.

Being vivivores they need homeostatic adaptations for preventing their hosts from being overexploited. Each host individual is a discrete habitat and has a sustainable carrying capacity which they exceed at their immediate peril. By definition they impose a trophic burden, but selection operates to minimize avoidable (e.g. toxic) forms of damage. In some life cycles succeeding hosts are predators on the preceding ones; but then the predator population must largely take over the group function of conserving the prey.

Population structure is two or more tiered, with its in-groups in individual hosts, and demes that approximate to those of the host. The populations benefit from homeostasis at both levels, and this function extends down to in-group level even in parasites where only 1 or 2 adults pre-empt a host and may be too few to constitute an effective genetic group. The in-group normally adjusts its demands to the host's nutritional state, and in the adult stage this generally means increasing or depressing egg-production. In the host's population (the parasite's deme) a high incidence of infected individuals is sustainable only if the average burden per host is sufficiently light; if the burden is heavy (as when the parasite castrates the host) the incidence level must obviously be restrained.

For feedbacks within in-groups, the production and monitoring of a parasite pheromone circulating in the host, plus monitoring the day-to-day nutritional plane, may be all that is necessary. Feedback from the deme regarding incidence rates is much more of a problem: my suggestion is that it is obtained by utilizing the host's immunity reaction as an indicator of how often the host is challenged by new arrivals of larvae seeking entry, and requires to suppress them. The adult parasites are evidently disguised and safe from molestation by host antibodies; but adults continue to keep the host's immune system alert, with the result that invading larvae trigger a systemic discharge of antibodies which seems likely enough to be detectable by the established members of the parasite in-group. The frequency and magnitude of challenges could then provide information from the outside world about the parasites' current transmission rate; and if established parasites innately sensed a norm, they could respond to aberrations from it by slowing or speeding their own reproduction.

Parasitic species would not long survive unless they, and not the hosts, were in command of the symbiosis, with at least one refuge host. In practice, selective pressures fall heavier on the dependent partner of the symbiont pair, which is doomed to perish if the mutualism fails; and this automatically vests the initiative in the parasite.

2   For illustrative examples we take first the schistosomes—immensely successful though debilitating parasites of man in the tropics and Orient. They are digenetic trematodes with a first parasitic stage in water snails (including *Biomphalaria*) and an adult stage (for present purposes) in the mesenteric and portal veins of man and other primates. *Schistosoma mansoni* is the one considered here though there are 4 species in all. The worms' tertiary aquatic larvae leave the snail and can penetrate the human skin from the water. After entry they are carried in the lymph and blood to the lungs where they spend several days, preparing themselves to move to their permanent habitat. Unusually among platyhelminths they have separate sexes and as adults form persistent pairs, with the female held in a groove on the male's ventral side. They may survive for years.

Animal hosts have been experimentally infected with either 100 or 600 larvae. Before long mature worms were producing eggs which the hosts evacuated, at first in similar numbers per worm pair from both light and heavy infections; but between the 3rd and 6th months 2/3 of the adult parasites in the 'heavy' in-groups died although there was no corresponding mortality in the 'light' ones. This confirmed a phenomenon previously inferred or recognized in other kinds of parasites, known as self-cure on the supposition that the host is responsible; but the experiment indicated that it was a density-dependent response on the part of the parasites, presumably to lighten an unsustainable burden on the host. If the host had the ability to rid itself of parasites it would do it completely, whether the infection were large or small.

A naive host can readily be infected with schistosome larvae because it takes several days to produce tailor-made antibodies for the first time; but if an already infected host is later exposed to another dose it responds promptly by killing the invaders. Smithers and Terry (1969) designated this host reaction as 'concomitant immunity'. It suggested that the established parasites which the host is powerless to attack must have become disguised and unrecognizable as foreign bodies ('non-self') by the host's immune system; and that larvae must lack the disguise on arrival and have to acquire it after entry.

In addition to monkeys they used mice as laboratory hosts; and to test this hypothesis they first showed it was possible to transplant adult worms surgically from one monkey to another uninfected one; the worms resumed egg-production in the new host with scarcely a check. But if the experimenters transferred them between, say, monkeys and mice, egg production was markedly depressed, and it took the worms 5 or 6 weeks to come back to normal output. Suspecting that the worms might be picking up the host's own antigen signature after entering the body, the experimenters proceeded to immunize a naive monkey against *mouse* antigens (by injecting them with homogenized mouse tissues), and then took worms from a mouse host and transplanted them into the treated monkey. Sure enough, the monkey now killed the worms with its mouse antibodies. By tritium labelling they even showed that the parasites do transfer the host's antigen molecules to their own teguments.

3   Further examples are taken from the tapeworms (class Cestoda). Ducks are hosts for cestodes that have pond-living copepods and ostracods as intermediate hosts, which the ducks ingest as food. The worms have annual life cycles. Avery (1969) experimented with two species in domestic ducks, and found some larvae had matured and produced eggs within 9 days of being fed to the hosts. Autopsies eventually showed that inoculations of 20 larvae generally produced in-groups of fewer than 20, and that there was always a wide range of individual worm sizes, i.e. a nutritional hierarchy that resulted in con-centrating resources and speeding the first onset of reproduction.

For his next experiments he placed 24 worm-free mallard ducklings in a pen for captive wildfowl containing a small pond that was visited by wild ducks, and he allowed the ducklings to become infected by natural feeding. In the early weeks the ducks were excreting eggs from 2 or more species of tapeworm; but within 9 weeks most hosts examined *post mortem* contained only stunted sterile survivors. After 3 months when the last hosts were killed their in-groups varied from 1 to 100 worms, and all were similarly stunted and sterile.

A year later he followed the progress of similar infections through the intermediate hosts as well, by monitoring the densities of the two crustaceans and the incidence rate of infections among them. Infected hosts held only one relatively huge cysticercoid larva and were rendered infertile by it. The ostracods trebled in density between February and July, and their infection rate rose from 0.5 to 30%. In the next few weeks it reached 50% and the whole ostracod population collapsed. Changes in the copepod host were parallel but less dramatic.

Ducks sampled in February averaged 120 fertile worms apiece; but long before July, egg-production stopped and only stunted worms were found. On 3 July another batch of young uninfected ducks were brought in to join the older birds. They picked up enormous worm burdens immediately, many of which matured, adding to the already teeming larval stocks in the pond hosts. But at the second sampling no fertile worms were found in the young ducks either.

In these semi-natural infections only the first-established in-groups ever produced fertile adults. I suggest that the worms very soon received warning feedbacks from the high and rising density of cysticercoids the ducks were ingesting, and responded homeostatically by suspending egg-production in order to ease the pressure on the crustacean host populations. The worms in the July ducks, initially naive and therefore fertile, showed a similar response, presumably as soon as the feedback from the pond had had time to switch them off. Note that all the ducks, old or young, shared the same food, and the tapeworm responses can have had no connection with nutrition. An additional feature to note is that six species of tapeworm were involved in the infections and all reacted in unison in this episode.

Helminths appear to possess no weapons, defences or perceptions for engaging in contests, except through pheromones and chemoreception. Nevertheless they managed to achieve effective hierarchies and exclusions.

4   The phenomena of over- and underdispersion are characteristic departures from random or Normal distributions found among members of living populations, and in animals even more conspicuously than in plants. Overdispersion means they are gregarious and clumped, underdispersion that they demand exclusive personal space. Both often occur at once on different scales, e.g. in flocks and schools (aggregates) the individuals often resent close encroachment. Living organisms frequently survive only through overcoming the accidents of chance.

Both types of dispersion occur in helminthic parasites, though aggregation is much the commoner, and allows a viable population of parasites per unit area without needing a high infection rate among hosts. It also allows an in-group of normal size to exist in a host.

Platyhelminths are almost all hermaphrodite, and self-fertilization is probably normal, though signs of cross-fertilization are not rare. Internal parasites occupy exceptionally changeless or seasonally predictable habitats, and depend for their long-term survival largely on the survival of their stocks of hosts. Inbreeding and impoverished gene pools may therefore entail less hazard for them than for most free-living species.

CHAPTER 19
GROUP SELECTION AT DIFFERENT
ORGANIZATIONAL LEVELS

19.1 Group selection can override individual selection

The view, widely held, that group selection can be dismissed as an effective evolutionary force is based on assumptions, not on evidence. It depends on the argument that cooperation for the common good makes it necessary for individuals to give way to the interests of others in matters that affect personal fitness, such as the exploitation of food resources and the control of population density; and that cooperators must thus lose in competition with self-seekers who are free to maximize their fitness by every possible means, including the consumption of the resources that the cooperators are trying to conserve. It is an argument uncritically derived from human experience, of cheaters, criminals and oppressors who live at other people's expense; and it ignores the fact that all viable kinds of exploiters in the animal world must be able when necessary to limit their own numbers. Living at a cost to others, whether of one's own or different species, is the vivivores' way of life, and it can be sustained only by keeping demand within the carrying capacity of one's food supply: otherwise consumer numbers will eventually rise to the point at which food resources are turned into wasting assets. For non-vivivores (such as the decomposer fauna) the position is effectively the same, because they too have to keep their daily or yearly needs within the food production rate.

Limiting numbers means, in practice, evolving the faculty of self-regulation, because relying on Darwin's other checks (predation, disease and 'climate') to act as limiting agents in perpetuity, is to court disaster and eventual extinction when they fail. Evidence that animal populations are homeostatic abounds, as previous chapters have attested, and the same is true of population structuring. Evidence to the contrary has been conspicuously absent. We have moved from one animal taxon to another, finding variants of the same basic programs, and from one aspect of sociality or structuring to another, noting that they exhibit products of an evolution that cannot be accounted for by selection acting at the individual level alone, but necessitates a group-selection process. It has emerged that group selection and individual selection are not generally contradictory to one another, but complementary instead. Because of the fact that life is vested in individuals and its transmission is hereditary, it is beneficial for individuals to be tested by selection and the less fit phenotypes eliminated from their ranks; and it is equally true that the fitness value of an individual can be measured by the contribution it makes to the viability of its group, and of the higher organizations of which the group is part.

The power that groups hold over individuals arises from the benefits that are attainable through cooperation, by raising individual fitness and group survivability to levels higher—and in the course of time enormously higher—than would be possible for non-cooperators each acting on its separate behalf. The advantages stem from the regulated management of trophic and genetic resources. Cooperative traits in consumers that lead to improvements in the productivity of their food resources will be preferentially selected, and benefits so obtained may continue to increase over the millennia, and can evidently spread upwards to the species-community level, and probably even higher, as we shall see (pp. 320, 326). Conversely, non-cooperative individuals are a drag on their group's productivity: the more there are of them the worse the handicap and the greater the selective penalty against the group in which they live. Parallel arguments apply to the genetic benefits that population structuring induces, through drift and innovative recombination, and through the diversification of groups and the group selection thereby activated. Whether it acts at individual or group level, natural selection is essentially a comparative process, and the rewarding of some units of selection usually implies the penalizing of others.

The 'mathematical tradition' of group selection, originating with Wright (1945) and recently discussed by Wilson (1983), is expressed in a single-locus model, in which there are two alleles A_1 and A_2, present in the group with frequencies p and $(1-p)$. The A_1 allele codes for a character that is socially advantageous, but at any given moment is disadvantageous to the individuals that carry it, because, for example, it obliges them to regulate their reproductive output at a conservative level. The A_2 allele does not restrain its possessors' fecundity. The fitness of the A_1's is thus diminished by a factor $(1-s)$ because of their sacrifice of reproductive potential. All members share the collective benefits arising from the sacrifices made by the A_1's, which increase their individual fitness by a factor of $(1+bp)$. But when the respective frequencies and selective values of the genotypes A_1A_1, A_1A_2 and A_2A_2 are tabulated, and the change in the frequency of the A_1 allele per generation (Δp) is calculated, it turns out that Δp is negative: that is to say the A_1 allele would fail to establish itself, even though the average individual would benefit if it did.

The model therefore does no more than formulate a problem, to which Wright did not provide an explicit solution. The one I am proposing is that, in a structured population, the inherent tendency for selection to maximize individual fecundity is continuously opposed by the penalty consumers pay, when the local demand for food resources is allowed to exceed the latter's carrying capacity. In these circumstances selection and counterselection seem likely to result in a balance, stabilizing fecundity and other individual-fitness factors at levels that maximize the persistence and productivity of the population unit as a whole. Obviously fecundity is not controlled at a single locus but is a metric character, and also one that is often manipulated physiologically, within the phenotype, from year to year and place to place, as a variable in population homeostasis. This implies that its level is adjustable above or below the mean, according to the current carrying capacity of the habitat. Other factors may affect it, notably the social status of the individual concerned. Because the inputs are diverse, and numerous gene loci must be involved, the basic setting of fecundity in relation to environmental inputs is presumably stable, and rather slow to change under natural selection, especially under individual selection.

In any species—one could think for instance of an aquatic single-celled organism—where there is a strong element of isolation by distance and where populations are further subdivided by physical discontinuities that create small islands of habitat, group selection could have come into existence without much apparent difficulty. All that would be necessary to allow it to start would be conditions that accidentally broke a population into largely but not totally isolated units small enough for intergroup variances and group phenotypes to develop. Provided the situation persisted or tended to recur the number of sub-groups existing at any one time need not have been large, because some advantage from selection for group adaptation would begin to be felt almost at once, through drift, accelerated adaptation to change, and increased productivity: the drawbacks would have been none. Its progress could have gathered momentum by favouring populations that were situated in naturally subdivided habitats; and it would eventually have been rendered irreversible through the promotion of self-structuring machinery, which enabled populations to subdivide themselves in any kind of habitat. It is this that makes one suspect that group selection first developed at an early stage in organic evolution. Once it had become self-perpetuating in a few primitive species it could in time have become universal, through extinctions, survivals, and adaptive radiation by phylogenetic descent from the originators. Present indications are that it does now occur in all types of organisms, from microbes to the higher plants and animals.

19.2 D. S. Wilson's model of group selection

The first theoretical model to give group selection explicit and rational support was David Wilson's (1975, 1977, 1980). In general it offered a possible answer to the missing link in the *Animal Dispersion* argument, and that was enough to get me started on the preparation of this book. In detail it revealed the importance of the 'dispersal phase' and the 'trait group' in the machinery of population structuring. It was only later, when my reading and writing were well advanced, that the compatible but more comprehensive solution offered now came into view.

His basic population structure is the one I have adopted (p. 219). In it, individuals spend most stages of their lives residing and interacting with a small group of companions or neighbours—their trait group. (I attached to the latter an important behavioural as well as a genetic significance, and for that reason decided to use the alternative name of in-group.) The group members have little or no contact with their conspecifics in other trait groups at such times and can be assumed to be temporarily isolated from them. At least once in every generation there is a dispersal stage, often of short duration, as a result of which the individuals get reassorted into new trait groups. Dispersal gives reality to the deme, as a genetical entity more permanent and of larger extent than the trait group. Wilson has, of course, been primarily interested in resolving the individualism versus altruism conflict, and thus in traits that benefit (or in any way affect) other members of the trait group, and particularly in traits resulting in 'weak altruism'. The latter term implies that the doer (donor) shares to some extent in the benefit he confers by his action on the other members of his group (recipients).

To strip the model of unnecessary complications Wilson postulated a population of individuals genetically identical except at one locus, where there are two alleles A and B. These make their possessors either social (cooperative) or non-social (non-cooperative) respectively. The individuals are conceived as being haploid and thus having genotypes that are either A or B, with no heterozygotes.

Trait groups are founded by dispersers, for instance by a small number of fertilized female insects laying eggs on a food plant, which hatch into a group of larvae. The group might be isolated on a single leaf. The alleles A and B are present in the deme in proportions p and q (where $p+q = 1$); but in small samples like those comprising single trait groups one must expect the actual proportions to vary about the deme mean, by chance if not for any other reason. One can assume for simplicity that there are equal numbers of A's and B's in the deme as a whole ($p = q = \frac{1}{2}$), and that the deme contains only four trait groups, in which the variations in allele proportions can be exaggerated for purposes of demonstration, as in Figure 19.1.

Fig. 19.1 *Four discrete trait groups of the same size but varying in their proportions of A- and B-types. From D. S. Wilson (1980).*

The population mean proportion

$$p = \frac{0.8+0.6+0.4+0.2}{4} = 0.5$$

and the same is true for q. But if one considers the mean 'subjective frequencies' the answers are different. The subjective frequencies are those that individual A's (or B's) actually experience in their own trait groups. Thus four of the A's are in a group where the frequency of A's is 0.8; three of them experience an A-frequency of 0.6, and so on:

$$p_A = \frac{4(0.8)+3(0.6)+2(0.4)+1(0.2)}{10} = 0.6$$

The A-individuals' mean subjective experience of B's in the same groups works out at:

$$p_B = \frac{4(0.2)+3(0.4)+2(0.6)+1(0.8)}{10} = 0.4$$

These differences between the overall and subjective frequencies show the same arithmetical relationship as Lloyd's (1967) 'mean crowding' effect, and they result in this example from the assumption that the mixing of A's and B's in the deme has not been homogeneous, because (to choose a non-adaptive explanation for the moment) there have been sampling errors in individual dispersal when the small trait groups were formed.

The standard illustration of the sort of altruistic trait to which the simple model might apply is the giving of a warning call by a watchful donor (A) in, say, a grazing flock of wild geese, on sighting an approaching predator. Small extra costs (of effort or danger) are incurred in remaining vigilant and giving the alarm, but the donor is more than compensated for these if he makes the flock take wing because he is safer from predation in the flock than he would be if he flew off alone. This satisfies Wilson's stipulation that the model, when based simply on random variations in the scatter of A's and B's, will account only for weak altruism. Strong altruism is defined as imposing a net loss of fitness on the donor through his action, and for this to evolve a larger than binomial trait-group variance in gene frequencies would be required.

The B's in the imaginary flock of geese neither watch nor give the signal, though they benefit by heeding it and taking flight. Flocks with a higher proportion of A's are automatically more alert than those with fewer, and suffer fewer casualties from predation. In the two better protected trait groups in Figure 19.1 there are altogether 7 A's and 3 B's, and in the two more vulnerable groups, 3 A's and 7 B's. By the time the next dispersal stage occurs, predation can therefore be expected to have taken a higher toll of B's than A's, so that the proportion of A's will have risen in the deme as a whole. A similar picture could be drawn for a breeding trait group in which, say, the cooperative mobbing of potential robbers of eggs and nestlings led to better protection and a higher productivity in terms of breeding success.

A necessary ingredient in the validity of Wilson's model is that the benefits arising from the donors' actions should be shared in person by the donors themselves; whereas many of the group advantages we have been postulating in this book have typically been delayed, and capable of entailing a lag of some years or generations before much effect was felt. The homeostatic programs that match population density with food production, for instance, need to include a safety margin for overexploitable resources, whereby this year's density is kept below the level that would maximize the consumers' reproduction, for the sake of maximizing the renewal of next year's crop. A longer delay between cause and effect could occur in some slow-breeding vertebrates such as deep-water seabirds, which may have an adolescent stage in their lives lasting some three to ten years (see p. 30). During if they presumably form a standing reserve of recruits, able to bolster production if any catastrophe were to befall the breeding population. Mutations might have to be suppressed that tempted such recruits to set up new colonies where no dominant seniors existed to prevent them; and if they did, it might be many years before the growing stock actually overtaxed the marine food supply.

These are the kind of situations I visualized in Wright's shifting balance process in Chapter 15 (p. 202), where differently balanced gene pools are perpetuated side by side, and those demes that manage their resources of food and genes better and are more productive in consequence, can be expected in the course of time to supersede the less productive ones. Wilson (1980: 41–2) points out that his model will not account for the evolution of group adaptations involving such delays as these. The reason is that in his plan of population structuring, the diversity that group selection has to work on is vested in sets of trait groups; and breeding trait groups are considered to be transient, so that their gene pools are broken up by dispersal, most commonly once a year, and merged into the common pool of the deme. It was mathematically too complex to carry the model up into the higher tier and include interdemic selection as well. But it is the deme that introduces the crucial element of philopatry and semi-permanence into structured populations; and it was the deme that Wright took as his selective unit, though he did not explain exactly how group selection and individual selection could be integrated.

Wilson noted that increasing the duration time of trait groups could improve the conditions for the evolution of altruism, and that shortening the dispersal distances of each generation of offspring so as to deposit them in a narrow belt around their parents, would help to prolong the effective selection period beyond the lifetime of one generation. (An element of expansion had to be introduced to enable the beneficial A allele to spread.) Wade's experiments (Chapter 15.3) have confirmed that the isolation of population units for several generations allows the gene-frequency variances among them to reach a maximum, and consequently strengthens group selection. Much of the value of Wilson's model, however, depends on its rigorous argument, and it was this that made it necessary to restrict his discussion to the short-term benefits amenable to mathematical modelling.

Some examples of undelayed group-benefits have figured prominently in this book: among the commonest are those that flow from the social hierarchy which is latent in the in-group, even at times when it can be ignored because there are superabundant resources. When food, for instance, becomes scarce the tempo of competition is immediately reduced by its revival, which enables known subordinates to be excluded at little cost to the establishment, compared with the fighting that might otherwise be necessary to achieve the same result.

Another important feature of the model is its confrontation and rejection of a tacit assumption of orthodox evolutionary theory, namely that the allele frequencies characterizing a particular population are uniformly spread throughout its spatial extent, and are identically experienced by every individual within it. The model recognizes that, on the contrary, gene pools cannot avoid being to some extent heterogeneous in dispersion, with equally unavoidable repercussions on the consequences of natural selection.

19.3 Community integration and selection

In the second half of his book Wilson (1980) discusses the possible application of group selection principles to whole biotic communities, and the mutual adaptation of their member species to form an integrated superorganism. Interspecific co-adaptation is clearly widespread in the biosphere. Some of it can readily be explained as resulting from individual selection: thus the fittest individuals in a typical plant population presumably include those that withstand the depredations of herbivores with least cost to themselves. Among predatory species the fittest individuals must include those that are most successful in finding and catching their prey. Very many interspecific relationships in a community pertain to this trophic category, where one species is using or is used by the other as a food resource. Another prevalent kind of relationship where individuals also come under selection occurs in species that are potential competitors for the use of the same resources.

The populations of component species in a climax community, however, are not normally found to suffer from unsupportable interspecific pressures. Their respective densities are regulated at or below the habitat carrying capacity, and competition between them for resources has been eliminated or stabilized. On the other hand, beneficial relationships can often be found, mutually enhancing the productivities of the species concerned. Familiar examples are the production of nectar and fleshy fruits by flowering plants, as attractants and rewards to the animals that pollinate their flowers and disperse their seeds; and transcending all others is the almost universal interlocking of species-guilds of plants, microbes and animals in nutrient cycling.

The evolution of these helpful relationships is sometimes impossible to account for wholly in terms of individual selection, as the following examples may serve to show. The four that are chosen here are arranged in order of increasing difficulty. The first is a well-known mutualism between two species, one a kind of hermit-crab, *Eupagurus prideauxii*, and the other its constant symbiont the 'cloak' sea-anemone, *Adamsia palliata*. Their partnership starts when the anemone settles on the sea-snail shell, which a young crab has adopted as its portable shelter. Like other arthropods the crab grows in size by rapidly swelling just after it has moulted, in the transient period while the new integument is still soft and elastic. Most hermit-crabs are obliged to find another larger shell to fit them each time they moult and while they are in the soft vulnerable state. The cloak anemone obviates this for *E. prideauxii* by living up to its name and growing at a speed that enables its wrap-around foot to secrete a horny extension, and thus enlarge the crab's chamber, for as long as the partnership lasts. In return the crab provides food for the anemone in the form of crumbs, scattered by the crab's oral appendages and floating or settling within reach of the anemone's tentacles. The anemone's stinging cells, to which the crab is immune, protect both of them from larger predators. Together they make a binary unit, each part increasing the fitness of the other; and no theoretical problem would arise about selecting the fittest units as if they were single individuals.

A second example can be found in the African birds known as ox-peckers (*Buphaga* spp.) which specialize in eating ticks and other ectoparasites from the coats of hoofed mammals. The birds free the host from much immediate irritation, and exploit their credit balance by helping themselves to scraps of tissue and blood from parasite-induced sores at the same time. This is a food-niche that affects the ecology of three types of animal—a host mammal, various arthropod parasites, and a delousing bird. The bird as usual requires density-limiting adaptations to prevent the overexploitation of its food resource, and that entails group selection for its evolution and maintenance. The mammal is acting solely in its own individual interest by not shaking the bird off so long as its ministration is acceptable. The parasites are also likely to be density-regulated, and the removal of some of their number must normally make room for more.

Next we may return to the tiny flowerpeckers and mistletoe-birds (Dicaeidae) of south Asia and Australia, previously referred to in Chapter 5 (p. 66). Most species feed largely on mistletoe berries (Loranthaceae); and they possess the capability of defaecating the extremely sticky seed within only minutes of eating the berry. It will be recalled that the seed frequently adheres to the twig or branch on which the bird is sitting, and germinates within a few days, taking root on the host tree. The birds thus propagate the crop on which they feed.

So far, the mutualism between bird and plant is, at least superficially, all attributable to individual selection. It is of individual advantage to the small birds to get rid of the heavy seed quickly and thus to acquire anatomical and physiological adaptations for the purpose. It is of individual advantage to the plant to make its fruit more attractive and its seed more glutinous, and also if possible to assist the bird to digest the fruit fast so that the seed will not be carried away and evacuated on some unsuitable tree elsewhere (a mutual disadvantage). But both birds and mistletoes are vivivores, ultimately dependent on the productivity of the host trees. As one might expect, the males of *Dicaeum erythrorhynchus*, the best known species, hold food territories (Ali & Ripley 1974, Vol. 10: 11), and those of other species probably do the same; in many species they are brightly coloured and aggressive, and keep up an almost incessant *chik-chik-chik* . . . call. The delay between depositing the seed and harvesting the crop, plus the lifespan of the *Loranthus* plant, may be long enough to prevent the individual donors receiving much direct benefit from their actions. If one has to accept the prospect of non-cooperators reaping much of the rewards, the premises of individual selection break down, and one is thrown back on group selection to explain how the mutualism could have evolved and be sustained. Birds that sow food plants for future use are simply carrying the practice of resource conservation a further step beyond the minimum, and it is predictable that the indirect benefits in fitness, if any, will as a rule accrue to succeeding generations of the same structured deme.

In some places flowerpeckers become a nuisance because they are so successful in disseminating the *Loranthus* parasite to cultivated orchard trees such as mangoes. Thus both plant and bird must benefit in natural habitats from homeostasis, and it is possible that the *Loranthus* species themselves are independently programmed against over-parasitizing their host trees. The inter-relationship is clearly ancient and highly evolved, because it has evoked profound anatomical and physiological responses in both *Dicaeum* and *Loranthus,* and both of them have speciated freely together over their wide geographical range.

The final illustration, taken from Springett (1968), was quoted previously by Wilson himself (1980: 116), and carries us a stage further into what Wilson terms dependent specialization. It concerns in the first place two species of burying beetles, *Necrophorus humator* and *N. investigator,* and a mite, *Poecilochirus necrophori,* all members of a ground community on the Farne Islands off the Northumberland coast in northern England. The mite is phoretic on the beetles, which means several are carried about (commonly 10–30) on the beetles as they fly in search of carrion. The beetle species have resolved their potential competition for resources like Box and Cox by competitive exclusion in time. Each has a single generation per year; but the former hibernates when an adult and the latter when a prepupa. The *humator* adults emerge from hibernation in April and breed in May–June. After guarding their larvae until the latter pupate *in situ,* the adults die in August. Pupal diapause lasts until September, when the new generation of adults appear and hibernate. *N. investigator,* on the contrary, does not complete its hibernation and pupal metamorphosis till July. Its adults are eclosed in July, breed in August, and die after their larval offspring have hibernated in September. In short, each has its active stage while the other species is in diapause underground. These are all adaptations that could have evolved through individual selection.

When a suitable corpse (e.g. that of a small rodent or bird) is found, adult beetles assemble and collaborate to bury it; after that it is monopolized by one pair, and when their larval offspring hatch they eat the carcass. The female remains in the underground chamber with them and feeds them each time they moult, probably to re-inoculate them with digestive microbes. The mites reproduce at the same time, and their progeny emerge with the next crop of beetles.

In their quest for carrion to serve as larval food, the beetles have formidable competitors in the blowflies or bluebottles (*Calliphora* sp.) which lay eggs on exposed putrifying flesh. Springett found that when a beetle discovers and alights on a corpse, the mites it is carrying dismount, and rapidly search for blowfly eggs. Any they find are eaten immediately.

He showed by experiment that within 4 hours of arrival the mites could normally destroy the 100 eggs he had placed on the dead and eviscerated body of a woodmouse (*Apodemus sylvaticus*). Thus the mites keep the carcass free from the main competitors until it has been safely interred by the beetles.

He carried out a series of culture experiments, each with a corpse plus one of six permutations of the three species involved in the triangle, and each replicated eight times, as follows. (1) With all three present, that is, one pair of beetles, 30 mites and 100 *Calliphora* eggs; (2) omitting the beetles; (3) omitting the mites; (4) omitting the blowfly eggs; (5) with beetles alone; and (6) with blowfly eggs alone. The results were clear cut, and showed (1) that *Necrophorus* could reproduce successfully when alone, or, if blowfly eggs were present, when there were also mites to destroy them; (2) that blowfly larvae always won the corpse unless there were mites to destroy their eggs and prevent their development; and (3) that the mites failed to reproduce in the absence of a beetle. A year later he repeated the experiments, except that he introduced blowfly larvae instead of eggs. The mites could only kill very young larvae, and, as expected when bigger ones were introduced, *Calliphora* always finished by excluding the beetles. Otherwise the results were the same as before.

Here two species, a beetle and a mite, belonging to different classes of Arthropoda, each perform tasks that are of immense advantage and perhaps indispensible to the survival of the other. It is not entirely certain whether *Poecilochirus necrophori* is phoretic on any other species of beetle, though present indications are against it. It was never found on any other beetle species caught in Springett's pitfall traps; and he carried out 'preference' experiments, presenting the mites with a range of other beetle species including another carrion beetle, only to find the mites shunned them all. He also discovered that a dark brown fluid regurgitated by *Necrophorus* is a powerful attractant to the mites.

There are at least two episodes in these interlocking life-cycles where group selection clearly intervenes. First, the individual beetle that finds the corpse is thought to signal by pheromone for help in burying it. After collaborators have assembled and finished doing this, dominance contests for possession follow, male versus male and female versus female, which result in selecting one pair to reproduce. Finally the female is left alone in the cell to lay her eggs (the average is about 15 in *Necrophorus vespillo*) and attend the larvae (details from Linssen 1959: 163–4, quoting Pukowski 1933). Under individual selection, on the contrary, the finder would be expected to bury the corpse alone; and if conspecific competitors were to arrive before the task was finished he would have either to drive them off or give place to them. The second collaborative act is performed by the mites that provide a scavenger force of appropriate size, normally 10–30 individuals, sufficient to devour the number of eggs that the (also predictably homeostatic) blowflies can be expected to lay. Springett found that, though they still ate the blowfly eggs, the mites would not reproduce in the absence of the beetles. The fact that the corpse provides only a finite amount of food for the decomposers means that the life cycles of any of these three principal actors—beetles, mites and blowflies—could come to grief unless each had in-groups that were capable of regulating their own reproductive rates.

Wilson proceeds to consider the effect such mutualisms can have on a community within which dependent specializations of many kinds exist. A species will tend to benefit from (1) fostering the other species on which it depends, and (2) discouraging those whose presence is inimical to itself. This accords with my postulate that members of consumer species must become adapted to conserve their living food organisms; and it accords also with the widely accepted principle of niche diversification plus competitive exclusion, which forces inimical competitors out of the community. By the time these adjustments are complete every species could be contributing something positive to the welfare of the whole, and every species consequently made more secure and productive than it would have been, had it lacked the interspecific support it receives. Even predators can perform useful or essential tasks for their prey, as exemplified by their preferential killing of social outcasts among the red grouse; or by the famous population of mule deer on the Kaibab plateau in Arizona, which 'exploded' and overtaxed its food supply after the pumas and wolves had been destroyed by man (p. 39).

The question of whether an ecological community can evolve as a corporate entity or superorganism is frequently evaded, and rarely receives an unequivocal answer. The mutual benefits that can arise through interspecific and intraspecific cooperation are, however, all essentially the same in kind and principle, and relate to the conservation of habitat and resources for the benefit of successors in generations to come. In as much as the benefits are indirect the donors receive no reward for their actions in their own lifetimes; and that clearly implies that community integration can evolve only through some group-selection process. Wilson (1980: 107, 134) has constructed a model based on the existence of variation in the species composition of communities from place to place. These supply the counterpart of the local variance in allele frequencies found among trait groups, which provides the raw material on which his basic group-selection model operates. Communities are often stable and persistent, but that does not preclude the possibility that there could be some presence-or-absence differences, or relative abundance differences, among the member species found in spatially separate, natural replicates of a particular community. If these were to result in productivity or survivability differences between local units, the better balanced ones could be expected to expand in the course of time and supplant the less good, in a typical process of group selection.

The model itself does not prove anything, but it encounters no obstacles and indicates that the development of community integration accords with the principles of group selection. The conclusion that biotic communities can become more than the sum of their parts is shared by many ecologists. It receives strong support from the fact already mentioned, that in climax communities, at least, all three of the basic industries are normally active and mutually integrated, and have evolved a stable balance which maintains the status quo as closely as the physical environment allows.

19.4 Biospheric homeostasis and Lovelock's 'Gaia' hypothesis

The main environmental media—sea-water and the atmosphere—vital as they are to organisms, are with minor exceptions global in their ambience and subject to continual circulation. As a result, ecosystems cannot be wholly enclosed and self-contained. The media are common property to all the communities that make use of them.

We tend to take very much for granted the almost unbelievable ingenuity inherent in living organisms, and it cannot come as any great surprise therefore to be told that there are reasons for suspecting that biotic homeostasis extends even to these biospheric media as well. The case has been strongly advocated by James Lovelock (1979) in his book *Gaia*. Gaia is the Greek poetic name for the earth goddess, or as we might say, Mother Earth. His thesis proposes that the physical and chemical states of the seas and atmosphere have been influenced or manipulated by the world's biota for virtually the whole of the 3.5 aeons (aeon $= 10^9$ years) that life has existed, so as to improve the conditions for sustaining it. The evidence is not easily dismissed.

Take the seas and oceans first. Their mean salinity is about 3.4%. Adequate estimates are available for the rate at which the world's rivers add salt to the oceans; and contrary to the geography that I was taught at school, it would only take 80 million years, starting with fresh water, to bring the ocean salinity to its present level. The evidence from palaeontology and physiology is clear, namely that the salinity has remained much the same as it is now for a period 30 or 40 times as long as that estimate suggests. The body-fluids of the majority of marine organisms are kept hypotonic to sea-water, at salinity equivalents of less than 1%; and only a small minority of specialists among them can survive an external salinity of more than 6%, which for the rest is enough to disintegrate their cell membranes. If the sea is really prevented from growing more saline, as seems almost certain, there must be mechanisms for removing sodium and magnesium cations, and chloride and sulphate anions (to name the most abundant kinds), as fast as they are added. What the mechanisms are is still a mystery. We do know, however, that living organisms intervene on a massive scale in the removal of silicon from the sea. It falls from the plankton to the ocean floor as a rain of dead skeletons of diatoms and radiolarian protozoa, and much of it is incorporated into sedimentary rock. A parallel process occurs with the calcareous shells of foraminiferan protozoa; and large concretions of calcareous rock are also sequestered in coral reefs.

Organisms actively remove toxic substances from the sea by forming methyl compounds, for example with mercury, lead, arsenic, sulphur and selenium; and they do the same with fluorine, chlorine, bromine and iodine. These are all made volatile by methylation and escape into the atmosphere. What homeostatic mechanisms, if any, exist for regulating the outputs have still to be investigated.

If one compares the surface of our planet with those of its lifeless neighbours Venus and Mars, the contrasts are striking. Venus has lost all its water and Mars has little left. Both are too acid to support life: whereas on Earth water exists in enormous quantities, and both waters and soils are close to neutrality. Lovelock characterizes Earth as the water planet; and ammonia, organically produced as a nitrogenous waste, is, in his opinion, almost certainly responsible for neutralizing the rain, which would otherwise fall with an acid pH of about 3.

The Earth's atmosphere is no less anomalous. In Lovelock's words (1979: 67), its composition, when he came to examine it experimentally, appeared to be 'so curious and incompatible a mixture that it could not possibly have arisen or persisted by chance. Almost everything about it seemed to violate the rules of equilibrium chemistry, yet amidst apparent disorder relatively constant and favourable conditions for life were somehow maintained'. The atmospheres of Venus and Mars have come to contain 98 and 95% carbon dioxide, and no, or very little, oxygen; whereas on Earth the atmosphere contains 0.03% carbon dioxide and 21% oxygen. The Earth has enormous turnovers of these two gases due to photosynthesis and respiration. The oxygen figure of 21% is interesting for another reason, namely that it appears to be no arbitrary proportion. Lovelock remarks that oxygen is a potentially dangerous substance, and this is about all it would be safe for the biosphere to carry. One needs at least 12% of oxygen in order to be able to light a fire; but 25% would lead to raging conflagrations, lit by lightning perhaps, in which even rain forests and damp detritus would go up in smoke. The 79% of nitrogen which accounts for almost all the rest of the reactive gases in the atmosphere, is thus essential as a fire-extinguisher. On a lifeless planet, on the contrary, the nitrogen would be dissolved or deposited, largely as nitrate.

Gaseous nitrogen is produced and removed from the air in part by bacteria. Microbial metabolism also results in the release of two other gases, nitrous oxide and methane (the latter produced at high energy cost). The nitrous oxide is split up by ultraviolet light and thus *adds* to the oxygen content of the air. Methane on the contrary is a hydrocarbon and its oxidation in the air *removes* oxygen from circulation. Both are biotically produced in quantities that could contribute appreciably to any homeo-static control of the oxygen level. One may reasonably ask, why other-wise should 'expensive' methane be produced, and how else is the atmospheric oxygen proportion kept from changing? If methane produc-tion stopped, carbonaceous material might simply be interred in the ground.

The Earth is equally the 'coloured' planet, due to the blue oceans, and on land to the presence or absence of vegetation. The shades of colour, combined with cloud, ice and snow, determine its total albedo, or reflec-tion of incident solar radiation. It amounts to 45%. Some of the gases in the atmosphere, including carbon dioxide and water vapour, are responsible for the 'greenhouse effect', by delaying the escape of this reflected energy. As a result, temperatures in the biosphere are kept tens of degrees higher than they would be if there were no such gases present. Bearing in mind (1) the large organic turnover in carbon dioxide, in relation to its actual concentration in the atmosphere, (2) the amelioration of climate when the concentration is right, and (3) the lethal overheating that too much of it would produce, this effect seems a particularly likely candidate for Gaian control. In comparison, Venus is now too hot and Mars too cold to support life.

The cybernetics of these and other biospheric variables, if they exist, have hardly begun to be investigated, so that the lack of demonstration and proof of their existence is not surprising at this stage. We know that some global relationships, for example between primary production as a whole and decomposition as a whole, must be close and cybernetically connected; and the same is true of primary production as a whole and its consumption as a whole by animal consumers (see Chapter 3). Positive controls are essential to keep these balances stable. Environmental homeostasis is recognized as occurring on a smaller scale, for instance in the production and maintenance of soils, and of coral reefs. Against the background perspective that group selection has given, the evolution of vital controls in the sea and atmosphere appears as an entirely rational proposition. Lovelock reports that in the past 3.5 aeons the sun's radiant energy is thought to have risen by 30%, apart from temporal fluctuations; and when one considers how vulnerable the thread of life must be to cosmic changes on that scale, and to the secular changes of aeons in the planet itself, it is clear that there will have been strong selective forces at work to promote homeostatic controls over the global environment.

The most challenging gaps in the Gaia hypothesis, as it stands at present, are perhaps (1) the nature of the feedbacks from the global environment to the control programs of guilds of operators extending worldwide, and collectively capable of correcting imbalances and restoring the norm; and (2) how group selection could be brought to bear on what at first sight seems to be an indivisible unit, the global biosphere. Selection requires the existence of a set of separate, virtually autonomous subdivisions, within which variances in homeostatic efficiency can develop, and can be put to the test in terms of the survival and productivity of their biotas. Without 'competing' regional subdivisions, environmental homeostasis could neither evolve nor be maintained. Increased knowledge may of course show that there *are* sufficiently permanent, discrete, and comparable entities, covering particular continents or seas perhaps, in the atmosphere and ocean waters; and that living organisms have had the same opportunity of evolving controls over the composition of these vital media as they have over the perpetuation of a rain-forest or a temperate-zone soil. A fuller understanding of the subject is urgently needed because of the central part it may play in keeping life going on Earth. For the latest available references, see Lovelock (1983) and Watson and Lovelock (1983).

19.5 Kin selection

Kin selection is the name given by Maynard Smith (1964) to a process suggested by W. D. Hamilton (1964) that might account for the origin and evolution of altruism; that is to say of behaviour in which the performer restrains or foregoes his (or her) own selfish pursuit of personal fitness while contributing to the fitness of others—in this context his own near of kin. Altruism exists in the natural world, and Hamilton's hypothesis seeks to justify the fact in accordance with the principles of classical selection theory.

His basic assumption is that altruistically motivated species have populations so 'viscously' structured that a significant proportion of an average individual's companions or neighbours are close relations; and they of necessity possess replicas of many of the same genes as his own, derived from their common parentage. It is a consequence of normal biparental reproduction that a parent shares half its autosomal genes with each son or daughter; and the latter, as sibs, share half with one another, and a quarter with each step-sib, uncle, aunt, nephew, niece, grand-parent or grandchild, an eighth with each first cousin, and so on. When all these replicas are added together they give the performer in effect an expanded or 'inclusive' fitness potential; and this is a quantity that should also tend to become maximized like individual fitness.

On that basis it would be justifiable to sacrifice part or all of one's personal fitness provided the sum of the benefits one's action conferred on replicas of one's genes held by the recipients, exceeded the cost to oneself. It will be recalled, for instance, that in the telemetry experiments with male red grouse (p. 172), several closely related birds were found to have established contiguous territories as adults, giving them an inclusive fitness valued at 2 or more, just among their closest neighbours. Hamilton suggested that if populations were kept sufficiently viscous, as opposed to being randomized by dispersal, appropriate conditions might often be created.

The concept seemed to have particular relevance to the evolution of sterile worker castes in the eusocial Hymenoptera—a group on which Hamilton is especially expert. The major sacrifice of fitness the workers make has been a notorious puzzle to evolutionary theorists ever since Darwin (1859) realized that it might actually be fatal to his whole theory, unless he were to concede that the family could become a unit of selection. Eusociality has evolved at least 11 times in the Hymenoptera (E. O. Wilson 1975: 415), and only once (in the termites) in any of the 20-odd other orders of insects, suggesting that the Hymenoptera have some special predisposition in its favour.

This might well be due to the rather uncommon, facultative method of sex-determination that is widely prevalent in the order. Thus when a virgin queen is inseminated on her nuptial flight she receives the sperm into a reservoir, the spermatheca, in which they can be kept alive; and later when she lays her eggs she has the power of releasing, or withholding, sperms to fertilize them. Fertilized eggs produce females, and unfertilized eggs males; and the latter are consequently haploid. All the sperms a male produces are therefore identical (apart from rare copying errors) and unless the female (queen) is inseminated by several males during her nuptial flight, this alters the genetic coefficients of relatedness among her progeny. Daughters of singly-inseminated queens are related to their mother by the usual coefficient of ½, but to their father by a coefficient of 1, to one another by ¾, and to their brothers by ¼. In short, a worker female engaged in feeding and tending her younger sisters in these circumstances is contributing more to her own inclusive fitness than if she were tending daughters of her own. Initially at least, this bizarre relationship appeared likely enough to have been the predisposing factor that had triggered so many parallel evolutions of sterile female castes.

When the hypothesis is pursued further, however, it encounters a good many difficulties and objections. Hamilton's (1972) discussion of the diversity of eusocial adaptations found in Hymenoptera shows that there are many problems of detail for which it provides no answer. Two inconsistencies may be noted here, the first being that the multiple insemination of eusocial queens is known to occur in some species and may be a common phenomenon; and the second that there are many eusocial species, especially among the wasps and ants, that habitually maintain multi-queened colonies. Both these practices must tend to diminish inbreeding and the coefficient of relatedness between sisters in the colonies concerned, possibly to or beyond the point where workers have to tend larvae to which they are related by a coefficient of $< \frac{1}{2}$, that is, less than they would share with their daughters if they were able to regain fertility.

As regards population viscosity and mating-group size, there seems little evidence for supposing that the eusocial species differ significantly from the norms found among non-eusocial species. The colony is equivalent to an in-group of potentially enormous size and complexity; but it can still occupy an exclusive feeding territory, or share a home range that overlaps in part with those of other conspecific colonies in a typical manner. Eusociality appears to offer as efficient a mechanism as any for the homeostatic control of population density and thus for achieving high survivability and productivity; and this seems likely to be its primary function.

Kin selection offers no advantage over individual selection as an explanation for the evolution of the group adaptations connected with sociality, population structuring, and the conservation and management of trophic and genetic resources. This is so because 'inclusive fitnesses' are coincident with individual lives, and just as transitory. A population parameter such as effective mating-group size, which affects the permanent conservation of genic diversity in a population, cannot be optimized through a short-term individual selection that is sure to foster close inbreeding. Rather it necessitates a structured, Wrightian popu-lation, in which random innovation combines with selection to create diversity among neighbouring demes, and the demes are persistent enough to prove their long-term worth through their differential survival and productivity.

With consequences no less illuminating than before, Wade (1980) has exposed kin selection to experimental investigation in the laboratory with populations of *Tribolium confusum*. Finding a form of social behaviour that could be tested quantitatively was critically important: the behaviour of the performing individuals had to have measurable effects on the fitness of other individuals; it had to be subject to genetic variation between performers; the degree of relatedness between performers and recipients had to be open to experimental manipulation; and the costs and benefits to the performers had to be at least partially understood. These are all properties known to pertain to *Tribolium* larvae in their habit of cannibalism on *Tribolium* eggs; and although at first sight this may seem to be the exact opposite of altruistic behaviour, it is actually an important density-dependent factor in the population homeostasis of *Tribolium*. That would qualify it as a social behaviour within the meaning of Wade's definition of 'an interaction between conspecifics which changes the fitness of the interactants'. Should that fail to placate the reader, then at least he can be fairly sure that cannibalism on the eggs of kin is the kind of behaviour that ought to be selected against under kin selection.

The base population of the flour beetles was the genetically varied stock that had been used for the group selection experiments, recounted in Chapter 15. There were six treatments (detailed below) each replicated 15 times and continued for eight generations. Each line was started in a separate vial of medium with a pair of virgin adults taken from stock, and kept under the usual standard conditions. After seven days the contents of each of the 90 vials were examined, and five of the first-instar larvae were removed and transferred as a group to a new vial, in which they were allowed to go on feeding and growing for another seven days; all surplus larvae were rejected. Cannibalism on eggs tends to increase with larval age, and it was at the end of this second period that the larvae were presented with eggs on which to prey, and the initial step was taken in streaming them into six parallel treatments. The eggs were given in three batches of 15 at two-day intervals (45 eggs in all).

Eggs of three different parentages were used, each type being assigned to two treatment streams that were going to be channelled apart at a later stage. The first 30 vials all received eggs that were full of sibs of the larvae in the vials, having been laid more recently by the same parents (Treatment R .50); the second 30 received half-sib eggs laid by the same mother after being re-mated with another male from within the set of replicates (R .25): and the third 30 received non-related eggs from the parents of a different line of larvae in the set (R .00). The larvae and surviving eggs were then allowed sufficient time to develop into pupae. Incidentally the viability rates of all the sets of eggs used were separately checked because they affect the rates of survival and cannibalism.

Thereafter the pupae in all groups were sexed and the sexes held separately, vial by vial, until the adults had emerged. Some groups were more populous, some less, as a result of the different intensities of cannibalism they had experienced; and in this way selection caused by the behaviour trait had taken effect, and resulted in changes in the mean frequencies of genes controlling cannibalism in the different sets of survivors.

In making up propagules for the second and all subsequent generations, two alternative mating systems were adopted, namely random mating and within-group mating, thus completing the separation of the treatments into six streams. The 30 replicates subjected to each larva-egg relationship were divided into two lots of 15. In those assigned to random mating all the young males were pooled and placed in one finger-bowl, and all the young females in another. When they had had time to mix, a male and a female were picked at random and placed in a new vial as parents; and this was repeated until 15 similar lines had been set up. Because the parents had been picked at random they provided a statistical sample of the gene frequencies surviving to maturity in their generation. In those replicates assigned to within-group mating, the male and female for each parental pair were picked from the same single vial of survivors (and were therefore all full sibs in the R.50 treatment). To make these parents also reflect the overall surviving gene frequencies, the relative size of each single set was taken into account and, by the use of a table of random numbers, each was given a probability of supplying 0, 1, or more of the 15 pairs required, in proportion to its size. These procedures were all repeated in the propagation of eight successive generations.

The three degrees of kinship between the interacting larvae and eggs and the two mating systems employed with each, can be regarded as comprising six differently structured populations. Interaction structure (e.g. altruism towards relatives) was considered by Hamilton to play a leading part in social and eusocial evolution; and mating structure (i.e. whether or not a Wrightian population was essential) was considered by Maynard Smith (1964) to provide a criterion for distinguishing group selection from kin selection. Wade included these mating structures in his experiments in part because he disagreed with Maynard Smith's view that kin selection was a form of individual selection. Instead it seemed to him, in its requirement of a viscous population, to depend on structuring and thus more to resemble group selection. Under either mating regime it was to be expected that larval cannibalism would be diminished by natural selection in the R .50 treatments, as compared with the R .00, and that the R .25 would come in between; but it was also predictable from Wade's earlier experience that the rate of evolution would be faster in the within-group rather than the random mating lines.

This is what the results showed. At the end of the first generation there were no significant differences in cannibalism rate between any of the six treatments; but at the third generation, after two episodes of selection had taken place, the within-group mating lines had attained significant differences between all three larva-egg classes, in the predicted order, and these were retained throughout the rest of the experiment. (After generation 3 there was no further significant divergence between them.) The random matings on the contrary produced no consistent or significant differences in any generation. What had emerged, in short, was that the effects predicted by kin selection did occur, but only in populations structured to retain the integrity of their mating lines from generation to generation—that is, only under group selection conditions. Where mating was random the incipient variances were immediately extinguished by outbreeding.

In a later paper Wade and Breden (1981) constructed a model to investigate the evolution of altruism by kin selection in a family-structured population, and by means of computer simulations brought out the powerful contribution that inbreeding would make to its achievement. They concluded that kin selection might be regarded as a potentially viable balance between the forces of group and individual selection. Obviously their model was not considered against the fuller natural perspective of group selection available in this book, which shows the enormous benefit that trophic resource management can bring to fitness levels. This benefit relies on, and assures the maintenance of, population structuring and its ancillary group adaptations, such as the retention of genetic variance, which can be seen to take place in nature and to contribute such unique potential for long-term survival. In this setting, close inbreeding and the homozygous straight-jacket it would bring would be distinctly dangerous, even though, in a living world that contains numerous apomictics, it might be rash to say it could not survive.

Summary of Chapter 19

1   The belief that group selection cannot be effective as an evolutionary force is based on a false assumption, that cooperation within a group requires sacrifices at the individual level, whereas it normally pays large dividends. Food resources can be managed through cooperation, and then productivity built up and sustained; and non-cooperating individuals, being a drag on group productivity, can be suppressed by group selection. Similar benefits arise from the demographic management of genetic resources through population structuring. The automatic tendency for selection to favour the more fecund individuals and thus continually to press for population increase, appears to be countered and overcome by the penalty group members would pay if their food demands were allowed to exceed the carrying capacity of their resources. A balanced compromise, made evident by the evolution of population homeostasis, has presumably resulted from these opposing forces. Individual fecundity has been made adaptable to changing resource levels so that overexploitation is avoided; but basically fecundity is a metric character, and endowed with a corresponding degree of stability.

Group selection may have originated in a population living in a naturally subdivided habitat, which imposed suitable limits on group size and the interchange of genotypes. Between-group variances would then have arisen spontaneously and led to local differences in group productivity and survival, thus fostering the seeds of cooperation that would soon bring appreciable advantages. Later they would have enabled the machinery to improve until structuring became self-imposed in any kind of habitat. The first steps were probably taken early in organic evolution, and all surviving forms of life might then have descended and radiated from the same few types of microbial pioneers. There is no reason to doubt that all contemporary organisms are geared to group selection, except for a small minority of apomictic species that have forsaken genetic interchanges between individuals and, with them, the main mechanism of evolutionary change.

2   The first hypothesis to show that some type of group selection might be theoretically simple was D. S. Wilson's (1975). It was founded on the observation that individual animals generally belong to neighbourhood groups (trait groups) in which they interact with one another. Wilson simplified his model to the bare essentials in order to facilitate its mathematics: thus individuals were assumed to be haploid (no heterozygotes), and all genetically identical except at one locus where there were two alleles, A and B. Bearers of A were cooperators and those of B non-cooperators. A trait group could be formed, say, by insects hatching from eggs laid singly by females on a leaf, or by the gathering of juvenile insectivorous birds into flocks. A-individuals promote the common good e.g. by warning of approaching predators, whereupon all A's and B's benefit by evasive action. The A 'donor' that gives the alarm incurs a small cost; and the elementary model applies only to social behaviour in which the benefit to the donor exceeds this cost, i.e. only to 'weak' altruism.

Since trait groups form at random in the habitat they are statistical samples of the population, and will differ among themselves in their exact proportions of A's and B's because of normal sampling errors. Groups with relatively more A's suffer fewer casualties than those with more B's. It is an arithmetical fact that in a set of such assorted groups a majority of the A's (or the B's) will belong to groups that contain more A's (or B's) than average. Consequently when, later, a dispersal stage intervenes and mingles the members of the groups together into a common deme, probability dictates that the frequency of A's will have risen and that of B's fallen.

If the A's and B's were not assorted at random, e.g. because A's tended to associate, the costs of strong altruism could similarly be borne without adverse selection (see Section 5 below). During its life cycle an individual may belong to several trait groups; and it is a limitation imposed on the model by its artificial simplification that the benefits of cooperation which the groups afford must be experienced directly by their members. Long-term benefits like those obtainable in philopatric demes through their continuing inheritance of group phenotypes and the conservation of food resources that this promotes, are not accounted for.

An important innovation in Wilson's model is his recognition that allele phenotypes are not likely to be uniformly mixed within a population, nor identically encountered by every member in its interactions with other members, as classical selection theory assumes.

3   Assuming that the real scope of group selection is much wider than this simple model justifies, Wilson (1980) carries the same principle beyond the single-species population, to the evolution, productivity and survival of multi-species communities, in which mutualisms are well known to occur. At this level the variance on which group selection can presumably work would be found among community replicates occupying separated islands of habitat, and would consist of variations in species composition and proportions, and degrees of interspecific integration. Some forms of mutualism or symbiosis can reasonably be accounted for by individual selection, e.g. beneficial partnerships between single individuals belonging to two different species. Some cases of mutual exclusion, e.g. of one species that must give way to another because they compete openly for the same resource, are similarly explainable. But group selection is necessary for building up mutual benefits that are indirect, as might be true of corvid birds is they hide nuts and cones in the soil in habitats where they can grow, and leave many of them in place to propagate the trees that will yield crops for a later generation. In the text, four comparative examples can be found, beginning with a hermit-crab/sea-anemone symbiosis and ending with a complex mutualism between burying beetles and mites. Incidentally, even predators can

provide group benefits for their prey. The more advantageous or indispensable interspecific bonds there are, the more the coherent community resembles a superorganism in its ability to respond as a whole to selection.

4   The case for considering that the biota of the biosphere as a whole is cybernetically manipulated, for the purpose of obtaining homeostasis in the physical environment, has been powerfully advocated by J. Lovelock (1979). The balanced nutrient cycling that proceeds on a vast scale through the integrated activities of primary producers, consumers and decomposers, more or less worldwide, has already been remarked on (Chapter 4). Lovelock's 'Gaia' hypothesis begins with the realization that during most of the 3.5×10^9 years that life has existed, its broad physiological demands on the environment can have changed very little, and that this has been possible because the environment itself has been extraordinarily static. The mean salinity of the oceans has apparently remained in the region of 3.4%, its present level, notwithstanding that the Earth's present rivers could contribute that much salt in 80 M years. Only a minority of marine organisms can survive a 6% salinity; yet no mechanisms are known for removing several of the most abundant ions (Na, Mg, SO_4). Silicates and carbonates, on the other hand, are precipitated by organisms on a colossal scale.

Earth's neighbours, Mars and Venus, have little or no water. Both are too acid for life to exist and both have atmospheres with 95% or more of CO_2. Venus is also too hot and Mars too cold. The Earth is strikingly different, for reasons quite unknown unless the Gaia hypothesis is correct. The atmosphere is very far from the steady-state chemical equilibrium that has been approached on the nearby planets. It is a remarkable mixture, of 21% oxygen (neither dangerously much nor suffocatingly little) and 79% non-dangerous nitrogen gas (not all dissolved or deposited as nitrates as expected). It has a CO_2 content of only 0.03%, constantly being abstracted by green plants and restored by respiration. The planet is coloured, by blue oceans, green vegetation, deserts, clouds, snow and ice, which combine to give it an albedo or reflective coefficient of about 45% of the incoming radiation. The atmospheric CO_2 and water vapour suffice to create a greenhouse effect by slowing the escape of heat, and keep the biosphere habitable, not frigid as it would be otherwise.

Plants of course possess the incomparable faculty of storing energy and recycling oxygen; but there are also microbes that can fix, and release, atmospheric nitrogen; and others that, at great cost, produce methane gas, most of which oxidizes after release and lowers the oxygen level in the air. Still others variously gasify dissolved toxic metals and halogens. The continuity of life proves that the biosphere has never become lethal, notwithstanding an estimated increase in solar radiation of 30%, with added fluctuations, over the period.

There is enough evidence to suggest that hypothesis is likely to be right. Two large gaps remaining are (1) how global feedbacks could be monitored by the myriad operators, and (2) how selection could have taken effect when there is only one unit of selection, the biosphere itself. Can it be effectively subdivisible?

5   W. D. Hamilton's (1964) hypothesis of kin selection, as a possible explanation for the evolution of altruism, is finally considered. Individuals who are related to each other hold genes in common, due to their common parentage, in proportions varying with the relationship. In classical theory one's fitness is measured by one's success in passing on genes to the next generation; but the kin-selection theory proposes we should take into account the 'inclusive fitness' that the sum of our genes achieve, in whatever individuals they are to be found. If this were done, and provided the relations can be identified and evaluated, then it might be possible to increase one's inclusive fitness by making sacrifices on their behalf. If there were enough of them, it could even be worth giving one's life to save a commensurate number of theirs.

Related neighbours are probably common in nature as a result of close philopatry; but the hypothesis finds its strongest support in the eusocial Hymenoptera (bees, wasps, ants). By definition they have a caste of more or less sterile female workers which normally retain no fitness except indirectly, through the queens and males they nurse and protect. Most species have colonies consisting of family units with a single queen. Hymenoptera appear to have an inherent predisposition to sociality, in as much as it has arisen in them in parallel about 11 times. Hamilton suggests this stems from their unusual 'haplodiploid' method of sex-determination. Fertile females are inseminated during a nuptial flight with a lifelong supply of sperms, and when they lay eggs they can choose whether the eggs are left unfertilized and become males, or are fertilized and become females. Males, being haploid, produce all identical sperm. Sisters with the same parents consequently share ¾ of the same genes instead of the normal ½, and are genetically closer akin to the younger sisters they nurse than to the daughters they would bear if they could have any. Their inclusive fitness benefits accordingly.

There are problems of detail among the eusocial species which this hypothesis does not resolve. Kinship ties are strengthened by inbreeding; yet the multiple insemination of queens by several males, and the habit of forming many-queened colonies, are both known to occur, the latter in numerous species. More seriously, inclusive fitness can only last for the lifetimes of individuals, and the hypothesis cannot therefore account for the evolution of adaptations such as resource conservation and the genetic structuring of populations, in so far as their benefits are deferred until later generations. Likewise the close inbreeding that would strengthen kin selection would lead to gene fixation and the loss of adaptability to environmental change.

Wade (1980) has experimented with kin selection, using *Tribolium* populations, and cannibalism by the larvae on conspecific eggs as the social interaction under test. This can be regarded as negative altruism or victimization, and the expectation is that it should be selected *against* when the larvae are presented with eggs closely related to them, and to a lesser extent with eggs coming from more distant or unrelated sources. The experiments ran for nine generations, using eggs that were either sibs, half-sibs or not related to the larvae; and using two mating systems, one outbred, the other kept within the same hereditary line. Only in the latter, which perpetuated the genetic structuring, as previously done in the group selection experiments (Chapter 15), were significant results obtained, in the expected rank order as regards larva-egg kinships.

This confirms the opinion that there is no evolutionary effect that kin selection could produce which group selection cannot produce more satisfactorily, particularly for benefits that are necessarily delayed, and for avoiding gene fixation; and only group selection can account for the evolution of adaptations for trophic and genetic management.

CHAPTER 20
SOME PROBLEMS SOLVED

20.1 Eusocial cooperation and group selection

Cooperation is an especially conspicuous attribute of the eusocial insects. In species where eusociality is well developed their populations are structured, at the lowest level, into superorganismal societies; and the individual members are specialists belonging to one or other of a set of mutually supportive castes. The castes may be polymorphic, and all of them are necessary for achieving full production. The three basic ones are those of male and female reproductives, and of workers. The latter construct, tend and guard the nest, defend the feeding territory, gather provisions, and feed both the reproductives (especially the egg-laying females) and the helpless young. Workers are typically sterile.

All the termites (Isoptera) are in the advanced category as regards their state of eusociality. Among the Hymenoptera the same applies to all the ants; but in the more heterogeneous groups generally known as wasps and bees there are still many 'solitary' species, and eusociality has arisen repeatedly among them, sometimes more than once in the same taxonomic family. As a result we have a range of stages of complexity of social organization, marking the evolutionary path (or paths) of eusociality from the rudimentary to the most sophisticated states (cf. E. O. Wilson 1975: 397; Andersson 1984; Brockmann 1984). This shows clearly that eusociality has developed from antecedent social states like those of the solitary bees and wasps, and broadly resembling the common forms that exist in most of the higher animals.

In an apparently primitive example, namely the species of *Polistes* (paper wasps) resident in temperate Europe and North America (the majority of the species being tropical), only fertilized females overwinter, in solitary hibernation. When they emerge in spring most start to build themselves an individual nest. But many quickly give up, perhaps because their food reserves have run too low, and join forces with another female, so that several usually assemble and cooperate in building the first small comb. Before any eggs are laid a hierarchy develops among the

341

females as a result of threat displays. The one that emerges as dominant from then on lays all the eggs, and the rest are relegated to serving as workers. Thus they invoke the normal social mechanism for apportioning communal food supplies and limiting quotas of breeders. Incidentally, sharing the same nesting site or the same nest is a common practice in Hymenoptera and has had much to do with the eusocial structuring they have evolved.

When the first cohort of eggs hatches, the adults feed the larvae on freshly chewed invertebrate food-items; but the queen increasingly tends to remain on the nest, where she is also fed. The emerging offspring all become workers, constructing the accommodation for further batches of eggs; and the reproductives are the last castes produced towards the end of the summer. In warmer, less seasonal climates the *Polistes* species are active the whole year, and form new colonies by dispatching swarms containing young fertilized queens and numerous workers (Richards 1953, E. O. Wilson 1975).

These and other observations indicate that eusociality is just a special type of social organization, though perhaps a uniquely elaborate one, and give no reason for doubting that it still assists in the normal functions of controlling population density, reproduction, and gene flow. Although it is a rare phenomenon, considering the animal kingdom as a whole, it can obviously perform efficiently as a group adaptation. Its special feature of employing sterile workers does in fact have partial parallels elsewhere, in the many bird and mammal species that employ non-breeding adults as helpers at the nest or den, or are organized to permit only one reproductive pair (or fertile female, as in some wolves and foxes) at a time in the in-group. There are less direct parallels too in the deferral of reproduction among young adults, resulting from the domination and exclusion practised by their seniors, which is widespread in long-lived vertebrates and often creates a standing reserve of potential breeders.

The reproductives among the eusocial Hymenoptera, during their development, and up to the time of their dispersal, insemination and establishment, are subject to individual selection in the normal way. If the females survive long enough to become the sole queen in a colony it seems probable that they will as a rule have been multiply inseminated. In that event they are probably best regarded as having been members of transient mating in-groups of males and females, and as now being the only survivors, each of them carrying a sample of the group's gene pool in their ovaries and sperm-receptacle. The relative fitnesses of the colonies they mother can then be accredited to local mating in-groups.

Colonies of termites differ notably from those of the eusocial Hymenoptera in as much as they are, with rare exceptions, founded by a single pair of reproductives which do not even mature and copulate until they have sought and found a suitable nest-site, excavated a nuptial chamber, and sealed themselves inside it. Their colonies tend to grow slowly at first, but, especially in the higher termites, the queen eventually becomes enormous and capable of laying tens of thousands of eggs a day. By that time the colony may be large enough for supplementary reproductives to be specially nourished and matured; but they will still be the progeny of the same founding pair. Thousands of alate (winged) young reproductives are released each year (the nests may survive up to a few decades), but the alates are weak fliers and are soon grounded. Little is known about the genetic structuring of their populations, although it seems clear that in-groups, if they exist at all, must be even more transient than those of the single-queened eusocial Hymenoptera. The coefficient of relatedness between each of the founders and their sons and daughters will be exactly one-half, and between the individual progeny it will be on average one-half.

Classical selection theory is unable to explain how selective forces can act to produce eusocial units which contain polymorphic, sterile individuals with mutually complementary functions, in addition to the reproductives. The same applies to those adults in some species of vertebrates that are rather similarly coerced into remaining subservient and assisting in the reproduction of their more dominant in-group companions. The intense interest and discussion that Hamilton's kin-selection hypothesis has aroused among evolutionists in recent years stems from the possibility that it could offer a way round this difficulty. In practice, however, it has not been able to solve all the problems, as already mentioned with regard to the eusocial Hymenoptera (p. 334).

When one turns to the vertebrates (for review papers see Macdonald and Moehlman 1982; S. T. Emlen 1982, 1984; Ligon 1982; Brown and Brown 1982; Woolfenden 1982), its explanatory value appears to vary from species to species, in the opinions of these various authors. In the grey-crowned babbler (*Pomatostomus temporalis*) the Browns found that helpers were for the most part closely related to the young they cared for; in the woodhoopoe (*Phoeniculus purpureus*) Ligon showed that they were usually but not necessarily offspring or sibs of the breeders; and in the Florida scrub jays (*Aphelocoma coerulescens*), many of them assisted breeders other than their own parents, and sometimes helped at such nests even though there were closer relatives at other nests nearby. Woolfenden, like the others, concluded that helping was a poor alternative to being a breeder, as a means of increasing personal fitness, though in a saturated habitat where vacant space was difficult to obtain it could contribute something to the helper's inclusive fitness. Its main advantage, to the more promising of the males at least, appeared to him to lie in the opportunity it gave for helping to gain extra ground and enlarge the group territory, to the point where the aspiring helper could carve off a section of it and take it as his own.

As to why young adults help instead of going to breed elsewhere, Emlen (1984), referring particularly to the white-fronted bee-eater (*Merops bullockoides*), discussed four possible types of ecological constraint that might tip the scales against breeding; these incorporate the one just mentioned above. They were (1) the high risk attached to dispersing, in comparison with staying on home ground and, with luck, eventually rising to breeder status; (2) a lack of vacant territories anywhere in high-quality habitat; (3) a dearth of sexual partners especially in the many species of cooperative breeder that have a chronic shortage of females; and (4) the prohibitive cost of attempting independent reproduction, especially in poor seasons in fickle environments, where changes in resource levels occur often, and happen too fast for the resident bird population to track them.

Except perhaps for number 3 these factors are of course general causes for restricting the recruitment of new breeders, applicable to almost any vertebrate consumer population and not peculiar to 'helper' species. An example appeared in an earlier chapter (p. 20) of a tawny owl population in which none, even of the established pairs, bred in a year when their rodent prey were unusually scarce.

We have seen that the kin-selection effect would rapidly weaken as the closeness of inbreeding diminished; and it is worth noting that in the helper species just discussed it tends to be the more closely philopatric male sex that predominates among the helpers. The females, as usual in birds, have presumably dispersed on average to a greater distance, and may have suffered a heavier mortality as well. Observed population structuring, though it confirms the prevalence of close philopatry, seems normally designed to prevent very close inbreeding rather than to promote it.

If one takes the group-selectionist viewpoint instead, one accepts in-groups and demes as being valid units of selection, and their survival and productivity as being the measures of their relative fitness. It is then obvious that to employ extra adults as nurses and guardians in seasons when the carrying capacity is not sufficient to allow them to breed independently, could increase the productivity of the established breeders, and thus of the group as a whole; and at the same time it will retain the extras in readiness as potential breeders to exploit a sudden upturn in resource levels, should that happen to occur. In group-selection theory there is no problem about sacrificing the fitness of some individuals if it benefits the fitness of the group as a whole to do so; and this applies not only to vertebrates in changeable habitats but to the special-duty sterile castes of insects as well.

Under group selection, in-groups and demes can gain these advantages and many others besides, without being particularly closely inbred; and the extra freedom this allows them in dispersal is actually consistent with the geometry of population structuring, as we commonly observe it in the field.

Employing helpers is only beneficial in the discharge of certain particular activities, one of which is the nourishment of callow young. The added presence of helpers in the vicinity of the nest may sometimes attract greater attention from predators or nest-parasites, but it is also true that helpers can assist in guarding the site against intruders. Most eusocial insects have formidable weapons for defence.

20.2 The sexes as cooperative castes

Few problems have puzzled evolutionary theorists more in recent years than the question of how reproduction involving male and female sexes, and the genetic recombination that results from it, came into existence, and have subsequently been maintained. It is easy to see that gene recombination is beneficial in the long run, because of the genetic variance it generates for natural selection to work on; but it is generally taken for granted that in the short term the individual could do better for itself if it could pass on its genome complete to each of its offspring, instead of having to go halves with an obligatory mate. A minority of animal and plant species exist in which the individuals do have this faculty of apomictic reproduction, either parthenogenetically through the female line or, in the lower phyla, by some form of fission. In many of the species nevertheless it is used just as a temporary expedient, and bi-parental generations are regularly interpolated.

Could this general commitment to the duality of sex have arisen as a result of individual selection, or does it demand the intervention of group selection instead? Maynard Smith (1978: 2) after writing a very illuminating book on the question, stated in the preface that he was still unable to make up his mind. I like to think this was due solely to an error made in his basic assumptions, namely that traits advantageous to the group in this context must be deleterious to the individual—the assumption I am trying here to falsify. For the same reason Hamilton *et al.* (1981: 376) concluded, in their paper on the problem of sexuality, that 'we still do not know what sex is for'.

The alternative view, namely that their basic assumptions are at fault because the group and not the individual has been the primary target for natural selection, is perhaps more convincingly supported by the genetic system than by any other major field of adaptation. To realize this one only needs to examine critically what sexual reproduction has been selected to achieve. Its central task is to produce novel genotypes in each generation, no two of them being alike except by monozygous twinning; and this endless diversity is generated by introducing a sufficient measure of randomization into the reproductive sequence of gameto-genesis and fertilization. The pairs of chromosomes that are present in the parent's diploid set are not identical to one another because many of their loci are heterozygous, and occupied by different alleles. There are two well-known factors in the process of separating the members of the pairs to make up haploid chromosome sets for the gamete nuclei. The first is crossing-over, which exchanges blocks of genes between one member and the other before the final separation takes place; and it results in new associations of genes. Because crossing-over is sufficiently randomized, every single dividing germ-cell can produce its own array of new associations. The second random factor is that when the paired chromatids are separated at the second maturation division and travel to opposite poles to form the gamete nuclei, the segregation is also a random one. Consequently, though any given gamete contains exactly half the parent's autosomal genome, the alleles on its chromatids have been randomized for the second time, making each gamete genetically different from the next.

Then, when it comes to fertilization there is a random element in how individuals come to be juxtaposed, and mate to become parents, although they may often be drawn from quite a small group. The result of this is to shuffle the gene combinations in the gene pool as well and thus produce a still wider variety of new genotypes for selection to work on. At the next higher level in the population, Wrightian structuring accomplishes a closely parallel process by producing another tier of unending innovation between local gene pools as units, again for selection to work on. At no stage, however, do any of the innovations mentioned cause, of themselves, changes in gene frequency in the population at large: these are changes to be determined by drift (i.e. small breeding in-groups comprising random samples of genotypes) and, above all, by selection.

Bisecting the parental genomes during gamete formation, and uniting two gamete nuclei to create a new genotype, are indispensible to the randomization. The comparative rarity of species that are permanently apomictic must presumably signify that the production of supremely fit genotypes, possessed by individuals endowed with ever longer life-spans, during which to pass them on intact to the next generation (which is what individual-selection logic rather leads one to expect), is not a generally reliable formula for securing the survival of the race. In practice it is as essential for the genomes to be halved as it is for their bearers to be mortal; and it is as important (in the long run) for the breeders to carry recessive alleles as it is for them to exhibit dominant ones. The system is unmistakeably geared to engendering variance, and not to attaining and fixing the ultimate in individual perfection.

As far as carrying out these complicated genetical manoeuvres is concerned there is no reason why the mated individuals should belong to separate male and female castes. In many invertebrate taxa the mating pairs are identical hermaphrodites, although even here there are polarized sexual roles which cause the mobile sperms to seek out the larger, immobile ova. Complete separation of the sexes has only become the almost invariable rule in the higher animals, and the alternatives of dioecy versus hermaphroditism are much more equivocally interspersed in the lower animals and in plants. The main reason is presumably that the higher animals are social, and it is in this context that the sexes have tended to develop a division of labour, which consigns females as a rule to bear the main burden of reproduction, and the males the main burden of mutual competition and homeostasis. In social species there is thus a non-genetic motive for turning the sexes into dimorphic castes, mutually complementary for demographic as well as genetic purposes.

It needs to be emphasized that adaptations which decree sex and

impose dimorphism (or polymorphism) on individuals in their morphology, physiology and behaviour, are just as essentially group adaptations
as those that segregate and recombine their genes for the perpetuation of
gene-pool variance. There seems no possibility of evolving either of these
results by means of individual selection alone.

20.3 The resource management syndrome

The selective advantages of possessing sterile castes and dimorphic sexes
are not the only theoretical perplexities that the group selection hypothesis appears to solve. Throughout the book it has shown its capacity for
suggesting how anomalous pieces of the evolutionary puzzle can be fitted
into place; and in this and the next sections of the chapter the more
important of these explanatory ideas are reassembled. Most of the group
adaptations we have identified are connected with one or the other of two
great adaptational syndromes, for the control of food-resource consumption and for gene-pool management respectively, and though they overlap somewhat it is not difficult to separate them and devote a section to
each.

Resource management should come first because it is the means of
attaining the superior group and individual fitnesses on which success in
other spheres depends. The management principle postulated here, as
the reader knows, is to control consumer demand by matching the
population density to existing levels or anticipated changes in the rate of
food supply, and admitting only as many consumers as can expect a
sufficient diet without overexploitation. Supernumeraries have to be
expelled. The response the individual has to make is no doubt automatic,
and depends on feedbacks received, regarding the quality of the habitat,
the current plane of nutrition, and the intensity of consumer competition.
The general result is to diminish an avoidable risk to the survival of the
group.

As it stands, the postulate just made is strictly hypothetical; but the
evidence behind it is actually very strong. It has been demonstrated many
times by experiment that changing the plane of nutrition can change the
consumer density, although the actual food is not at the time in short
supply. As a hypothesis, the postulate explains and unifies the function
of the systems of property tenure that exist, some of which have become
conventionalized into token holdings, or sublimated into abstract rankings in a hierarchy which need not involve holding any individual

property at all. When I originally made the suggestion in 1962 that these were complementary exclusion mechanisms there was no pre-existing concensus of opinion on the general functions of territory, nor had any link been recognized between territories and the (then) equally confusing phenomenon of hierarchies.

At the same time sociality itself was seen to be coincident with the conventionalizing of intraspecific competition and with cooperation for the common good, and its evolution by group selection was postulated. A corollary has been presented in the preceding section of this chapter that the predominance of males in social affairs arose long ago as a division of duties between the sexes, and that this division was the motive for evolving sexual dimorphism, which is of general occurrence only among social animals.

The improvement that food-resource conservation has generally brought to the carrying capacity of vivivore habitats results at any instant from the good management practised by consumers in the past—partly the recent past though often the more distant also, through prudential exploitation by previous generations. This potentiality for improvement may have given an early stimulus to selection for philopatry, which holds the gene pools that programmed the conservation in place until the deferred rewards mature. Philopatry is however also an important component of the genetic structuring of populations, and is mentioned again in the next section. It has a third type of advantage as well, when one interprets it to mean the permanence of individual domicile, in allowing social acquaintances and relationships between individuals to develop. This saves competitive effort by enabling neighbours to live on stable terms, and it also enables mated pairs that have parted after one breeding season to reunite before the next. Incidentally, canalizing competition by holding special events for the purpose is a common social practice.

In the longer term philopatry makes it possible to hand down traditions from one generation to the next, mostly relating to geographical locations such as individual breeding sites, colonies, lek arenas, dormitories, and hibernating or aestivating places. These serve, through competitive events of the kind just noted, as foci for the acquisition of epideictic feedbacks, or for direct controls over population density through conventional contests for sites and the exclusion of losers. The advantages gained are shared by the established members of the group in their joint capacity as bearers of the local gene pool.

The high intensity of social competition normally seen in the process of determining the appropriate density or quota of breeders each year, and the normal role the male sex plays in it, lead one to the conclusion that individual feeding territories, nest-sites in colonies, and stances or courts in an arena, are alternative types of prizes for males to win, in deciding which of them are to breed and how many females are to be inseminated per unit area. In the non-breeding season sleeping sites in a dormitory, for example, are likely to be held by males and females on equal terms.

In Chapters 17 and 18 an examination was made of the population dynamics of a selection of hermaphrodite molluscs and flatworms, as representatives of animals remaining in a presocial condition as regards cooperation and competition and picked out because they had been well enough researched to supply the necessary data. They too were found to possess homeostatic adaptations for the regulation of consumer demand in the interest of conserving their trophic resources. They gave evidence of group selection at work, no less convincingly than one finds among vertebrates and arthropods. Their population homeostasis depends for its inputs, as usual, on monitoring the levels of food supply and consumer density; and, for its output, on making correlated changes in consumer demand: that is, on accepting or repelling immigrant recruits, on increasing or decreasing reproduction, and on getting rid of, or incapacitating, part of the resident in-group when necessary. It was remarkable to find all these functions being discharged by helminth parasites living inside vertebrate hosts, considering that the only kinds of information they can receive from outside their bodies are chemical and tactile. Their feedbacks nevertheless allow them to monitor the abundance of their own free-living larvae, and, in some species, to make the adults aware of and respond to, a heavy mortality occurring in their intermediate hosts. Their in-groups, isolated in individual hosts, can moderate their collective trophic burden on the host, and also their egg-production rate, through a hierarchical mechanism that inhibits or expels some members of the group if need be. Population homeostasis, complex process though it is, appears indeed to be one of the most universal faculties in the animal kingdom; and, of course, it is one that could not evolve without group selection.

20.4 The genetic management syndrome

We have seen indications of the spatial structuring of populations sufficiently widespread to suggest that its occurrence may also be general:

clearly it is so in the vertebrates. The best evidence comes from birds, which are relatively easy to mark individually and to rediscover subsequently, sometimes without needing to be re-caught or killed. I have repeatedly stressed the significance of the sex differences that have recently been discovered by marking techniques in the mean dispersal distances of many bird and mammal species, and equally the significance of the statistical distributions of the individual dispersal distances, because these imply that dispersal must have been subjected to natural selection and therefore must be genetically controlled. In view of the array of individual distances that is generated in each dispersing year-cohort, it is clear that the selective advantage of controlling dispersal does not accrue to individuals separately, but is realized through its statistical effect on the population as a whole.

There is only one obvious way to interpret the survival value of the structuring that results from controlled dispersal, which is that it performs the function perceived by Sewall Wright, of maintaining a system of local gene pools, small and isolated enough to become differentiated one from another by drift, as well as by natural selection. He was of course able to predict that this would greatly increase the genic resources and frequency variances that a population could contain, and increase the chance that somewhere a local unit would exist that was already pre-adapted to cope with any general change in the environment whenever it occurred. Such a fortunate unit could then rapidly expand, while the adversely affected units declined. His type of structuring would keep the gene-pool units in a perpetual state of flux, and evolutionary creativity would thus be 'raised to the second power'. Rather than all conforming with a single optimal model, different species appear in fact to differ in the particular geometries that their dispersals produce; and this may be due in part to the other functions which dispersal has to perform, such as the recolonizing of vacant habitats, which are likely to differ from species to species in scale and frequency.

Selection falls on structured populations at both individual and group levels. Individuals are exposed to risks of injury or death from extraneous agents including predators, infections, and extremes of 'weather', and they differ genetically in their ability to evade or survive them. Within their in-groups, individuals face still greater risks from the effects of intraspecific competition. The competition normally consists of ritual duels between in-group members to establish, at relatively low cost, which of the two contestants shall assume dominance over the other.

Once decided, the relationship tends to last, and the eventual disadvantage to the loser can sometimes be long deferred. Hierarchic dominance plays an essential part in population homeostasis, as the means of cutting out subordinates whenever they happen to become superfluous either to the carrying capacity or to the breeding establishment. The contest for breeding rights may keep them waiting till their seniority is sufficient to admit them; and that applies with special force to the more or less polygamous males of strongly dimorphic species, which tend to develop symbols of greater power as they grow older—in body-size, colouring, plumes, horns, crests, wattles and the like.

It seems clear that these anatomical symbols are in fact cooperative traits, advertising accurately to the females and other males which of the males are the healthiest and best-tested survivors in the mating group. The value of age-related status (which also presumably affects the opposite choice of females by males) in assuring that the selected breeders are of high quality is probably a common component factor in determining the optimal balance between a lengthened lifespan and a shortened generation time, as statistical group traits. Every kind of manifestation which allows individuals to make an accurate appraisal of one another's genetic quality in status confrontations and to respond appropriately, will be improved through group selection to make it as reliable a reflection as possible of the genotype behind the mask, and keep the costly self-inflicted process of social selection efficient in maintaining group prosperity.

In a seasonal climate reproduction is generally possible only for a limited part of the year, when the carrying capacity remains high enough to support both the adults and the progeny they produce. If the parents are iteroparous and many of them are going to survive till the following year, it may be necessary to start submitting the young of the year to social selection early in their lives in order to weed out the less promising ones within the first few months. The social hierarchy probably has much influence on which ones succumb to predators or to inadequate diet during the juvenile mortality period. Juvenile mortality no doubt carries off the more handicapped bearers of double recessive genes, which are the price of preserving alternative alleles in the gene pool; and in some species juvenile hierarchies are known to develop among sibs before they become separated by dispersal. But it needs time for phenotypes to reveal all their hidden potential, and that may be one of the reasons for starting the annual breeding process at an early date, in order to give as long as possible before the annual flush of carrying capacity comes to an end and exclusions become imperative.

Our conclusion has been that the size of the effective interbreeding units is controlled by programming the pattern of dispersal, and is normally small; and this incidentally carries the collateral implication that the classical assumption of panmixis is false. Two advantages appear to arise from such a control. The first comes from isolating the local population units sufficiently to allow them to develop their own different group phenotypes, as predicted in Wright's model. In large continuous habitats where every individual is the centre of its own 'virtual' in-group, the group phenotypes would simply grade more or less steeply into one another. The effect in any case would be to keep many simultaneous 'experiments' running, with different permutations of the population gene pool, and thus to provide genetic versatility, with which to meet environmental challenges, either present or future. The second advantage is to perpetuate philopatric gene pools long enough to let them reap the results of their own resource conservation. The value of these advantages and the importance of philopatry to both must be enormous, if one can judge from the many parallel evolutions of precise navigation, apparently in most or all the two-way migrants in all the vertebrate classes, for the purpose of enabling them to retain their membership in philopatric breeding groups. (There are other migrants, including various bats and certain Lepidoptera, which home seasonally to traditional hibernating and aestivating caves or other refuges. These movements may in some cases have nothing to do with philopatric mating or breeding, but merely give access to scarce resorts where the dormant inmates are safe.)

The combination of programmed dispersal to achieve philopatry, and of migrants returning from great distances to conform with their dispersal and structuring programs, gave us a clue to the solution of a very old mystery, in the precocious spawning of the male parr of the Atlantic salmon (p. 259). These parr, never having left their natal streams apparently serve to lower the mean dispersal distance (i.e. increase the philopatry) of their breeding in-groups, whose remaining members are sea-run 'adults'. The dispersal pattern of the adults is known to be considerably more diffuse. We were even able to support this hypothesis and explain the probable evolutionary history of the habit, by reference to known sex-differences in the dispersal behaviour of other species of Salmonidae.

The fact that sex-differences in dispersal distance are common makes it probable that populations are generally viscous enough to sustain philopatry, but that it is achieved without incurring a harmful frequency of incestuous matings. Structuring does not therefore appear to be directed towards promoting kin selection, which would become ineffective unless matings occurred between close relations. What we know about structuring does, on the other hand, lend confirmation to Wright's model, on which he based his 'shifting balance' hypothesis of evolution. Wright's structuring, in so far as it facilitates group selection, has also received strong support from the experiments and theoretical conclusions of Wade and others (pp. 208, 332).

Another adaptive combination, this time of optimal mating-group size coupled with social selection for high-quality phenotypes as breeders, has repeatedly led to the evolution of polygamy, that is, of inequalities in the sex-ratio of breeding in-groups. For example, in the eusocial honeybee, single queens are normally inseminated by several males. The males on the other hand eject all their semen into the one queen they fertilize (E. O. Wilson 1975: 141), so that the multiple matings are polyandrous (and not promiscuous). In the vertebrates it is fairly common to find males that are polygynous; and if the adult sex-ratio is near 1:1, this means that there are more surplus males than females being excluded from breeding as a result of social selection. Polygyny is generally the more practicable type of polygamy to adopt if the participants of one sex are narrowed down, so as to employ fewer sires of higher quality; and since quality, when it is determined through social competition, depends on a summation of the whole phenotype, the individuals that achieve polygamy are likely to possess quite different phenotypes from one in-group to the next, and even from one member to another when they are in the same in-group, so that little gene fixation need result. Polygyny is usually practised on a small, and sometimes only a bigamous scale, by having several active males working together. It is especially simple for mammals to adopt polygyny because in many species the males play only a minor or negligible part in tending the young. Polygamy is thus another phenomenon easy to rationalize as a group trait for refining the gene pool, in species where the females are able to bring up single-parent or one-and-a-bit-parent families.

The likelihood that close inbreeding is generally avoided except as a last resort, because of the lowered mean viability of the offspring and the gene-fixations that result, suggested an explanation for the Allee effect, or undercrowding response, shown by various laboratory-culture

molluscs and arthropods (p. 275). It presumably exists also in wild populations of the same and allied species. The rates of growth and maturation, and especially of reproduction itself, are slowed in cultures seeded
with only a single hermaphrodite or a single dioecious pair, as compared
with those that have two or several potential pair-bonds. This occurs even
though all the cultures contain the same absolute quantities of living
space and food. No explanation has come forth in the half-century since
Allee discovered the phenomenon. It may be, however, that it is another
group adaptation; and that just as some species accept polygamy with the
exclusion of subordinate adults it entails, apparently to gain the genetic
advantage of elite parenthood, so other species are programmed to delay
reproduction when it would necessitate, or lead the next generation into,
incestuous matings. The delay would last as long as there was some
possibility that immigrants could arrive in time to avert the crisis. The
somewhat analogous inhibition of reproduction by individuals when it
would lead to overproduction and resource depletion if they did so, is one
of the commonest of group adaptations.

Population cycling (see p. 185) can be taken as a last illustration of the
heuristics of group selection. It will be recalled that cycling is best developed in three independent taxa, the grouse, the microtine rodents and
the snowshoe hare. All show very much the same characteristics, namely
a minimum periodicity of 8–10 generations between peaks, the peaks
being dissipated by the dispersal of the overstock produced and having a
tendency to recur irregularly, due to the lengthening of the quiescent
phase. Though the onset of the growth phase is frequently synchronized
over a large area due to a widespread climatic build-up of favourable
conditions, there are commonly only local 'islands' of growth, leaving
surrounding or nearby populations unaffected; and the peaks also vary
much in height. Typically there are two phases of the cycle—a run of 4–6
or more generations of normal philopatry and homeostasis, succeeded by
a rapid proliferation and continuous exodus phase, which eventually
runs the affected population down to a deep low before it resumes the
norm. In other words, there is an alternation of structuring between the
phase of static isolated in-groups, and the phase of increase and dispersal.

My tentative explanation, previously given, was that the normal
phase lasts sufficiently long to allow maximum variance in gene-
frequencies to develop between small in-groups by drift; and the growth-
and-dispersal phase assays, and then disseminates, the respective productivities of the groups, with proportionate effects on gene frequencies

ic pool. In strongly variable climates such as those of the
nd continental interiors, good and bad years appear in rapid
and new cycles are often triggered as soon as the previous one
has run its course; whereas in less extreme climates there may even be a
long wait, with the homeostatic phase running on, before the growth-
and-dispersal phase can start. Extension beyond the minimum period is
not critical to the genetic function I have proposed.

20.5 What constitutes fitness?

Individual-selectionist theory owes its present acceptance to the fact that
individual selection can readily be seen to occur, and that as a process it
can account quite credibly for a great many of the marvels of evolution—
the near-magical brains and eyes, the conquest of sea, land and air, even
the genetic code perhaps. That is enough to satisfy uncritical minds, and
the fact that there are other adaptations that it fails to explain worries few
but the evolutionary specialists.

It may bring the deficiencies nearer home to point out that we ourselves possess group adaptations that are readily recognisable as such.
The most familiar set of them centre round the neural mechanism we call
conscience. Its function is to steer and reinforce our moral conduct.
Reflection shows that moral standards themselves are not innate, but
acquired, normally in youth, from the group among whom we happen to
live. In certain sections of society the standards can become depraved, but
the mechanism is there just the same, coercing individuals 'who know no
better' to act morally according to their beliefs. That means promoting the
accepted interests and principles of their group, being selfless, loyal,
diligent, militant against opposing groups, generous and helpful to their
fellows—in short being cooperative members of a cooperative society.

In support of this we have innate brain programs to reward the efforts
we make on behalf of others, and to punish us with guilt and remorse
when we sidestep our duty. We belong to social hierarchies and are
delighted if we excel in the presence of spectators, disappointed or
ashamed if we let ourselves down. But hierarchies too have their cooperative side. If they bring us to positions of responsibility, conscience
bids us accept the burdens if we can. Conscience shows a wide range of
individual intensity, and its very weakness or strength can be seen to be
hereditary. If well tutored, socially reinforced by keeping enlightened
company, and obeyed, it goes a long way towards mitigating and exclud-

ing the selfish struggle for existence, upon which individual-selection theory inclines one to focus too much.

The classical theory has necessarily been preoccupied with individual fitness, as offering the only mechanism for changing the frequency of genes through time, and thus achieving evolution. Its foremost advocates have concluded that selection is bound to maximize the ability of individuals to bequeath their genes to the pool of adults in the next generation. Whichever theory one adopts, however, individuals are found in practice to be equipped to reproduce at their own characteristic rate, which may vary with environmental conditions; and whether this is the maximum rate that selection could elicit or not is generally obscured by other factors which preclude proof or disproof. It is only perhaps when one recognizes the general correlation that exists between fecundity and the risk of premature death, and compares helminth parasites or marine fish that have planktonic larvae, with those mammals, birds or fish which exercise the best parental care, that a shadow of doubt falls on the classical proposition.

Classical theorists have dismissed group selection in the past with two arguments. The first is that the fitness of groups would necessarily suffer through the need to regulate the fitnesses of their individual members. Not only would the two types of selection be in conflict, therefore, but group selection could never 'win' because, almost by definition, free individuals would fare better than regulated ones. Furthermore, the advocates of group selection had not explained how selective forces with opposite impacts could co-exist. The second argument is that, until recently, group selection could not be demonstrated in action.

Now both or all three of these objections have rather suddenly collapsed. In fact higher and more permanent fitness can be obtained by cooperation and regulation than by leaving individuals to act in competition, as human exploitation of renewable natural resources has repeatedly proved. Laboratory experiments have demonstrated that group selection is a potentially powerful force, given a particular type of population structure. This type of structure has been looked for and found to exist in a wide variety of animal species. Lastly, group-selection theory has been shown to dispel many previous anomalies, long recognized by individual-selectionists, and to reveal the unexpected existence of other very important group adaptations. I do not believe any valid objections to group selection now exist.

Individual and group selection can be successfully combined because individual selection is only permitted as long as it benefits the group. (If it were to persist, notwithstanding, it would soon reduce the existing fitness levels and become self-defeating.) It is used solely for the purpose of showing which individual animals are in the most viable and socially dominant class, and then according them the rights to feed and breed. The actual breeding density, and the due quota of progeny, are regulated by group-selected programs; and individual selection itself is largely artificial, in the sense that intraspecific competition is ruled by group-selected conventions. Only the extrinsic agents that deal injury and death impinge in the raw, and even these may fall heaviest on social outcasts.

Living organisms of every kind exist because of the ability of their stocks to survive. Clearly survival is their *summum bonum*. Since survival of the stock evidently benefits from the spatial structuring of populations in a manner that retains genetic variance between the component groups, it would be undesirable for the groups to pursue each other's extinction, unless new groups were certain to be created as fast as they were lost. Probably they are; but there could nevertheless be some delicacy in inter-unit relationships, moderating the spread of units enjoying temporary advantages, which would exterminate others in their path. If these are valid arguments, successful groups would move slowly, not aggressively, in displacing their neighbours, leaving the main threats to group survival to come from outside, from competing or exploiting species, from failure of recurrent resources, and from the physical environment.

There are two further problematical aspects of production among in-groups. First, it must be true that if two groups produce the same quantities of offspring, the average quality of the latter, in terms of survivability, will be the prime factor that determines the respective fitnesses of the groups. Higher productivity will allow a more stringent application of social selection, for the purpose of increasing or upholding the genetic quality of the quota of breeders; and there will necessarily be a best compromise between quantity production for recolonization and extension, and quality production for better survival through elite selection. The circumstances will therefore exist in which many progeny, which might superficially have been regarded as capable of adding to group fitness through expansion, are being sacrificed for the sake of increasing the quality of the survivors and hence the survivability of the group.

The second problem arises from the possibility that a vivivorous in-group could be assembled that was lax and incompetent in its practice of homeostasis. For a few years it might out-produce its neighbours simply by overexploiting the resources to which it had fallen heir. If that allowed its collective territory to expand, a chain reaction of resource destruction might conceivably spread outwards, leaving devastation in its wake. This possible source of instability is presumably countered by the certainty that the F_1 progeny would out-cross, and that this would quickly restore the norm. Inefficient groups must generally be penalized by their falling productivity.

The purpose of this speculative discussion is to suggest that ruthless productivity in terms of numbers of progeny, and booming expansion, may rather rarely serve to promote the survival of the stock; and the fact that group productivity is generally an inconspicuous and elusive phenomenon may not therefore be very surprising. Indeed it may be some time yet before the means by which fitness is realized are sharply characterized.

Summary of Chapter 20

1 Cooperation among the eusocial Hymenoptera and termites has led to a division of labour between male and female reproductives on the one hand, and one or more castes of sterile workers on the other. Eusociality has evolved repeatedly among the Hymenoptera, and many 'solitary' taxa still exist. This helps to make it clear that eusociality has been derived from normal sociality, and that it still promotes the usual functions of controlling population density and gene flow. In the termites all species are highly eusocial, and the polymorphism between their royal males and females and their two sexes of workers reach the greatest diversity. Eusocial insects are always structured into colonies (normally centred on a nest) which, as units of selection, are integrated superorganisms.

Under a regime of group selection the evolution of polymorphic castes and worker infertility presents no theoretical problems. The royal reproductives, especially in some termites, become enormously fecund as specialist reproductives, just as the complementary workers become specialists as helpers.

Non-reproductive helpers at the nest or den provide a well-known parallel among various birds and mammals. Authors have tended to ascribe their altruistic behaviour to kin selection, but the exceptions and anomalies observed make this explanation inadequate and unacceptable. Once again, if group selection is assumed, there are no theoretical difficulties.

2   The evolution of sexual reproduction has presented one of the most intractable of puzzles to the classical theorist. The process is primarily adapted to producing novel genotypes. The mechanisms evolved for the purpose include gamete formation and fertilization: the former dismembers the parental genotypes and the latter marries two half-genotypes together, showing that individual genotypes, however fit, have no permanent value. There are three randomizing events in the course of breaking up old and rebuilding new genotypes, which account for the endless stream of novel genotypes that results. The whole drastic process is placed in a still more curious light by the fact that a minority of animal and plant species exist which have 'opted out' of it, either facultatively or completely, and can actually pass on their individual genotypes intact from generation to generation instead. The individual-selectionist could be forgiven for wondering why this is not the generally preferred alternative.

No one doubts that in the long term the recombination of genes is of immense value, in continually exposing new genotypes to the test of selection. The Wrightian population structure is presumed to extend the same function to the next higher structural tier, by endlessly varying the constitutions of local gene pools and exposing them likewise to group selection.

Sexual reproduction involves the union of gametes produced by two individuals; but for that purpose there is no need to have dimorphic sexes. Two identical hermaphrodites can perform the union just as well, and even do it in both directions simultaneously. It is evidently for social and not genetic reasons that the sexes have evolved into two different, cooperative castes: indeed they are found consistently only among social species of animals. The two castes normally divide their labours by allotting the competitive and homeostatic role primarily to the males and the reproductive, family role to the females. Allocating individuals to alternative castes in this way is necessarily a group adaptation.

3   As a hypothesis, group selection has shown much versatility in explaining the enigmas of evolution, and more examples of this, gleaned from earlier chapters, are assembled in this section and the next. This section presents those connected with the management of food resources, which thereby raise the prosperity of cooperative consumer populations above what would be attainable by non-cooperators.

The evidence for population homeostasis, mediated by competition for property and status and by the exclusion of subordinate extras, is very strong. When it was proposed in 1962 that this was the function of territorial systems and hierarchies, and that the conventionalizing of the competition was the key element in sociality, much obscurity was dispelled and a broad segment of animal behaviour rationalized. The deduction just mentioned above about dimorphic sexes extends the segment further. Similarly we have rationalized philopatry and recognized the various functions it serves. Surprisingly, our survey showed that population homeostasis can operate efficiently even in presocial invertebrates, namely molluscs and platyhelminths, and thus that sociality only strengthens a more primitive adaptation, which makes it probable that the faculty is common to all animal phyla.

4   The 'genetic management syndrome' embraces another complex of adaptive traits, dependent on dispersal and the population structuring that results from it. We have found dispersal to be tightly controlled, at least in the vertebrates, apparently for the purpose of creating and perpetuating a Wrightian structure of small, almost isolated breeding units. This, according to Wright's prediction, greatly improves the population's chances of being able to respond quickly to environmental change. Such populations are incidentally heteromictic and not panmictic.

In structured populations selection falls both on individuals and groups. Much of the selection of individuals is 'internalized' within the in-groups, as social (hierarchic) selection. Even the incidence of mortality from predators, disease and malnutrition can be largely predetermined by the social status of the victims. Age tends to augment the individual's status symbols and power in accordance with the growing evidence it gives of good survivorship. Status symbols, generally most prominent in senior males, are essentially group-advantageous traits.

Other examples of apparent genetic management not previously rationalized include the evolution of polygamy (from Chapter 16.2), the Allee effect (from Chapter 17.2), the precocious spawning of fingerling parr in the Atlantic salmon (from Chapter 16.3), and population cycling (from Chapter 14).

5   Man possesses one very familiar group trait, in the innate brain mechanism we call conscience. Its function is to reinforce moral (which for present purposes can be taken as social and collaborative) conduct. Moral standards on the other hand are acquired in youth, usually from older companions, and may in fact be much depraved. This does not inhibit the innate mechanism, which continues to inspire loyalty, diligence, obedience to principle, militance against opposing groups, and

o associates, in terms of the owner's convictions whatever
There are innate brain programs to reward compliance and
:t of duty. Conscience varies individually in strength, and
some degree of heritability. In the best circumstances it
suppresses the selfishness on which individual-selection theory has
sometimes tended to concentrate.

Group selection has been dismissed by most theorists, chiefly on the
mistaken assumption that it would oppose individual selection and
would be unable to defeat it; and to a lesser extent for lack of evidence,
and tests, and clear explanations of how it might work. Recently all these
objections have rather suddenly collapsed, with the empirical discovery
of population structuring, with Wade's experiments, and with the realization that habitats improve under cooperative management.

Organisms owe their continued existence to the ability of their stocks
to survive, and thus stock-survival must be the central pillar of fitness.
Conspecific groups should not pose threats to one another's survival; and
this and other considerations appear to account for the obscurity that still
surrounds the way in which group productivity finds its practical
expression.

REFERENCES

Ali, S.A. (1932) The role of sunbirds and flower-peckers in the propagation and distribution of the tree-parasite, *Loranthus longifolius* Dest., in the Konkan (W. India). *Journal of the Bombay Natural History Society* **35**, 144–149.

Ali, S. & S.D. Ripley (1974) *Handbook of the Birds of India and Pakistan*. Volume 10. Bombay: Oxford Univ. Press.

Allee, W.C. (1931) *Animal Aggregations*. Chicago: Chicago University Press.

Allee, W.C. (1938) *The Social Life of Animals*. London and Toronto: Heinemann.

Allee, W.C., A.E. Emerson, O. Park, T. Park, & K.P. Schmidt (1949) *Principles of Animal Ecology*. Philadelphia and London: Saunders.

Altum, B. (1868) *Der Vogel und sein Leben*. Münster. (See E. Mayr 1935).

Anderson, P.K. (1970) Ecological structure and gene flow in small mammals. *Symposium of the Zoological Society of London* No. **26**, 299–325.

Andersson, M. (1984) The evolution of eusociality. *Annual Review of Ecology and Systematics* **15**, 165–189.

Andrewartha, H.G. (1961) *Introduction to the Study of Animal Populations*. London: Methuen.

Andrewartha, H.G. & L.C. Birch (1954) *The Distribution and Abundance of Animals*. Chicago: Chicago University Press.

Ashworth, H.P.C. (1895) The dispersal of mistletoe. *Indian Forester* **22**, 2–4.

Avery, R.S. (1969) The ecology of tapeworm parasites in wildfowl. *Wildfowl* **20**, 59–68.

Baer, J. & C. Joyeux (1961) Classe des trématodes. *In:* P.P. Grassé, *Traité de Zoologie*. Tome **4**, 561–692. Paris: Masson.

Baker, H.G. & I. Baker (1975) Studies of nectar-constitution and pollinator-plant coevolution. *In:* L.E. Gilbert & P.H. Raven (eds.). *Coevolution of Animals and Plants*. Austin and London: University of Texas Press.

Baker, M.C. & P. Marler (1980) Behavioral adaptations that constrain the gene pool in vertebrates. Pp. 59–80 *In:* H. Markl (ed.) *Evolution of Social Behavior, Hypotheses and Empirical Tests*. Weinheim: Verlag Chemie.

Banfield, A.W.F. (1974) *The Mammals of Canada*. Toronto: Toronto University Press.

Barriga, O.O. (1981) *The Immunology of Parasitic Infections*. Baltimore: University Park Press.

Beacham, T.D. (1980a) Demography of declining populations of the vole, *Microtus townsendii*. *Journal of Animal Ecology* **49**, 453–464.

Beacham, T.D. (1980b) Dispersal during population fluctuations of the vole, *Microtus townsendii*. *Journal of Animal Ecology* **49**, 867–877.

Bent, A.C. (1932) *Life histories of North American gallinaceous birds*. U.S. National Museum Bulletin 162 (Smithsonian Institution). Washington: Government Printing Office.

Bierman, W.H. & K.H. Voous (1950) Birds observed and collected during the whaling expeditions of the 'William Barendsz' in the Antarctic, 1946–47 and 1947–48. *Ardea* **37**, Extra-Nummer, 123 pp.

Birdsell, J.B. (1978) Spacing mechanisms and adaptive behaviour of Australian Aborigines. Pp. 213–244 *In:* F.J. Ebling & D.M. Stoddart (eds.) *Population Control by Social Behaviour*. London: Institute of Biology.

Blackman, F.F. (1905) Optima and limiting factors. *Annals of Botany* **19**, 281–295.

Bradbury, J.W. (1981) The evolution of leks. Pp. 138–169 *In:* R.D. Alexander & D. Tinkle (eds.) *Natural Selection and Social Behavior*. Newton, Mass.: Chiron Press.

Brand, C.J., L.B. Keith & C.A. Fischer (1976) Lynx responses to snowshoe hare densities in central Alberta. *Journal of Wildlife Management* **40**, 416–428.

Brockmann, H.J. (1984) The evolution of social behaviour in insects. Pp. 340–361 *In:* J.R. Krebs & N.B. Davies (eds.) *Behavioural Ecology: An Evolutionary Approach*. Oxford: Blackwell Scientific Publications.

Brown, J.L. & Esther R. Brown (1982) Kin selection and individual selection in babblers. Pp. 244–256. *In:* R.D. Alexander & D.W. Tinkle (eds.) *Natural Selection and Social Behavior*. New York: Chiron Press.

Bruns, H. (1960) The economic importance of birds in forests. *Bird Study* **7**, 193–208.

Buck, R.J.G. & D.W. Hay (1984) The relation between spawning stock and progeny of Atlantic salmon (*Salmo salar* L.) in a Scottish stream. *Journal of Fish Biology* **24**, 1–11.

Buck, R.J.G. & A.F.Youngson (1982) The downstream migration of precociously mature Atlantic salmon, *Salmo salar* L. parr in autumn; its relation to the spawning migration of mature adult fish. *Journal of Fish Biology* **20**, 279–288.

Campbell, B. & E. Lack (1985) *A Dictionary of Birds*. Calton: Poyser, and Vermillion: Buteo Books.

Campbell, J.S. (1977) Spawning characteristics of brown trout and sea-trout *Salmo trutta* L. in Kirk Burn, River Tweed, Scotland. *Journal of Fish Biology* **11**, 217–229.

Cassie, R.M. (1962) Frequency distribution models in the ecology of plankton and other organisms. *Journal of Animal Ecology* **31**, 65–92.

Chabrzyk, G. & J.C. Coulson (1976) Survival and recruitment in the herring gull *Larus argentatus*. *Journal of Animal Ecology* **45**, 187–203.

Chapman, R.N. (1928) The quantitative analysis of environmental factors. *Ecology* **9**, 111–122.

Charles, J.K. (1972) Territorial behaviour and the limitation of population size in the crow, *Corvus corone* and *Corvus cornix*. Unpublished Ph.D. thesis, Aberdeen University, Aberdeen, Scotland.

Cheever, A.W. & K.G. Powers (1972) *Schistosoma mansoni* infection in rhesus monkeys: comparison of the course of heavy and light infections. *Bulletin of World Health Organization* **46**, 301–309.

Chitty, D. (1967) The natural regulation of self-regulatory behavior in animal populations. *Proceedings of the Ecological Society of Australia* **2**, 51–78. (*Reprinted in* I.A. McLaren (ed.) *Natural regulation of animal populations.* New York: Atherton Press (1971) pp. 136–170.)

Chitty, D. (1970) Variation and population density. *In:* R.J. Berry & H.N. Southern (eds.) *Variation in Mammalian Populations.* Symposia of the Zoological Society of London **26**, 327–333.

Clark, W.C. (1974) Interpretation of life history pattern in the Digenea. *International Journal for Parasitology* **4**, 115–123.

Cole, F.R. & G.O. Bazli (1979) Nutrition and population dynamics of the prairie vole, *Microtus ochrogaster,* in central Illinois. *Journal of Animal Ecology* **48**, 455–470.

Collett, R. (1895) *Myodes lemmus,* its habits and migrations in Norway. *Christiania Videnskabs-Selskabs Forhandlinger 1895 No. 3,* 63 pp.

Cooch, F.G. (1963) Recent changes in distribution of color phases of *Chen c. caerulescens.* Proceedings XIII International Ornithological Congress, 1182–1194.

Cooke, F. & F.G. Cooch (1968) The genetics of polymorphism in the goose *Anser caerulescens. Evolution* **22**, 289–300.

Cooke, F., C.D. MacInnes & J.P. Prevett (1975) Gene flow between breeding populations of lesser snow geese. *Auk* **92**, 493–510.

Corbet, G.B. & H.N. Southern (1977) *The Handbook of British Mammals.* Oxford: Blackwell Scientific Publications.

Craig, D.M. (1982) Group selection versus individual selection. *Evolution* **36**, 271–282.

Cramp, S., W.R.P. Bourne & D. Saunders (1974) *The seabirds of Britain and Ireland.* London: Collins.

Criddle, N. (1930) Some natural factors governing fluctuations of grouse in Manitoba. *Canadian Field-Naturalist* **44**, 79.

Criddle, S. (1938) A study of the snowshoe rabbit. *Canadian Field-Naturalist* **52**, 31–40.

Crombie, A.C. (1943) The effect of crowding on the natality of grain-infesting insects. *Proceedings of the Zoological Society of London, ser. A,* **113**, 77–98.

Crompton, D.W.G. & S.M. Joyner (1980) *Parasitic Worms.* London: Wykeham Publications.

Darwin, C. (1859) *The Origin of Species by Means of Natural Selection.* London: Murray (Citations refer to 6th edition with additions and corrections, 1872.)

Darwin, C. (1874) *The Descent of Man.* 2nd ed., 2 vols. London. (Page numbers from John Murray's new ed., 1901.)

Davidson, J. & H.G. Andrewartha (1948a) Annual trends in a natural population of *Thrips imaginis* (Thysanoptera). *Journal of Animal Ecology* **17**, 193–199.

Davidson, J. & H.G. Andrewartha (1984b) The influence of rainfall, evaporation and atmospheric temperature on fluctuations in the size of a natural population of *Thrips imaginis* (Thysanoptera). *Journal of Animal Ecology* **17**, 200–222.

de Kock, L.L. & A.E. Robinson (1966) Observations on a lemming movement in Jämtland, Sweden, in autumn 1963. *Journal of Mammalogy* **47**, 490–499.

Diamond, J.M. (1975) Assembly of species communities. Pp. 342–444 *In:* M.L. Cody & J.M. Diamond (eds.) *Ecology and Evolution of Communities.* Cambridge, Mass. and London: Harvard University Press.

Dixon, A.F.G. (1970) Quality and availability of food for a sycamore aphid population. Pp. 271–286. *In:* A. Watson (ed.) *Animal populations in Relation to Their Food Resources.* Oxford: Blackwell Scientific Publications.

Docters van Leeuwen, W.M. (1954) On the biology of some Loranthaceae and the role birds play in their life-history. *Beaufortia* **4**, 105–208.

Dodd, A.P. (1936) The control and eradication of prickly pear in Australia. *Bulletin of Entomological Research* **27**, 503–517.

Duncan, J.S., H.W. Reid, R. Moss, J.D.P. Phillips & A. Watson (1978) Ticks, louping ill and red grouse on moors in Speyside, Scotland. *Journal of Wildlife Management* **42**, 500–505.

Dunnet, G.M. (1955) The breeding of the starling *Sturnus vulgaris* in relation to its food supply. *Ibis* **97**, 619–662.

Dunnet, G.M., J.C. Ollason & A. Anderson (1979) A 28-year study of breeding fulmars *Fulmarus glacialis* in Orkney. *Ibis* **121**, 293–300.

Eisenberg, R.M. (1966) The regulation of density in a natural population of the pond snail, *Lymnaea elodes. Ecology* **47**, 889–906.

Elton, C.S. (1924) Periodic fluctuations in the numbers of animals: their causes and effects. *British Journal of Experimental Biology* **2**, 119–163.

Elton, C. (1942) *Voles, Mice and Lemmings.* Oxford: Clarendon Press.

Emlen, J.M. (1984) *Population Biology: the Coevolution of Population Dynamics and Behavior.* New York and London: Macmillan.

Emlen, S.T. (1982) Altruism, kinship, and reciprocity in the white-fronted bee-eater. Pp. 217–230. *In:* R.D. Alexander & D.W. Tinkle (eds.) *Natural Selection and Social Behavior.* New York: Chiron Press.

Emlen, S.T. (1984) Cooperative breeding in birds and mammals. Pp. 305–339 *In:* J.R. Krebs & N.B. Davies (eds.) *Behavioural Ecology: an Evolutionary Approach.* Oxford: Blackwell Scientific Publications.

Esch, G.W., J.W. Gibbons & J.E. Bourque (1975) An analysis of the relationship between stress and parasitism. *American Midland Naturalist* **93**, 339–353.

Falconer, D.S. (1981) *Introduction to Quantitative Genetics.* 2nd ed. London and New York: Longman.

Finerty, J.P. (1980) *The Population Ecology of Cycles in Small Mammals. Mathematical Theory and Biological Fact.* New Haven and London: Yale University Press.

Fisher, H.I. (1976) Some dynamics of a breeding colony of Laysan albatrosses. *Wilson Bulletin* **88**, 121–142.

Flowerdew, J.R. (1974) Field and laboratory experiments on the social behaviour and population dynamics of the wood mouse (*Apodemus sylvaticus*). *Journal of Animal Ecology* **43**, 499–511.

Flowerdew, J.R. (1978) Residents and transients in wood mouse populations. Pp. 49–66 *In:* F.J. Ebling & D.M. Stoddart (eds.) *Population Control by Social Behaviour.* London: Institute of Biology.

Foerster, R.E. (1936) The return from the sea of sockeye salmon (*Oncorhynchus nerka*) with special reference to percentage survival, sex proportions and progress of migration. *Journal of the Fisheries Research Board of Canada* **3**, 26–42.

Frank, F. (1957) The causality of microtine cycles in Germany. *Journal of Wildlife Management* **21**, 113–121.

Freeland, W.J. (1974) Vole cycles: another hypothesis. *American Naturalist* **108**, 238–245.

Fretwell, S.D. (1972) *Populations in a Seasonal Environment*. Princeton: Princeton University Press.

Fryer, G. (1966) Habitat selection and gregarious behaviour in parasitic crustaceans. *Crustaceana* **10**, 199–209.

Fryer, G. & T.D. Iles (1972) *The Cichlid Fishes of the Great Lakes of Africa: their Biology and Evolution*. Edinburgh: Oliver & Boyd.

Gasaway, W.C. (1976a) Seasonal variation in diet, volatile fatty acid production and size of cecum of rock ptarmigan. *Comparative Biochemistry and Physiology* **53A**, 109–114.

Gasaway, W.C. (1976b) Voltile fatty acid and metabolizable energy derived from cecal fermentation in the willow ptarmigan. *Comparative Biochemistry and Physiology* **53A**, 115–121.

Gasaway, W.C. (1976c) Cellulose digestion and metabolism by captive rock ptarmigan. *Comparative Biochemistry and Physiology* **54A**, 179–182.

Gasaway, W.C., D.E. Holleman & R.G. White (1976) Digestion of dry matter and absorption of water in the intestine and cecum of rock ptarmigan. *Condor* **78**, 77–84.

Gee, A.S., N.J. Milner & R.J. Hemsworth (1978) The effect of density on mortality in juvenile Atlantic salmon (*Salmo salar*). *Journal of Animal Ecology* **47**, 497–505.

Gelting, P. (1937) Studies on the food of the East Greenland ptarmigan. *Meddelelser om Grønland* **116**(3), 196 pp.

Gentry, J.B. (1968) Dynamics of an enclosed population of pine mice, *Microtus pinetorum*. *Research in Population Ecology* **10**, 21–30.

Gibb, J.A. (1958) Predation by tits and squirrels on the eucosmid *Ernarmonia conicolana* (Heyl.). *Journal of Animal Ecology* **27**, 375–396.

Gibb, J.A. (1960) Populations of tits and goldcrests and their food supply in pine plantations. *Ibis* **102**, 163–208.

Gilmour, J.S.L. & J.W. Gregor (1939) Demes: a suggested new terminology. *Nature*, Lond. **144**, 133.

Gimingham, C.H. (1972) *Ecology of Heathlands*. London: Chapman & Hall.

Glen, D.M., N.F. Milsom & C.W. Wiltshire (1981) The effect of predation by blue tits (*Parus caeruleus*) on the sex-ratio of codling moth (*Cydia pomonella*). *Journal of Applied Ecology* **18**, 133–140.

Greenwood, P.J. (1980) Mating systems, philopatry and dispersal in birds and mammals. *Animal Behaviour* **28**, 1140–1162.

Greenwood, P.J. (1983) Mating systems and the evolutionary consequences of dispersal. Pp. 116–131 *In:* I.R. Swingland & P.J. Greenwood (eds.) *The Ecology of Animal Movement*. Oxford: Clarendon Press.

Greenwood, P.J., P.H. Harvey & C.M. Perrins (1979a) The role of dispersal in the great tit (*Parus major*): the causes, consequences and heritability of natal dispersal. *Journal of Animal Ecology* **48**, 123–142.

Greenwood, P.J., P.H. Harvey & C.M. Perrins (1979b) Kin selection and territoriality of birds? A test. *Animal Behaviour* **27**, 645–651.

Gudmundsson, F. (1960) Some reflections on ptarmigan cycles in Iceland. *Proceedings XII International Ornithological Congress, Helsinki,* 259–265.

Gudmundsson, F. (1972) The predator-prey relationship of the gyrfalcon (*Falco rusticolus*) and the rock ptarmigan (*Lagopus mutus*) in Iceland. *Proceedings XV International Ornithological Congress, Leiden,* p. 649.

Hairston, N.G., F.E. Smith & L.B. Slobodkin (1960) Community structure, population control, and competition. *American Naturalist* **94**, 421–425.

Halvorsen, O. & H.H. Williams (1968) *Gyrocotyle* (Platyhelminthes) in *Chimaera montrosa* from Oslo Fjord, with emphasis on its mode of attachment and a regulation in the degree of infection. *Nytt Magasin for Zoologi* **15**, 130–142.

Hamilton, W.D. (1964) The genetical evolution of social behavior. *J. Theoret. Biol.* **7**, 1–51. (*Reprinted* in G.C. Williams (ed.) 1971 *Group Selection.* New York and Chicago: Aldine Atherton. Pp. 23–89.)

Hamilton, W.D. (1972) Altruism and related phenomena, mainly in social insects. *Annual Review of Ecology and Systematics* **3**, 193–232.

Hamilton, W.D., P.A. Henderson & N.A. Moran (1981) Fluctuation of environment and coevolved antagonist polymorphism as factors in the maintenance of sex. Pp. 363–381 *In:* R.D. Alexander & D.W. Tinkle (eds.) *Natural Selection and Social Behavior.* New York: Chiron Press.

Heinrich, B. (1975) The role of energetics in bumblebee-flower interrelationships. *In:* L.E. Gilbert & P.H. Raven (eds). *Coevolution of Animals and Plants.* Austin & London: Texas University Press.

Hewson, R. (1962) Food and feeding habits of the mountain hare *Lepus timidus scoticus* Hitzheimer. *Proceedings of the Zoological Society of London* **139**, 515–526.

Hewson, R. (1976a) A population study of mountain hares (*Lepus timidus*) in north-east Scotland from 1956–1969. *Journal of Animal Ecology* **45**, 395–414.

Hewson, R. (1976b) Grazing by mountain hares *Lepus timidus* L., red deer *Cervus elaphus* L. and red grouse *Lagopus l. scoticus* on heather moorland in north-east Scotland. *Journal of Applied Ecology* **13**, 657–666.

Hilden, O. & S. Vuolanto (1972) Breeding biology of the red-necked phalarope (*Phalaropus lobatus* L.) *Ornis Fennica* **49**, 57–85.

Houston, D.C. (1979) The adaptations of scavengers. Pp. 263–286 *In:* A.R.E. Sinclair & M. Norton-Griffiths (eds). *Serengeti: Dynamics of an Ecosystem.* Chicago: University of Chicago Press.

Howard, H.E. (1920) *Territory in Bird Life.* London: Murray.

Hutchinson, K.J. (1971) Productivity and energy flow in grazing/fodder-conservation systems. *Herbage Abstracts* **41**, 1–10.

Hutton, J.A. (1924) *The Life-history of the Salmon.* Aberdeen: University Press.

Imms, A.D. (1957) *A General Textbook of Entomology.* 9th ed. revised by O.W. Richards & R.G. Davies. London: Methuen.

Irving, L., G. West & L. Peyton (1967) Winter feeding program of Alaska willow ptarmigan shown by crop contents. *Condor* **69**, 69–77.

Janzen, D.H. (1966) Coevolution of mutualism between ants and acacias in Central America. *Evolution* **20**, 240–275.

Janzen, D.H. (1969) Seed-eaters versus seed size, number, toxicity and dispersal. *Evolution* **23**, 1–27.

Janzen, D.H. (1971) Seed predation by animals. *Annual Review of Ecology and Systematics* **2**, 465–492.

Janzen, D.H. (1974) Tropical blackwater rivers, animals and mast fruiting by the Dipterocarpaceae. *Biotropica* **6**, 69–103.

Jenkins, D. (1963) Population control in red grouse (*Lagopus lagopus scoticus*). *Proceedings XIII International Ornithological Congress* **2**, 690–700.

Jenkins, D. & A. Watson (1962) Fluctuations in a red grouse (*Lagopus scoticus* Latham) population, 1956–9. Pp. 96–114 *In:* E.D. Le Cren & M.W. Holdgate (eds) *The Exploitation of Natural Animal Populations*. Oxford: Blackwell Scientific Publications.

Jenkins, D., A. Watson & G.R. Miller (1963) Population studies on the red grouse, *Lagopus lagopus scoticus* (Lath.) in north-east Scotland. *Journal of Animal Ecology* **32**, 317–376.

Jenkins, D., A. Watson & G.R. Miller (1967) Population fluctuations in the red grouse *Lagopus lagopus scoticus*. *Journal of Animal Ecology* **36**, 97–122.

Jenkins, D., A. Watson & N. Picozzi (1965) Red grouse chick survival in captivity and in the wild. *Transactions VI Congress, International Union of Game Biologists*, 63–70.

Jenni, D.A. & G. Collier (1972) Polyandry in the American jaçana (*Jacana spinosa*). *Auk* **89**, 743–765.

Jones, J.W. (1959) *The Salmon*. London: Collins.

Jones, J.W. & J.H. Orton (1940) The paedogenetic male cycle in *Salmo salar* L. *Proceedings of the Royal Society B* **128**, 485–499.

Joyeux, C. & J.G. Baer (1961) Classe des cestodes. *In:* P.P. Grassé, *Traité de Zoologie*. Tome 4, 347–560. Paris: Masson.

Kalela, O. (1957) Regulation of reproduction rate in subarctic populations of the vole *Clethrionomys rufocanus* (Sund.). *Annales Academiae Scientiarum Fennicae, ser. A IV*, No. **34**, 60 pp.

Kalela, O. (1962) On the fluctuations in the numbers of arctic and boreal small rodents as a problem of production biology. *Annales Academiae Scientiarum Fennicae, ser. A, IV. Biologica*, No. **66**, 38 pp.

Keast, A. (1958) The influence of ecology on variation in the mistletoe bird, *Dicaeum hirundaceum*. *Emu* **58**, 195–206.

Kennedy, J.S. & Lorna Crawley (1967) Spaced-out gregariousness in sycamore aphids *Drepanosiphum platanoides* (Schrank) (Hemiptera, Callaphididae). *Journal of Animal Ecology* **36**, 147–170.

Kistchinski, A.A. (1975) Breeding biology and behaviour of the grey phalarope *Phalaropus fulicarius* in east Siberia. *Ibis* **117**, 285–301.

Krebs, C.J. (1964) The lemming cycle at Baker Lake, Northwest Territories, during 1959–62. *Arctic Institute of North America, Technical Paper No. 15*, 104 pp.

Krebs, C.J. (1978) A review of the Chitty hypothesis of population regulation. *Canadian Journal of Zoology* **56**, 2463–2480.

Krebs, C.J., M.S. Gaines, B.L. Keller, J.H. Myers & R.H. Tamarin (1973) Population cycles in small rodents. *Science* **179**, 35–41.

Krebs, C.J. & J.H. Myers (1974) Population cycles in small mammals. *Advances in Ecological Research* **8**, 267–399.

Krebs, J.R. & C.M. Perrins (1978) Behaviour and population regulation in the great tit (*Parus major*). Pp. 23–47 *In:* F.J. Ebling & D.M. Stoddart (eds.) *Population Control by Social Behaviour*. London: Institute of Biology.

Kruuk, H. (1967) Competition for food between vultures in East Africa. *Ardea* **55**, 171–193.

Lack, D. (1944) Ecological aspects of species-formation in passerine birds. *Ibis* **86**, 260–286.

Lack, D. (1954) *The Natural Regulation of Animal Numbers*. Oxford: Clarendon Press.

Lack, D. (1966) *Population Studies of Birds*. Oxford: Clarendon Press.

Lack, D. (1968) *Ecological Adaptations for Breeding in Birds*. London: Methuen.

Lack, D. (1971) *Ecological Isolation in Birds*. Oxford: Blackwell Scientific Publications.

Lancaster, D.A. (1964) Biology of the brushland tinamou. *Bulletin of American Museum of Natural History* **127**, 269–314.

Lance, A.N. (1978a) Territories and the food plant of individual red grouse. II. Territory size compared with an index of nutrient supply in heather. *Journal of Animal Ecology* **47**, 307–313.

Lance, A.N. (1978b) Survival and recruitment success of individual young cock red grouse *Lagopus l. scoticus* tracked by radio-telemetry. *Ibis* **120**, 369–378.

Leopold, A.S. (1943) Deer irruptions. *Wisconsin Conservation Department Publication* **321**, 1–11.

Leopold, A.S. (1953) Intestinal morphology of gallinaceous birds in relation to food habits. *Journal of Wildlife Management* **17**, 197–203.

Leopold, A. & J.N. Ball (1931) British and American grouse cycles. *Canadian Field-Naturalist* **45**, 162–167.

Leslie, A.S. (1912) *The Grouse in Health and Disease, being the popular edition of the Report of the Committee of Inquiry on Grouse Disease*. London: Smith, Elder & Co.

Leuthold, W. (1966) Variations in territorial behaviour of Uganda kob *Adenota kob thomasi* (Neumann 1896). *Behaviour* **27**, 215–258.

Levins, R. (1970) Extinction. Pp. 75–108 *In:* M. Gerstenhaber (ed.) *Some Mathematical Problems in Biology* (Lectures on Mathematics in the Life Sciences, vol. 2). Providence: American Mathematical Society.

Leyhausen, P. (1965) The communal organization of solitary mammals. *Symposia of the Zoological Society of London* **14**, 249–263.

Ligon, J.D. (1982) Demographic patterns and communal breeding in the green woodhoopoe, *Phoeniculus purpureus*. Pp. 231–243 *In:* R.D. Alexander & D.W. Tinkle (eds) *Natural Selection and Social Behavior*. New York: Chiron Press.

Linssen, E.F. (1959) *Beetles of the British Isles*, First series. London: Warne.

Lloyd, M. (1967) Mean crowding. *Journal of Animal Ecology* **36**, 1–30.

Lloyd, M. & H.S. Dybas (1966a, b) The periodical cicada problem: I, population ecology. II, evolution. *Evolution* **20**, 133–139; 466–505.

Lockie, J.D. (1956) Winter fighting in feeding flocks of rooks, jackdaws and carrion crows. *Bird Study* **3**, 180–189.

Lovelock, J.E. (1979) *Gaia—a new look at Life on Earth*. Oxford: Oxford University Press.

Lovelock, J.E. (1983) Gaia as seen through the atmosphere. Pp. 15–25 *In:* P. Westbroek & E.W. de Jong (1983) *Biomineralization and Biological Metal Accumulation*. Dordrecht, The Netherlands: D. Reidel Publishing Co.

MacArthur, R.H. & E.O. Wilson (1967) *Island Biogeography*. Princeton: Princeton University Press.

McBee, R.H. & G.C. West (1969) Cecal fermentation in the willow ptarmigan. *Journal of Wildlife Management* **17**, 197–203.

McCauley, D.E. & M.J. Wade (1980) Group selection: the genetic and demographic basis for the phenotypic differentiation of small populations. *Evolution* **34**, 813–821.

McClure, H.E. (1966) Flowering, fruiting and animals in the canopy of a tropical rain forest. *Malayan Forester* **29**, 182–203.

Macdonald, D.W. & Patricia D. Moehlman (1982) Cooperation, altruism, and restraint in the reproduction of carnivores. Pp. 433–467 *In:* P.P.G. Bateson & P.H. Klopfer (eds.) *Perspectives in Ethology, vol. 5, Ontogeny*. New York: Plenum Press.

McGovern, M. & C.R. Tracy (1981) Phenotypic variation in electromorphs previously considered to be genetic markers in *Microtus ochrogaster*. *Oecologia* **51**, 276–280.

MacLagan, D.S. (1932) The effect of population density upon rate of reproduction with special reference to insects. *Proceedings of the Royal Society of London B* **111**, 437–454.

Marr, J.W.S. (1962) The natural history and geography of the antarctic krill (*Euphausia superba* Dana). *Discovery Reports* **32**, 33–464.

Marrett, R.R. (1914) *Anthropology*. London: Williams & Norgate.

Mather, K. & J.L. Jinks (1971) *Biometrical Genetics: the Study of Continuous Variation*. (2nd ed.) London: Chapman & Hall.

May, R.M. & R.M. Anderson (1983) Epidemiology and genetics in the coevolution of parasites and hosts. *Proceedings of the Royal Society B* **219**, 281–313.

Maynard Smith, J. (1964) Kin selection and group selection. *Nature* **201**, 1145–1147.

Maynard Smith, J. (1978) *The Evolution of Sex*. Cambridge: Cambridge University Press.

Mayr, E. (1935) Bernard Altum and the territory theory. *Proceedings of the Linnean Society of New York*, Numbers **45** and **46**, 24–38.

Medway, Lord (1972) Phenology of a tropical rainforest in Malaya. *Biological Journal of the Linnean Society, London* **4**, 117–146.

Menzies, W.J.M. (1931) *The Salmon, its Life Story*. New Edition, Edinburgh: Blackwood.

Miller, G.R. (1968) Evidence for selective feeding on fertilized plots by red grouse, hares and rabbits. *Journal of Wildlife Management* **32**, 849–853.

Miller, G.R. (1979) Quantity and quality of the annual production of shoots and flowers by *Calluna vulgaris* in north-east Scotland. *Journal of Ecology* **67**, 109–129.

Miller, G.R., D. Jenkins & A. Watson (1966) Heather performances and red grouse populations. I. Visual estimates of heather performance. *Journal of Applied Ecology* **3**, 313–326.

Miller, G.R. & A. Watson (1978a) Heather productivity and its relevance to the regulation of red grouse populations. Pp. 277–285 *In:* O.W. Heal & D.F. Perkins (eds.) Ecological Studies 27, *Production ecology of British moors and montane grasslands.* Berlin: Springer.

Miller, G.R. & A. Watson (1978b) Territories and the food plant of individual red grouse. I. Territory size, number of mates and brood size compared with the abundance, production and diversity of heather. *Journal of Animal Ecology* **47**, 293–305.

Miller, G.R., A. Watson & D. Jenkins (1970) Response of red grouse populations to experimental improvement of their food. pp. 323–335 *In:* A. Watson (ed.) *Animal Populations in relation to their Food Resources.* British Ecological Society Symposium No. 10. Oxford: Blackwell Scientific Publications.

Milne, H. (1965) Seasonal movements and distribution of eiders in northeast Scotland. *Bird Study* **12**, 170–180.

Milne, H. & G.M. Dunnet (1972) Standing crop, productivity and trophic relations of the fauna of the Ythan estuary. Pp. 86–106 *In:* R.S.K. Barnes & J. Green (eds.) *The Estuarine Environment.* Barking, England: Applied Science Publishers.

Milne, H. & F.W. Robertson (1965) Polymorphisms in egg albumen protein and behaviour in the eider duck. *Nature, Lond.* **205**, 367–369.

Moffat, C.B. (1903) The spring rivalry of birds. Some views on the limit to multiplication. *Irish Naturalist* **12**, 152–166.

Morton, E.S. (1973) On the evolutionary advantages and disadvantages of fruit eating in tropical birds. *American Naturalist* **107**, 8–22.

Moss, R. (1967) Probable limiting nutrients in the main food of red grouse (*Lagopus lagopus scoticus*). Pp. 369–379. *In:* K. Petrusewicz (ed.) *Secondary Productivity of Terrestrial Ecosystems,* Vol. 1. State Scientific Publishing House, Warsaw & Krakow.

Moss, R. (1969) A comparison of red grouse (*Lagopus l. scoticus*) stocks with the production and nutritive value of heather (*Calluna vulgaris*). *Journal of Animal Ecology* **38**, 103–122.

Moss, R. (1972a) Effects of captivity on gut lengths in red grouse. *Journal of Wildlife Management* **36**, 99–104.

Moss, R. (1972b) Food selection by red grouse (*Lagopus lagopus scoticus* (Lath.)) in relation to chemical composition. *Journal of Animal Ecology* **41**, 411–428.

Moss, R., G.R. Miller & S.E. Allen (1972) Selection of heather by captive red grouse in relation to the age of the plant. *Journal of Applied Ecology* **9**, 771–778.

Moss, R. & J.A. Parkinson (1972) The digestion of heather (*Calluna vulgaris*) by red grouse (*Lagopus lagopus scoticus*). *British Journal of Nutrition* **27**, 285–298.

Moss, R., A. Watson & R. Parr (1975) Maternal nutrition and breeding success in red grouse (*Lagopus lagopus scoticus*). *Journal of Animal Ecology* **44**, 233–244.

Moss, R., A. Watson, R. Parr & W. Glennie (1971) Effects of dietary supplements of newly growing heather on the breeding of captive red grouse. *British Journal of Nutrition* **25**, 135–143.

Moynihan, M. & A.F. Rodaniche (1982) *The Behavior and Natural History of the Caribbean Reef Squid* Sepioteuthis sepioidea. Berlin: Parey.

Murdoch, W.W. (1966) "Community structure, population control, and competition"—a critique. *American Naturalist* **100**, 219–226.

Myers, J. & C.J. Krebs (1971) Genetic, behavioral and reproductive attributes of dispersing field voles *Microtus pennsylvanicus* and *Microtus ochrogaster*. *Ecological Monographs* **41**, 53–78.

Myllymäki, A., J. Aho, E.A. Lind & J. Tast (1962) Behaviour and daily activity of the Norwegian lemming, *Lemmus lemmus* (L.), during autumn migration. *Annales Zoologici Societatis 'Vanamo'* **24**, No. 2, 31 pp.

Nelson, J.B. (1978) *The Sulidae: Gannets and Boobies*. Oxford: Oxford University Press.

Nethersole-Thompson, D. (1973) *The Dotterel*. London: Collins.

Noordwijk, A.J. van, & W. Scharloo (1980) Inbreeding of an island population of the great tit. *Evolution* **35**, 674–688.

Ollason, J.C. & G.M. Dunnet (1983) Modelling annual changes in numbers of breeding fulmars, *Fulmarus glacialis*, at a colony in Orkney. *Journal of Animal Ecology* **52**, 185–197.

Österdahl, L. (1969) The smolt run of a small Swedish river. Pp. 205–215 *In*: T.G. Northcote (ed.) *Symposium on Salmon and Trout in Streams* (H.R. MacMillan lectures in Fisheries). Vancouver: University of British Columbia.

Paine, R.T. (1966) Food web complexity and species diversity. *American Naturalist* **100**, 65–75.

Park, T. (1933) Studies in population physiology. II. Factors regulating initial growth of *Tribolium confusum* populations. *Journal of Experimental Zoology* **65**, 17–42.

Patterson, I.D. (1982) *The Shelduck: a Study in Behavioural Ecology*. Cambridge: Cambridge University Press.

Payne, R.B. (1984) *Sexual Selection, Lek and Arena behavior, and Sexual Size Dimorphism in Birds*. Ornithological Monographs No. 33. Washington: American Ornithologists' Union.

Pearson, T.H. (1968) The feeding biology of sea-bird species breeding on the Farne Islands, Northumberland. *Journal of Animal Ecology* **37**, 521–552.

Perrins, C.M. (1965) Population fluctuations and clutch-size in the great tit, *Parus major* L. *Journal of Animal Ecology* **34**, 601–647.

Perrins C.M. (1979) *British Tits*. London: Collins.

Picozzi, N. (1968) Grouse bags in relation to the management and geology of heather moors. *Journal of Applied Ecology* **5**, 483–488.

Pitelka, F.A. (1973) Cyclic pattern in lemming populations near Barrow, Alaska. Pp. 199–215 *In*: M. & E. Britton (eds.) *Alaskan Arctic Tundra*. Technical paper no. 25, Arctic Institute of North America, Washington.

Preble, E.A. (1908) *A biological investigation of the Athabaska-Mackenzie region.* North American Fauna No. 27. (See pp. 199–205.) U.S. Dept. of Agriculture, Washington.

Pukowski, E. (1933) Ökologische Untersuchungen an *Necrophorus* F. *Zeitschrift für Morphologie und Ökologie der Tiere* **27**, 518–586.

Raner, L. (1972) Förekommer polyandri hos smallnäbad simsnäppa (*Phalaropus lobatus*) och svartsnappa (*Tringa erythropus*)? *Fauna och Flora* **67**, 135–138.

Rappaport, R.A. (1967) *Pigs for the Ancestors: ritual in the ecology of a New Guinea people.* New Haven: Yale University Press.

Read, C.P. & J.E. Simmons, jr. (1963) Biochemistry and physiology of tapeworms. *Physiological Reviews* **43**, 263–305.

Rhijn, J.G. van (1983) On the maintenance and origin of alternative strategies in the ruff *Philomachus pugnax. Ibis* **125**, 482–498.

Rich, S.S., A.E. Bell & S.P. Wilson (1979) Genetic drift in small populations of *Tribolium. Evolution* **33**, 579–584.

Richards, O.W. (1953) *The Social Insects.* London: Macdonald.

Ricker, W.E. (1950) Cycle dominance among the Fraser sockeye. *Ecology* **31**, 6–26.

Ricker, W.E. (1972) Hereditary and environmental factors affecting certain salmonid populations. Pp. 27–160 *In:* R.C. Simon & P.A. Larkin (eds.) *The Stock Concept in Pacific Salmon* (H.R. MacMillan lectures in Fisheries). Vancouver: University of British Columbia.

Ricklefs, R.E. (1973) *Ecology.* Portland, Oregon: Chiron Press, 1972. (London: Nelson, 1973).

Robertson, R.A. & G.E. Davies (1965) Quantities of plant nutrients in heather ecosystems. *Journal of Applied Ecology* **2**, 211–219.

Royama, T. (1970) Factors governing the hunting behaviour and selection of food by the great tit (*Parus major* L.) *Journal of Animal Ecology* **39**, 619–688.

Ryman, N. & G. Ståhl (1981) Genetic perspectives of the identification and conservation of Scandinavian stocks of fish. *Canadian Journal of Fisheries and Aquatic Sciences* **38**, 1562–1575.

Savory, C.J. (1975) Seasonal variation in the food intake of captive red grouse. *British Poultry Science* **16**, 471–479.

Savory, C.J. (1977) The food of red grouse chicks *Lagopus l. scoticus. Ibis* **119**, 1–9.

Savory, C.J. (1978) Food consumption of red grouse in relation to the age and productivity of heather. *Journal of Animal Ecology* **47**, 269–282.

Schad, G.A. (1977) The role of arrested development in the regulation of nematode populations. Pp. 111–167 *In:* G. Esch (ed.) *Regulation of Parasite Populations.* New York: Academic Press.

Schamel, D. & D. Tracy (1977) Polyandry, replacement clutches and site tenacity in the red phalarope (*Phalaropus fulicarius*) at Barrow, Alaska. *Bird-Banding* **48**, 314–324.

Scheffer, V.B. (1958) *Seals, Sea Lions and Walruses, a Review of the Pinnipedia.* Stanford: Stanford University Press.

Schjelderup-Ebbe, T. (1922) Beiträge zur Sozialpsychologie des Haushuhns. *Zeitschrift für Psychologie* **88**, 226–252.

Schorger, A.W. (1955) *The Passenger Pigeon, its Natural History and Extinction.* Madison: University of Wisconsin Press.

Semeonoff, R. & F.W. Robertson (1967) A biochemical and ecological study of plasma esterase polymorphism in natural populations of the field vole, *Microtus agrestis* L. *Biochemical Genetics* **1**, 205–227.

Seton, E.T. (1909) *Life-histories of Northern Animals, an Account of the Mammals of Manitoba.* 2 vols., New York: Scribner.

Seton, E.T. (1912) *The Arctic Prairies.* London: Constable.

Shaw, J. (1836) Account of experimental observation on the development and growth of salmon fry, etc. *Transactions of the Royal Society of Edinburgh* **14**, 547–566. (Largely reprinted on pp. 172–204 in W. Yarrell (1859) *A History of British Fishes.* 3rd ed., 2 vols. London: Van Voorst.)

Siivonen, L. (1957) The problem of the short-term fluctuations in numbers of tetraonids in Europe. *Papers on Game Research, Helsingfors* **19**, 1–44.

Skaife, S.H. (1961) *The Study of Ants.* London: Longmans.

Slobodkin, L.B., F.E. Smith & N.G. Hairston (1967) Regulation in terrestrial ecosystems, and the implied balance of nature. *American Naturalist* **101**, 109–124.

Smith, F.E. (1961) Density dependence in the Australian thrips. *Ecology* **42**, 403–407.

Smith, G. (1909) Crustacea. *In: The Cambridge Natural History,* vol. 4, 1–217. London: Macmillan.

Smith, R.F.C. & L.C.H. Wang (1977) Arctic hares on Truelove Lowland. Pp. 461–466 *In:* L.C. Bliss (ed.) *Truelove Lowland, Devon Island, Canada.* Edmonton: University of Alberta Press.

Smithers, S.R. & R.J. Terry (1969) Immunity in schistosomiasis. *Annals New York Academy of Sciences* **160**, 826–840.

Smithers, S.R. & R.J. Terry (1976) The immunology of schistosomiasis. *Advances in Parasitology* **14**, 399–422.

Smithers, S.R., R.J. Terry & D.J. Hockley (1969) Host antigens in schistosomiasis. *Proceedings of the Royal Society B* **171**, 483–494.

Snedecor, G.W. & W.G. Cochran (1967) *Statistical methods.* 6th Ed. Ames, Iowa: Iowa State University Press.

Snow, B.K. (1970) A field study of the bearded bellbird in Trinidad. *Ibis* **112**, 299–329.

Snow, B.K. & D.W. Snow (1971) The feeding ecology of tanagers and honeycreepers in Trinidad. *Auk* **88**, 291–322.

Snow, D.W. (1961–2) The natural history of the oilbird, *Steatornis caripensis*, in Trinidad, W.I. Part I. General behavior and breeding habits. *Zoologica* **46**, 27–48; Part II. Population, breeding ecology and food. *Zoologica* **47**, 199–221.

Snow, D.W. (1976) *The Web of Adaptation: Bird Studies in the American Tropics.* New York: Quadrangle/New York Times.

Solomon, M.E., D.M. Glen, D.A. Kendall & N.F. Milsom (1976) Predation of overwintering larvae of codling moth (*Cydia pomonella* (L.)) by birds. *Journal of Applied Ecology* **13**, 341–352.

Soper, J.D. (1921) Notes on the snowshoe rabbit. *Journal of Mammalogy* **2**, 101–108.

Soper, J.D. (1928) *A faunal investigation of Southern Baffin Island.* National Museum of Canada Bulletin No. 53, 143 pp. Ottawa.

Southern, H.N. (1970) The natural control of a population of tawny owls (*Strix aluco*). *Journal of Zoology* **162**, 197–285.

Southern, H.N., R. Carrick & W.G. Potter (1965) The natural history of a population of guillemots (*Uria aalge* Pont.). *Journal of Animal Ecology* **34**, 649–665.

Springett, B.P. (1968) Aspects of the relationship between burying beetles, *Necrophorus* spp. and the mite, *Poecilochirus necrophori* Vitz. *Journal of Animal Ecology* **37**, 417–424.

Syers, J.K. & J.A. Springett (1983) Earthworm ecology in grassland soils. Pp. 67–83 *In:* J.E. Satchell (ed.) *Earthworm Ecology*. London: Chapman & Hall.

Taitt, Mary J. & C.J. Krebs (1981) The effect of extra food on small rodent populations. *Journal of Animal Ecology* **50**, 125–137.

Thomas, J.D. (1973) Schistosomiasis and the control of molluscan hosts of human schistosomes with particular reference to possible self-regulatory mechanisms. *Advances in Parasitology* **11**, 307–394.

Thomas, J.D. (1982) Chemical ecology of the snail hosts of schistosomiasis: snail-snail and snail-plant interactions. *Malacologia* **22**, 81–91.

Thomas, J.D. & M. Benjamin (1974) The effects of population density on growth and reproduction of *Biomphalaria glabrata* (Say) (Gasteropoda: Pulmonata). *Journal of Animal Ecology* **43**, 31–50.

Thomas, J.D., G.J. Goldsworthy & R.H. Aram (1975) Studies on the chemical ecology of snails: the effect of chemical conditioning by adult snails on the growth of juvenile snails. *Journal of Animal Ecology* **44**, 1–27.

Thompson, D.Q. (1955) The 1953 lemming emigration at Point Barrow, Alaska. *Arctic* **8**, 37–45.

Thomson, A.L. (1923) The migration of some British ducks: results of the marking method. *British Birds* **16**, 262–276.

Thorpe, J.E. (1977) Bimodal distribution of length of juvenile Atlantic salmon (*Salmo salar* L.) under artificial rearing conditions. *Journal of Fish Biology* **11**, 175–184.

Thorpe, J.E. & K.A. Mitchell (1981) Stocks of Atlantic salmon (*Salmo salar*) in Britain and Ireland. *Canadian Journal of Aquatic Science* **38**, 1576–1590.

Townsend, C.R. (1974) Mucus trail following by the snail *Biomphalaria glabrata* (Say). *Animal Behaviour* **22**, 170–177.

Varley, G.C. (1970) The concept of energy flow applied to a woodland community. Pp. 389–404 *In:* A. Watson (ed.) *Animal Populations in Relation to their Food Resources*. Oxford: Blackwell Scientific Publications.

Vaurie, C. (1965) *The Birds of the Palearctic Fauna, Non-Passeriformes*. London: Witherby.

Vickery, P.J. (1972) Grazing and net primary production of a temperate grassland. *Journal of Applied Ecology* **9**, 307–314.

Voous, K.H. (1960) *Atlas of European Birds (English edition)*. London: Nelson.

Wade, M.J. (1977) An experimental study of group selection. *Evolution* **31**, 134–153.

Wade, M.J. (1978) A critical review of the models of group selection. *Quarterly Review of Biology* **53**, 101–114.

Wade, M.J. (1980) An experimental study of kin selection. *Evolution* **34**, 844–855.

Wade, M.J., (1982) Group selection: migration and the differentiation of small populations. *Evolution* **36,** 949–961.

Wade, M.J. (1984) Changes in group-selected traits that occur when group selection is relaxed. *Evolution* **38,** 1039–1046.

Wade, M.J. & F. Breden (1981) Effect of inbreeding on the evolution of altruistic behavior by kin selection. *Evolution* **35,** 844–858.

Wade, M.J. & D.E. McCauley (1980) Group selection: the phenotypic and genotypic differentiation of small populations. *Evolution* **34,** 799–812.

Watson, A. (1964) Aggression and population regulation in red grouse. *Nature, Lond.* **202,** 506–507.

Watson, A. (1965) A population study of ptarmigan (*Lagopus mutus*) in Scotland. *Journal of Animal Ecology* **34,** 135–172.

Watson, A. (1970) Territorial and reproductive behaviour in the red grouse. *Journal of Reproductive Fertility, Supplement* **11,** 3–14.

Watson, A. (1972) The behaviour of the ptarmigan. *British Birds* **65,** 6–26, 93–117.

Watson, A. & D. Jenkins (1964) Notes on the behaviour of the red grouse. *British Birds* **57,** 137–170.

Watson, A. & D. Jenkins (1968) Experiments on population control by territorial behaviour in red grouse. *Journal of Animal Ecology* **37,** 595–614.

Watson, A. & G.R. Miller (1971) Territory size and aggression in a fluctuating red grouse population. *Journal of Animal Ecology* **40,** 367–383.

Watson, A. & R. Moss (1971) Spacing as affected by territorial behavior, habitat and nutrition in red grouse (*Lagopus l. scoticus*). Pp. 92–111 *In:* A. Esser (ed.) *Behaviour and Environment: the Use of Space by Animals and Men.* New York: Plenum Press.

Watson, A. & R. Moss (1980) Advances in our understanding of the population dynamics of red grouse from a recent fluctuation in numbers. *Ardea* **68,** 103–111.

Watson, A., R. Moss & R. Parr (1984) Effects of food enrichment on numbers and spacing behaviour of red grouse. *Journal of Animal Ecology* **53,** 663–678.

Watson, A., R. Moss, J. Phillips & R. Parr (1977) The effects of fertilizers on red grouse stocks on Scottish moors grazed by sheep, cattle and deer. Pp. 193–212 *In:* P. Pesson (ed.) *Écologie du Petit Gibier et Amenagement des Chasses.* Paris: Gauthier-Villars.

Watson, A., R. Moss, P. Rothery & R. Parr (1984) Demographic causes and predictive models of population fluctuations in red grouse. *Journal of Animal Ecology* **53,** 639–662.

Watson, A. & P.J. O'Hare (1973) Experiments to increase red grouse stocks and improve the Irish bogland environment. *Biological Conservation* **5,** 41–44.

Watson, A. & P.J. O'Hare (1979a) Red grouse populations on experimentally treated and untreated Irish bog. *Journal of Applied Ecology* **16,** 433–452.

Watson, A. & P.J. O'Hare (1979b) Spacing behaviour of red grouse at low density on Irish bog. *Ornis Scandinavica* **10,** 252–261.

Watson, A. & R. Parr (1969) Red grouse behaviour. *13th Progress Report, Grouse Research in Scotland* (Nature Conservancy), p. 11–16.

Watson, A.J. & J.E. Lovelock (1983) Biological homeostasis of the global environment: the parable of Daisyworld. *Tellus* **35B,** 284–289.

Watts, C.H.S. (1969) The regulation of wood mouse (*Apodemus sylvaticus*) numbers in Wytham Woods, Berkshire. *Journal of Animal Ecology* **38**, 285–304.

Wetmore, A. (1914) The development of the stomach in the Euphonias. *Auk* **31**, 458–461.

Willis, E.O. (1966) Competitive exclusion and birds at fruiting trees in Western Colombia. *Auk* **83**, 479–480.

Wilson, D.S. (1975) A theory of group selection. *Proceedings of National Academy of Sciences USA* **72**, 143–146.

Wilson, D.S. (1977) Structured demes and the evolution of group-advantageous traits. *American Naturalist* **111**, 157–185.

Wilson, D.S. (1980) *The Natural Selection of Populations and Communities*. Menlo Park, California: Benjamin-Cummings.

Wilson, D.S. (1983) The group selection controversy: history and current status. *Annual Review of Ecology and Systematics* **14**, 159–187.

Wilson, E.O. (1975) *Sociobiology—the New Synthesis*. Cambridge, Mass.: Harvard University Press.

Windberg, L.A. & L.B. Keith (1976) Experimental analyses of dispersal in snowshoe hare populations. *Canadian Journal of Zoology* **54**, 2061–2081.

Woolfenden, G.E. (1982) Selfish behavior by Florida scrub jay helpers. Pp. 257–260 *In:* R.D. Alexander & D.W. Tinkle (eds.) *Natural Selection and Social Behavior*. New York: Chiron Press.

Wright, C.A. (1960) The crowding phenomenon in laboratory colonies of freshwater snails. *Annals of Tropical Medicine and Parasitology* **54**, 224–232.

Wright, S. (1931) Evolution in Mendelian populations. *Genetics* **16**, 97–159.

Wright, S. (1932) The roles of mutation, inbreeding, crossbreeding and selection in evolution. *Proceedings 6th International Congress of Genetics* **1**, 356–366.

Wright, S. (1940) The statistical consequences of Mendelian heredity in relation to speciation. pp. 161–183 *In:* J. Huxley (ed.) *The New Systematics*. Oxford: Oxford University Press.

Wright, S. (1943) Isolation by distance. *Genetics* **28**, 114–138.

Wright, S. (1945) 'Tempo and mode in evolution': a critical review. *Ecology* **26**, 415–419.

Wright, S. (1946) Isolation by distance under diverse systems of mating. *Genetics* **31**, 39–59.

Wright, S. (1949) Adaptation and selection. pp. 365–389 *In:* G.L. Jepsen, E. Mayr & G.C. Simpson (eds.) *Genetics, Paleontology, and Evolution*. New Haven: Princeton University Press.

Wright, S. (1968–78) *Evolution and the Genetics of Populations. Vol. 1 Genetic and biometric foundations, 1968; Vol. 2 The theory of gene frequencies, 1969; Vol. 3 Experimental results and evolutionary deductions, 1977; Vol. 4 Variability within and among natural populations, 1978*. Chicago: University of Chicago Press.

Wright, S. (1980) Genic and organismic selection. *Evolution* **34**, 825–843.

Wynne-Edwards, V.C. (1939) Intermittent breeding of the fulmar (*Fulmarus glacialis* (L.)), with some general observations on non-breeding in sea-birds. *Proceedings of the Zoological Society of London* **109A**, 127–132.

Wynne-Edwards, V.C. (1955) The dynamics of animal populations. *Discovery* 1955, 433–436.

Wynne-Edwards, V.C. (1959) The control of population density through social behaviour. *Ibis* **101,** 436–441.

Wynne-Edwards, V.C. (1962) *Animal Dispersion in Relation to Social Behaviour.* Edinburgh: Oliver & Boyd.

Wynne-Edwards, V.C. (1963) Intergroup selection in the evolution of social systems. *Nature* **200,** 623–626.

Wynne-Edwards, V.C. (1964) Self-regulating systems in populations of animals. *Science* **147,** 1543–1548.

Wynne-Edwards, V.C. (1972) Ecology and the evolution of social ethics. pp. 49–69 *In:* J.W.S. Pringle (ed.) *Biology and the Human Sciences: The Herbert Spencer lectures 1972.* Oxford: Clarendon Press.

Yom-Tov, Y. (1974) The effect of food and predation on breeding density and success, clutch-size and laying date of the crow (*Corvus corone* L.) *Journal of Animal Ecology* **43,** 479–498.

Young, J.Z. (1978) *Programs of the Brain.* Oxford: Oxford University Press.

INDEX